Statistics for Social and Behavioral Sciences

Advisors:
S.E. Fienberg
W.J. van der Linden

T0137685

For other titles published in this series, go to
http://www.springer.com/series/3463

Wicher Bergsma · Marcel Croon
Jacques A. Hagenaars

Marginal Models

For Dependent, Clustered, and Longitudinal Categorical Data

 Springer

Wicher Bergsma
London School of Economics
Department of Statistics
Houghton Street
London, WC2A 2AE
United Kingdom
w.p.bergsma@lse.ac.uk

Jacques A. Hagenaars
Tilburg University
Fac. of Social & Behavioural Sciences
Department of Methodology
PO Box 90153
5000 LE Tilburg
The Netherlands
jacques.a.hagenaars@uvt.nl

Marcel Croon
Tilburg University
Fac. of Social & Behavioural Sciences
Department of Methodology
PO Box 90153
5000 LE Tilburg
The Netherlands
m.a.croon@uvt.nl

Series Editors:
Stephen Fienberg
Department of Statistics
Carnegie Mellon University
Pittsburgh, PA 15213-3890
USA

Wim J. van der Linden
CTB/McGraw-Hill
20 Ryan Ranch Road
Monterey, CA 93940
USA

ISBN   978-1-4419-1873-4   e-ISBN   978-0-387-09610-0
DOI 10.1007/978-0-387-09610-0
Springer Dordrecht Heidelberg London New York

Printed on acid-free paper

Springer is part of Springer Science+Business Media (www.springer.com)

Preface

This book has been written with several guiding principles in mind. First, the focus is on marginal models for *categorical* data. This is partly done because marginal models for continuous data are better known, statistically better developed, and more widely used in practice — although not always under the name of marginal models — than the corresponding models for categorical data (as shown in the last chapter of this book). But the main reason is that we are convinced that a large part of the data used in the social and behavioral sciences are categorical and should be treated as such. Categorical data refer either to discrete characteristics that are categorical by nature (e.g., religious denominations, number of children) or result from a categorical measurement process (e.g., using response categories such as *Yes* versus *No*, *Agree* versus *Neither Agree nor Disagree* versus *Disagree*). Treating categorical data as realizations of continuous variables and forcing them somehow into models for continuous data always implies making untestable assumptions about the measurement process and about underlying continuous distributions. In general, an explicit a priori model about the specific way in which the respondents 'translate' the continuous information into a choice among discrete categories must be postulated, and the underlying latent variable must be assumed to have a specific form, e.g., a normal distribution. Violation of these largely untestable assumptions can seriously distort the conclusions of an analysis. Therefore, it is very fortunate that during the last few decades enormous progress has been made in the development of models that take the categorical nature of the data as given without having to make strong untestable assumptions. This monograph will apply and extend these insights into the area of marginal modeling.

Our second guiding principle seems an obvious one: *marginal models* are needed to answer important research questions. However, we state this principle explicitly because there is some controversy about it. This monograph will elaborate our position extensively (throughout the book and more explicitly in the next and last chapters), but a concise illustration of it runs as follows. We have two variables A and B, which are measurements of the characteristic *Political Preference* at two instances, e.g., political preference at time one (variable A) and two (B) for the same sample, or the political preferences of husband (A) and wife (B) for a sample of married couples. To answer questions like *Are there any net differences between time one and two?* or *Are there any overall differences between the opinions of husbands and wives?*, one needs marginal modeling. Because the marginal distributions of A and B do not come from independent samples, standard estimation and testing techniques involving independent observations cannot be used. Comparison of the marginal distributions, taking the dependencies of the observations into account, is the specific purpose of marginal modeling. This approach is often referred to as 'unconditional' or 'population-averaged' and contrasted with the conditional or subject-specific approach. In the latter one, the scores on B are regressed on A (or vice versa) and the interest lies in the (expected) individual scores on B, conditional upon and as a function of the scores on A. In this sense, the interest in conditional analyses lies explicitly in the nature of the dependencies in the data. In terms of longitudinal studies, one focuses on the gross changes and transitions from A to B. The unconditional and the conditional analyses will generally show different outcomes. It does not make much sense to argue about what approach is better — they just answer different research questions.

An important and widely used alternative way of handling dependencies in the data is by means of random coefficient models, also referred to as multilevel or hierarchical models. The (dis)similarities between the marginal and the random coefficient approaches may be rather complex. Just a few main remarks will be made here. In the next and especially the last chapter, a more extensive discussion will be presented. In marginal modeling, the marginal distributions of A and B are being compared, taking the dependencies in the data into account in the estimation and testing procedures, without making any restrictive assumptions about the nature of the dependencies. Many researchers tend to use random coefficient models for essentially the same purposes. In terms of our little example, one can test and estimate the differences between A and B while introducing random intercepts for each individual (in the longitudinal example) or for each couple (in the husband-wife example), thus explicitly taking the dependencies in the data into account. Generally (and certainly for loglinear models), the outcomes of random coefficient models and corresponding marginal models will not be the same. In the random coefficient model for the example above, one investigates the differences between the marginal distributions of A and B conditioning on subject (or couple) differences. In this sense, random coefficient models belong to the class of conditional or subject-specific models. Further, when introducing random coefficients, one introduces certain assumptions about the distribution of the random coefficient and about the nature of the dependencies in the data, assumptions that usually constrain the dependencies in the data and that may or may not be correct. Finally, it is usually not so easy and straightforward to constrain random-effect models in such a way that these constraints yield the intended marginal restrictions by means of which the intended hypotheses about the marginal distributions can be tested.

The estimation method that will be used in this book is *Maximum Likelihood* (ML) estimation, directly building upon the work of Lang and Agresti (Lang, 1996a; Lang & Agresti, 1994). ML estimation for marginal modeling has many advantages, but ML estimates are sometimes difficult to obtain. The algorithms proposed here at least partly overcome these problems. In an alternative general approach towards the testing and estimation of marginal models, weighted least squares (WLS) procedures are used. This approach is often called the GSK method, named after its developers Grizzle, Starmer and Koch (Grizzle, Starmer, & Koch, 1969; Landis & Koch, 1979). In general, WLS estimates are computationally simpler to obtain than ML estimates, but have some statistical disadvantages. To overcome some of the computational difficulties of ML estimation without the disadvantages of WLS, Liang and Zeger developed a quasi-likelihood procedure — the GEE (Generalized Estimating Equations) method — for marginal-modeling purposes (Liang & Zeger, 1986; Diggle, Heagerty, Liang, & Zeger, 2002). GEE provides consistent parameter estimates, but faces problems regarding the efficiency and accuracy of the estimated standard errors. A more extensive comparison between ML on the one hand, and WLS and GEE on the other hand will be discussed in the last chapter. A third alternative to ML might be Bayesian inference. Despite the important recent developments in Bayesian methods, at least for categorical data analysis, its accomplishments and promises for marginal modeling of categorical data are still too unclear to treat them here. It would also go beyond the intended scope of this book.

This book has been written for *social and behavioral scientists* with a good background in social science statistics and research methods. Familiarity with basic loglinear modeling and some basic principles of matrix algebra is needed to understand the contents of this book. Nevertheless, the emphasis is on an intuitive understanding of the methodology of marginal modeling and on applications and research examples. Parts that are statistically more difficult are indicated by ***: they are included to provide a deeper insight into the statistical background but are not necessary to follow the main argument.

The real world examples presented in this book are on the book's website:

www.cmm.st

In addition, our Mathematica and R programmes for fitting marginal models can be found there, as well as the user-friendly code needed to fit the models discussed in the book. This is further discussed in Chapter 7, along with a presentation of some other user-friendly programs for marginal modeling.

In the first chapter, we will explain the basic concepts of marginal modeling. Because loglinear models form the basic tools of categorical data analysis, loglinear marginal models will be discussed in Chapter 2. However, not all interesting research questions involving marginal modeling can be answered within the loglinear framework. Therefore, in Chapter 3 it will be shown how to estimate and test nonloglinear marginal models. The methods explained in Chapters 2 and 3 will then be applied in Chapter 4 to investigate changes over time using longitudinal data. Data resulting from repeated measurements on the same subjects probably form the most important field for the application of marginal models. In Chapter 5, marginal modeling is related to causal modeling. For many decades now, Structural Equation Modeling (SEM) has formed an important and standard part of the researcher's tool kit, and it has also been well developed for categorical data. It is shown in Chapter 5 that there are many useful connections between SEM and marginal modeling for the analysis of cross-sectional or longitudinal data. The use of marginal models for the analysis of (quasi-)experimental data is another important topic in Chapter 5. In all analyses in Chapters 2 through 5, the observed data are treated as given, in the sense that no questions are asked regarding their reliability and validity. All the analyses are manifest-level analyses only. Marginal models involving latent variables are the topic of Chapter 6. In the final Chapter 7, a number of important conclusions, discussions and extensions will be discussed: marginal models for continuous data, alternative estimation methods, comparisons of marginal models to random and fixed-effect models, some specific applications, possible future developments, and very importantly, software and the contents of the book's website.

The very origins of this book lie in the Ph.D. thesis *Marginal Models for Categorical Data* (Bergsma, 1997), written by the first and supervised by the second (as co-promotor) and third author (as promotor). Each of us has written in one form or another all lines, sections, and chapters of this book, and we are all three responsible for its merits and shortcomings, but at the same time we acknowledge the fundamental work done in the Ph.D. thesis. Grants from the Netherlands Organization for Scientific Research NWO have made the Ph.D. project possible, as well as a subsequent postdoc project (NWO grant 400-20-001P) that contributed enormously to the birth of this monograph.

Finally, we would like to thank several people and acknowledge their often critical but always constructive, helpful, and important contributions. Jeroen Vermunt, Joe Lang, Antonio Forcina and Tamas Rudas contributed in many ways, in discussions, in answering questions, in working together both on the aforementioned Ph.D. thesis and on this book. John Gelissen, Steffen Kühnel, J. Scott Long, Ruud Luijkx, and Michael E. Sobel commented on (large) parts of the manuscript. Andries van der Ark contributed to the R routines for fitting the models. Matthijs Kalmijn provided us with the Dutch NKPS data (Netherlands Kinship Panel; http://www.nkps.nl) used in Chapters 2, 5, and 6. Marrie Bekker allowed us to use the data on body satisfaction which are analyzed in Chapters 2 and 3. Finally, we thank Bettina Hoeppner for providing us with the smoking cessation data from the Cancer Prevention Research Center, University of Rhode Island (Kingston, RI), used in Chapter 5 (and we will not insult the readers' intelligence by assuming that they might think the aforementioned people are in any way responsible for the book's errors and shortcomings).

If it were not for the patience and enduring support of John Kimmel, editor at Springer Verlag, these pages would never have appeared between covers.

Wicher Bergsma, Marcel Croon, Jacques Hagenaars

Contents

1

Introduction

The starting point for the marginal modeling of categorical data is a multidimensional table representing the joint probability distribution of two or more categorical variables. Over the last decades, many techniques have been developed to analyze the relationships in such a table, excellent overviews of which have been provided by Agresti at several levels of difficulty (Agresti, 1990, 1996, 2002). Very often, however, researchers are not interested in the analysis of the complete joint distribution but in a marginal analysis, i.e., in the comparison of two or more marginal tables that may be formed from the original complete table by collapsing cells.

In the first section of this chapter, the concept of marginal modeling of categorical data will be further explained, partly and briefly in contrast to alternative approaches that deal with more or less related research problems (as already mentioned in the Preface). Thereafter, some historical background will be provided, identifying previous statistical work on testing marginal homogeneity, and some statistical approaches that can be seen as forms of marginal modeling but are usually not explicitly identified as such. In the final section of this chapter, several descriptive statistics are introduced that are used in the remainder of this book to describe and test particular differences between marginal distributions.

1.1 Marginal Models for Categorical Data

A very simple, but typical, example of marginal modeling is the investigation of the net changes in a particular characteristic for two points in time by means of panel data. For example, information on the respondents' vote intentions is obtained in the beginning and at the end of a political campaign, interviewing the same persons twice. The pertinent categorical variables are denoted as X (*Vote intention at time 1*) and Y (*Vote intention at time 2*). The joint)probability that a randomly chosen person belongs to a particular cell (i, j) of table XY is indicated as π_{ij}^{XY}. These joint probabilities are the entries of turnover table XY and may be used to study gross changes and transitions over time in political preference. *Gross changes* in this example refer to changes on the individual level. The proportion of individuals, for example, whose

W. Bergsma et al., *Marginal Models: For Dependent, Clustered, and Longitudinal Categorical Data*, Statistics for Social and Behavioral Sciences,
DOI: 10.1007/978-0-387-09610-0_1, © Springer Science+Business Media, LLC 2009

vote intention did not change from time 1 to time 2 equals $\sum_i \pi_{i\,i}^{XY}$, and the proportion of changers is $1 - \sum_i \pi_{i\,i}^{XY}$. *Transitions* refer to the transition probabilities from time 1 to time 2, i.e., the conditional probabilities $\pi_{j\,|\,i}^{Y\,|\,X}$ of $Y = j$, given $X = i$. However, the research question of interest may not concern the transitions or the gross changes, but the net changes over time. *Net changes* in a characteristic refer to the comparison of marginal distributions of the characteristic at the different time points and represent the overall trend in the data. The marginal probabilities of turnover table XY are indicated as $\pi_i^X (= \sum_j \pi_{i\,j}^{XY} = \pi_{i\,+}^{XY})$ and $\pi_j^Y (= \sum_i \pi_{i\,j}^{XY} = \pi_{+j}^{XY})$. For example, when studying the overall net impact of a political campaign, the researcher should investigate and compare the marginal distributions π_i^X and π_j^Y of table XY rather than focus on the joint probabilities $\pi_{i\,j}^{XY}$. If there is no gross change, there cannot be net change, but gross change itself does not necessarily lead to net change since the changes at the individual level may neutralize each other in such a way that the marginal distributions remain the same over time.

If the X and Y observations have been obtained from two different independent samples at time 1 and time 2, respectively, standard statistical procedures may be used to test and estimate the net change models one is interested in. One sets up a table *Time* × *Vote intention* in which the observed sample distribution of *Vote intention* at the first time point corresponds to p_i^X and the observed sample distribution of *Vote intention* at the second time point to p_j^Y and then applies a standard chi-square test for statistical independence to this table to test the no net change hypothesis $\pi_i^X = \pi_j^X$ for $i = j$. However, in our example above, the observations on X and Y are in principle not independent of each other, as they are based on the same sample of respondents; they come from a panel study, and not from a trend (repeated cross sections) study. The over time observations tend to be dependent on each other, as reflected in the full table XY. If the researcher is only interested in the overall net changes, this dependency as such is not important and is just a nuisance. But it is a nuisance that has to be taken into account when comparing and modeling the marginal distributions of X and Y. Ignoring the dependencies between X and Y generally leads to nonoptimal estimators in terms of desirable properties like efficiency. Moreover, it often leads to similarly nonoptimal parameter estimates for the (restricted) differences between the two marginal distributions. Why then should one ever use dependent rather then independent observations? The answer is simple: sometimes only dependent observations are available to answer the research questions. Moreover, in many cases, tests based on dependent samples will have more power than tests based on two independent samples (for an illustration, see Hagenaars, 1990, p. 208-210). Further, the need for marginal-modeling procedures may arise not only from the (longitudinal) design, but also from the particular nature of the research question, even in a one-shot cross-sectional design with independent observations. For example, imagine a researcher using cross-sectional (independent) data interested in comparing the marginal distributions of *Political Party Preference* and *Political Candidate Preference* using the same categories for both variables. Because the same respondents are involved in both marginal distributions, their equality cannot be investigated by means of the standard chi-square test, but marginal modeling is required.

Marginal modeling is not restricted to the investigation of one-variable marginals of a two-dimensional joint table, as in the above example. At one extreme, the joint table may represent the distribution of just one variable and the marginal tables to be compared may be particular partially overlapping cumulative probabilities formed from this one variable distribution. For example, the 'joint' distribution is π_i^A with $i = 1, \ldots, 5$, and the 'marginals' to be compared are $(\pi_1^A + \pi_2^A + \pi_3^A)$ and $(\pi_1^A + \pi_3^A + \pi_5^A)$. Several examples of this kind will be presented in this book. At the other extreme, the joint table may be high-dimensional, e.g., pertaining to ten variables, and the marginal tables to be compared may be several five-dimensional tables obtained by collapsing the ten-dimensional table in several different ways. In this way, for example, one can investigate for panel data whether a particular causal model for five characteristics is identical in some or all respects at time 1 compared to time 2. In general, marginal modeling refers to the simultaneous analysis of marginal tables coming from the same joint table. There are many ways to form marginal tables, but as several examples in this book will show, it is not always necessary to apply the marginal methods advocated here. Sometimes simpler and more direct methods are available. A more formal definition of marginal modeling is given by Bergsma, 1997, Section 4.1.

Dependency of observations may be seen as a clustering effect and often arises in multistage or cluster sampling. Marginal modeling can be used, as many examples in this book will show, to take this clustering effect into account. However, these kinds of dependencies can also be handled in several ways, not just by means of marginal modeling. In longitudinal studies, a distinction is often made between conditional and unconditional approaches (Goldstein, 1979; Plewis, 1985; Hagenaars, 1990). In conditional models, the interest lies explicitly in the nature of the dependency of the observations at time t on the observations from previous points in time. In terms of our simple two-wave panel example above, *Vote intention* at time two is regressed on *Vote intention* at time one, modeling the joint probabilities π_{ij}^{XY}. These models are also called transition models or subject-specific models, because the pertinent research question is about the transitions from time 1 to time 2, i.e., the conditional probabilities $\pi_{i\,j}^{Y|X}$, and about what a respondent's expected vote intention is at time 2, conditional upon her or his individual score at time 1. In comparison, marginal models are unconditional models, also called population-averaged models, because they just compare, in terms of our little panel example, the marginal distributions of X and Y, substantively ignoring the dependencies of the observations (Diggle et al., 2002; Lindsey, 1999; Agresti, 2002). In marginal modeling, the dependencies are treated as a nuisance that has to be taken into account in the statistical testing and estimation procedures, but is of no substantive interest as such. Whether to use conditional or unconditional methods depends essentially on the nature of the research question. Because in many situations, especially with categorical data and nonlinear models, conditional and unconditional analyses may yield very different outcomes, one should be very careful to choose the right approach. Also note that the term 'unconditional analysis' for marginal models may be confusing in that it can be investigated by means of marginal modeling whether, in our panel example, the marginal distribution of X and Y differ from each other, both for men and women, that is, conditional upon gender.

Nowadays, perhaps the most popular way to handle dependencies and cluster effects in the data are random-effect models, often referred to as multilevel models or hierarchical models (Hsiao, 2003; Goldstein, 1979; Raudenbush & Bryk, 2003). Recently, there has been enormous progress in the development of random-effect models for categorical data. Comprehensive overviews of these developments are available, often including extensive comparisons with the conditional and unconditional models discussed above (Agresti, 2002; Diggle et al., 2002; Lindsey, 1999; Vermunt, 2003; Agresti, Booth, Hobert, & Caffo, 2000; Vermunt & Hagenaars, 2004; Molenberghs & Verbeke, 2005; Verbeke & Molenberghs, 2000). Precisely because of the importance of these developments and the opportunities they offer, researchers should make deliberate decisions on what techniques to choose and not automatically follow the fads and foibles in social science research methodology. In Chapter 7, a more extensive comparison of marginal and random-effect models will be presented, but for now it is good to keep a few main things in mind. In general, random-effect models make specific assumptions about the nature of the dependencies in the data. If these hypotheses are not true, the substantive conclusions about the effects of particular variables may be very wrong. In line with this, the random terms in the equations (which are essentially latent variables) are assumed to have a particular, often normal, distribution, and are in most applications assumed to be independent of the other observed variables in the equation. Again, if this is not true, all conclusions from the data might be wrong. Finally, random-effect models are a form of conditional or 'subject specific' analysis. In our panel example, the effect of one variable on another is investigated conditioned on the individuals. Again, such conditional analyses may differ very strongly from marginal unconditional analysis. This is especially true for categorical data and nonlinear models. Moreover, it is usually not straightforward to derive the necessary restrictions on the relevant marginals from (random effect) models for the full joint table.

As is clear from the title of this book, we confine ourselves to the application of marginal models to categorical data; marginal models for continuous data will only be briefly discussed in the last chapter. Perhaps without being aware of it, researchers are often fairly well-acquainted with marginal modeling of continuous data using linear models. The t-tests for differences between two means for matched pairs are a standard part of basic statistics courses, as are (M)ANOVA designs for repeated measures, etc. In general, the standard linear techniques for independent observations are more readily transferred to the dependent case than is true for categorical data and nonlinear techniques. Actually, the development of marginal models for categorical data is much more recent (Diggle et al., 2002; Lindsey, 1999; Agresti, 2002) and less well-known. This is unfortunate as marginal models are needed to answer the kinds of important research questions discussed in this book, and because categorical data are part and parcel of the social science research practice. Forcing categorical data into linear models for continuous data is, in our view, a practice not to be recommended in general. For example, assigning ('estimating') scores to the categories of a particular categorical variable on the basis of its relationships with one or more other variables or on the basis of particular (often normal) underlying continuous distributions may make the outcomes of the measurement process strongly dependent on the other variables chosen or on the normality assumption which can have a strong influence on the outcomes of the analyses. In that respect, we favor a Yulean as opposed to a Pearsonian approach (Agresti, 2002, 619-621; Aris, 2001; Hagenaars, 1998) and advocate the development of truly categorical marginal models for unordered and ordered categorical data.

1.2 Historical and Comparable Approaches

There has been a continuing interest in categorical marginal modeling at least since the 1950's, mainly focusing on simple marginal homogeneity models for square tables. Simple marginal homogeneity models are often used for the investigation of net changes over time in categorical characteristics, often in a (quasi-)experimental context. Comprehensive overviews of early and more recent work are provided by Agresti (2002), Bishop, Fienberg, and Holland (1975), Haberman (1979), Hagenaars (1990), and Bergsma (1997). The approach to marginal modeling that will be explained and extended in this book is based on the full Maximum Likelihood (ML) estimation procedure with constraints on the parameters, first developed in a very general way by Aitchison and Silvey (1958, 1960). This approach was applied to categorical data by Lang and Agresti, and further elaborated by Bergsma and other authors (Haber, 1985; Lang & Agresti, 1994; Becker, 1994; Balagtas, Becker, & Lang, 1995; Lang, 1996a; Bergsma, 1997; Vermunt, Rodrigo, & Ato-Garcia, 2001). Its main characteristics are that any model for the marginal tables is formulated directly in terms of restrictions on the cell probabilities of these marginal tables, and that no assumptions are made or needed for the nature of the dependencies in the joint table. The only requirement is that the conditions for multinomial sampling are fulfilled.

Several other approaches exist for obtaining estimates for restricted models for marginal tables. Perhaps the best known is the GSK approach, named after Grizzle, Starmer and Koch (Grizzle et al., 1969; Landis & Koch, 1979; Kritzer, 1977). They developed a very flexible and general approach for the analysis of categorical data, based on Weighted Least Squares (WLS) estimation. It includes marginal-modeling methods that can be used to solve essentially the same kinds of substantive questions and problems dealt with in this book. User-friendly computer programs for applying their methods have been available for a long time, including modules in generally available computer packages such as SAS. WLS estimates are often much easier to compute than ML estimates. Moreover, just as ML, WLS estimates have desirable asymptotic properties. In general, for both small and large samples, ML tends to have superior properties to WLS, as is made clear in the discussion section of Berkson (1980).

More recently, to overcome the sometimes big computational difficulties encountered when trying to obtain the maximum likelihood estimates for complex marginal models and avoid the problems with WLS, Liang and Zeger developed an extended quasi-likelihood approach called Generalized Estimating Equations (GEE) (Wedderburn, 1974; Liang & Zeger, 1986; Liang & Zeger, 1986; Diggle et al., 2002). Although GEE methods can be used outside marginal modeling, its main uses have been in this area. The main difficulty in finding the maximum likelihood estimates for the parameters of marginal models is the often complex structure of the dependencies among the observations. Essentially, in the GEE approach, this complex dependence structure is replaced by a much more simple one, e.g., statistical independence or uniform association, and consistent estimates are obtained. In this way, computations become much easier. If the dependency structure is misspecified, GEE may yield nonoptimal estimates, especially of the standard errors. Because in more complex marginal models dependencies are often complex involving higher-order interaction effects, we will rely on ML in this book because it allows us to take these complexities explicitly into account. Nevertheless, it is good to remember that if the ML procedures proposed here break down for computational reasons, WLS and GEE might provide an alternative. In the last chapter, ML will be compared more extensively with WLS and GEE.

There exist a few other approaches that also use full maximum likelihood methods. They can be divided into two broad families that contain many variants. They will not be used or discussed further in this book because of some practical shortcomings (in our view), but they are mentioned here for the sake of completeness. The first one is essentially a testing approach, more precisely a conditional testing approach and goes back to important fundamental work by Caussinus (1966). The starting point is the symmetry model for a square table in which it is assumed that the cell frequency for cell (i, j) is the same as the cell frequency for cell (j, i) for all values of $i \neq j$. Maximum likelihood estimates for the cell frequencies under this restriction can be obtained, and a statistical test for symmetry is possible, using the maximum likelihood chi-square G_S^2 with degrees of freedom df_S. The next model is the quasi-symmetry model, in which it is assumed that the square table is symmetric as far as the differences between the marginal distributions of this table allow. The corresponding test statistic is denoted as G_{QS}^2 with df_{QS}. Quasi-symmetry in combination with marginal homogeneity, implies and is implied by symmetry. Therefore, a (conditional) test of marginal homogeneity is possible by means of the conditional test statistic $G_{MH}^2 = G_S^2 - G_{QS}^2$ and $df_{MH} = df_S - df_{QS}$. If the symmetry model is valid, this conditional test statistic has asymptotically a chi-square distribution, conditional upon the quasi-symmetry model being true in the population. However, even when the quasi-symmetry model is only roughly approximately true, the chi-square distribution for the marginal homogeneity test is still adequately approximated. These models can be formulated within the loglinear framework and readily extended to three or more variables. In this way, Duncan and Hagenaars have tested several of the marginal models discussed in this book (Duncan, 1979,1981; Hagenaars, 1990,1992). However, for practical work, it has a serious shortcoming in that it remains a testing procedure that does not provide expected frequencies or parameter estimates under the assumed marginal model. These have to be determined in an ad hoc way. Furthermore, it is not clear whether the methodology can be extended to more complex models.

In another family of general marginal approaches, a model is formulated for the joint probabilities in the full table, and restrictions are imposed on the parameters in the model for the full table in such a way that the desired restrictions in the marginal tables are defined. Or alternatively, a model for the multi-way table is formulated by defining a set of parameters concerning the one-way marginals, a non-overlapping set of parameters for the two-way marginals, then for the three-way marginals, some-times up to the full joint probabilities (Glonek & McCullagh, 1995; Glonek, 1996). For an overview and more references, see Agresti (2002, p. 466) and Heagerty and Zeger (2000). Fitting these models is often computationally very involved. To avoid such computational complexities in the end, often a non-saturated model is postu-lated for the dependencies at the higher multiway level, but misspecification of the nature of the dependencies might have a serious influence on the conclusions about the marginal models. Moreover, extensions of this approach to nonloglinear models may be very cumbersome.

There exist many more models and methods commonly used in the social and behavioral research that are, in fact, marginal models for more specific purposes and apply marginal-modeling principles, but are often not recognized as such. For example, cumulative odds and odds ratios are widely used in the analysis of or-dered categorical data (Agresti, 2002). Modeling such odds (ratios) usually requires marginal-modeling principles. A simple example may be one in which the indepen-dent variable is *Sex* (= 1 for men and 2 for women) and the categorical dependent variable is *Education* (with scores running from 1 = only elementary school to 7 = university degree). In ordinary logit or loglinear modeling, one investigates how the conditional odds $\pi_{i\,s}^{E|S}/\pi_{j\,s}^{E|S}$ of belonging to category i rather than to category j of *Education* are different for boys and girls. However, given the ordinal character of *Education*, one may form conditional cumulative odds of belonging to category 1 rather than to a category larger than 2, or of belonging to category 1 or 2 rather than to a category larger than 2, etc. Each of these cumulative odds might be separately investigated for differences between boys and girls. However, if it is assumed that the *Sex* differences are the same for all successive cumulative odds, marginal-modeling methods are necessary because the successive cumulative odds are no longer inde-pendent of each other under the simultaneous restrictions.

Another instance is causal modeling of categorical data by means of modified path models, also referred to as directed graphical models, or directed loglinear mod-els (Goodman, 1973; Hagenaars, 1998; Vermunt, 1997b; Cox & Wermuth, 1996). Assume five categorical variables A, B, C, D, and E where the alphabetical order in-dicates the postulated causal order, A being the first independent and E being the last dependent variable. For testing and estimating a structural equation (causal) model for these variables, a series of marginal tables of increasing complexity must be set up. Treating the observed marginal distribution of A as given, the first interesting marginal table is Table AB, which is obtained by collapsing the complete joint table over C, D, and E. This marginal table is used to estimate the logit parameters for the effects of A on B. Next, we have marginal table ABC to estimate the logit effects of A and B on C, etc. In principle, these subtables (marginal tables) and submodels can be analyzed separately. However, as soon as restrictions are imposed on probabilities or parameters that belong to different subtables or submodels, some form of marginal modeling is needed (Croon, Bergsma, & Hagenaars, 2000). For example, a researcher may want to test whether the logit effect of A on B in marginal table AB is identical to the partial or to the marginal effect of B on C, modeling marginal tables AB and ABC, or AB and BC simultaneously. In Chapter 5, these modified path models will be discussed extensively.

Markov chain models are prime examples of the combination of marginal and regular causal modeling. Let variables A through E introduced above now refer to five successive points in time. The order of the markov chain can be determined in the standard causal modeling way. If a first order chain is assumed, in which the observations at time t are only influenced by the observations at time $t-1$ and not by $t-2, t-3$, etc., only B, and not A, has an influence on C in the marginal table ABC. Similarly, in the logit equation for table $ABCD$, variable D is only influenced by C and not by A and B. A second order chain would imply, for example, that in the logit equation for table $ABCD$, both B and C influence D, but not A. Although marginal tables are involved, so far no marginal modeling is required in the restricted sense indicated above. However, the introduction of a markovian homogeneity constraint would require marginal modeling. The simple homogeneity restriction implies that all marginal transition tables from t to $t+1$ are identical: the transitions in marginal tables AB, BC, CD, and DE are all the same, e.g., $\pi_{i\ j}^{B|A} = \pi_{i\ j}^{C|B}$. This restriction creates a dependency between the observations that has to be taken into account. In that sense, first order homogenous markov chains are a form of marginal modeling. On the other hand, the estimation and testing of the first order homogenous markov chain without further restrictions does not necessarily require the use of the full joint table (as in our general marginal-modeling approach), and more efficient ML procedures are possible for particular models (Bishop et al., 1975, Chapter 7).

From all these introductory remarks, the reader should have gained a good informal and intuitive insight into the nature and purposes of marginal modeling and may now be ready for a more extended and precise treatment of the topic. But before presenting this more precise treatment, one more aspect needs to be introduced. Only rarely does a researcher want to compare complete marginal distributions. More often than not, the interest focuses on particular aspects of these distributions, such as comparing the averages or variances of one-dimensional margins or the associations of several bivariate marginal tables. The main statistics that will be used in this book will be introduced in the remainder of this chapter.

Note that in this and all following chapters, all estimates will be based on maximum likelihood procedures, and all estimates with a caret (ˆ) on top are the maximum likelihood estimates of the corresponding population parameters, although not always the other way around, e.g., the maximum likelihood estimate for the product moment correlation coefficient ρ in the population will be denoted by r and not by $\hat{\rho}$. Similarly, we will often use the symbol p in stead of $\hat{\pi}$ to denote the ML estimate of the population cell probability π.

1.3 Coefficients for the Comparison of Marginal Distributions

Imagine a researcher conducting a monthly panel study about the support for a particular economic reform proposal, e.g., freezing wages for the next two years. The support is measured on a five-point rating scale. Among other things, the researcher is interested in the net changes over time of the overall support for this proposal. Most probably, this researcher will not be satisfied with the conclusion that the marginal response distributions have changed. Instead, she wants to know whether the average response has increased or decreased, whether the variance of the response has remained the same, or whether the correlation or the association of the response variables with other explanatory variables has changed over time, etc.

Several descriptive statistics will be presented in this section that are useful for the comparison of one-dimensional marginals and for investigating similarities in bivariate marginal distributions. For the comparison of univariate distributions, several coefficients will be given that can be used to detect differences in location and dispersion. For the comparison of bivariate marginal distributions, several association coefficients will be presented. Because categorical data may be measured at different levels of measurement, the coefficients will be classified according to their applicability to interval, ordinal or nominal-level data.

With a few exceptions, most coefficients presented here are well-known. In later chapters, their usefulness for marginal modeling will be illustrated and ML methods will be developed for applying these coefficients in a marginal-modeling context with dependent observations.

1.3.1 Measuring Differences in Location

Probably the most common perspective from which univariate distributions are compared is from the perspective of location or central tendency: one computes for each marginal distribution of an interval or ordinal-level variable a measure of central tendency and compares the values found for the different distributions.

Interval Data

For interval-level measurements, the arithmetic mean is the most common central tendency measure. In more formal terms, for a real valued random variable X, the mean is defined as its expected value:

$$\mu(X) = E(X) .$$

If X is categorical with I categories and category scores $(x_1, \ldots, x_I)$, we can write

$$\mu(X) = \sum_{i=1}^{I} x_i \pi_i^X$$

where $\pi_i^X = P(X = i)$. Its sample value is

$$\hat{\mu}(X) = \sum_{i=1}^{I} x_i p_i^X$$

where p_i^X is the observed proportion of observations in cell i. The sample value of μ coincides with the ML estimate, which is used in this book.

Ordinal Data

If the ordinal data are treated as truly ordinal without numerical scores assigned to the response categories, the median is often recommended as a measure of central tendency. It is defined as a point at which at least half the observations are greater than or equal to it, and at least half the observations are less than or equal to it. One serious drawback of the median for our purposes here is that it is not a continuous and differentiable function of the response probabilities. By way of example, suppose that X is a dichotomous variable with two categories 1 and 2. If we have 11 observations in Category 1 and 10 observations in Category 2, the median score is equal to 1. If a single observation is moved from category 1 to category 2, the median score becomes 2. The jump in the median score illustrates that the median is not a continuous function of the data. This discontinuous character of the median excludes it from being discussed further in this book, since the methods that will be developed require that the statistics are continuous and differentiable functions of the data. We might have tried to overcome the discontinuous character by some approximation using common interpolation methods for determining the median, but this does not really solve the problem and, moreover, good alternatives are available.

These alternatives are based on the fact that more often than not, and certainly in marginal modeling, researchers are interested in *comparing* the locations of two or more (ordinal-level) variables. A comparison of the locations of two distributions brings us actually into the realm of association. Nevertheless, we discuss these alternatives here, and not in the next subsection on association measures, because our specific goal is to compare central tendencies and not to compare associations.

Suppose a researcher wishes to compare the locations of K ordinal random variables $X_1, \ldots, X_K$. An intuitive coefficient for comparing the locations of X_i and X_j is to see what happens if we draw independently from the marginal distribution of X_i and from the marginal distribution of X_j: is there a tendency that the X_i observations are larger than the X_j observations, or vice versa? The coefficient L_{ij} expresses these tendencies (ignoring possible ties):

$$L_{ij} = P(X_i > X_j) - P(X_i < X_j) .$$

Since the inequalities in this definition are strict, pairs of tied observations do not contribute to L_{ij}. It is easily seen that $L_{ij} = -L_{ji}$. Further, L_{ij} is bounded: $-1 \leq L_{ij} \leq 1$, where the extreme values 1 and -1 indicate that X_i is with probability 1 greater or smaller than X_j, respectively. $L_{ij} = 0$ implies that the locations of X_i and X_j are the same. Positive values $L_{ij} > 0$ imply that the X_i values tend to be larger than the X_j values, i.e., that the location of X_i is larger than that of X_j; negative values imply the opposite. For example, an outcome of $L_{ij} = .30$ means that if we have two independent draws, one from the marginal distribution of X_i and one from the marginal distribution of X_j, then the probability of obtaining a higher score on the first draw is .30 larger than the probability of the first draw having a smaller value than the second. In a sports context, where X_i represents the performance of player i, $L_{ij} > 0$ means that player i is more likely to beat player j than vice versa. A value of $L_{ij} = .30$ would mean that if players i and j played 100 independent games against each other, player i would be expected to win 65 games and lose 35. This coefficient, which was already described in Agresti (1984, Section 9.3), was intensively studied by Cliff (1993, 1996), who discussed some inferential procedures for this coefficient and its possible extensions to multigroup and factorial designs.

The dissimilarity of the two probabilities $P(X_i > X_j)$ and $P(X_i < X_j)$ can also be expressed by taking their (log)ratio. (see also Agresti, 1984, Section 9.3). This log-multiplicative alternative to L_{ij} will be denoted by L'_{ij} and is defined by

$$L'_{ij} = \log \frac{P(X_i > X_j)}{P(X_i < X_j)}$$
$$= \log P(X_i > X_j) - \log P(X_i < X_j) .$$

This coefficient L'_{ij} fits naturally into the framework of loglinear modeling. In our applications, both coefficients will be used.

Coefficients L_{ij} and L'_{ij} have a natural interpretation for the comparison of two (marginal) distributions. However, if there are more than two variables, an interpretational problem of the two coefficients as difference scores arises. In the following discussion, we focus on L_{ij}, but all our remarks and suggestions are also valid for coefficient L'_{ij}. If the mean or the median is used as a location or central tendency measure, knowledge of the difference in location of variables X_i and X_j and of X_j and X_k will determine exactly the location difference between X_i and X_k. However, this is not true in general for the L_{ij}'s, since for given i, j and k

$$L_{ij} + L_{jk} \neq L_{ik}.$$

Furthermore, there may be an intransitivity problem when comparing two or more coefficients. The paradoxical situation that $L_{ij} > 0$, $L_{jk} > 0$ and $L_{ki} > 0$ may occur at the same time, i.e., X_i is located to the right of X_j (cf. $X_i > X_j$); X_j is located to the right of X_k (cf. $X_j > X_k$); but at the same time X_k is located to the right of X_i (cf. $X_k > X_i$). In a sports analogy, players higher on the ranking list will usually beat opponents that rank lower, but 'interacting' playing styles may be such that a lower-ranked player is more likely to win against the higher-ranked player. However, what is not a problem in sports becomes a nuisance when comparing central tendencies in a set of variables, and leads to interpretation problems in most cases. To overcome this intransitivity problem, we took our lead from a model in another area, viz., the Bradley-Terry model for pairwise choices (Bradley & Terry, 1952; Fienberg, 1980, Section 8.4; Agresti, 2002, Section 10.6), which is the loglinear model

$$\log \frac{P(X_i > X_j)}{P(X_i < X_j)} = \lambda_i - \lambda_j .$$

If the Bradley-Terry (BT) model is valid, the choices are by definition transitive in the sense that if $L_{ij} \geq 0$ and $L_{jk} \geq 0$, then $L_{ik} \geq 0$.

As a starting point, L_{ij} is decomposed as follows:

$$L_{ij} = l_i - l_j + l_{ij}$$

where the main parameters l_i and the interaction effects l_{ij} have to be constrained to obtain an identified model. Although other identification schemes such as dummy coding are also frequently applied, we opt here in favor of effect coding which means that the main effects are subject to the restriction $\sum_i l_i = 0$ and the interaction terms l_{ij} to $\sum_j l_{ij} = \sum_j l_{ji} = 0$. If the interaction effects l_{ij} are all zero, there are no longer intransitivities of the nature described above. Although the l_i are still not absolute measures of location, such as means or the medians, they now do allow a transitive comparison of the relative locations of two or more variables. As follows directly from the above, in the restricted model with $l_{ij} = 0$,

$$L_{ij} = l_i - l_j, \tag{1.1}$$

it is true that

$$L_{ij} + L_{jk} = L_{ik}$$

since $(l_i - l_j) + (l_j - l_k) = (l_i - l_k)$. This means that in the restricted model, $L_{ij} = l_i - l_j$ is a true difference score for the locations of X_i and X_j, and the intransitivity problem disappears. However, if the interaction effects l_{ij} are non-zero, the ordering of the variables X_i on the basis of the main effects l_i may not adequately reflect the information given by the L_{ij} and intransitivity problems may be present.

As is done in the next chapter, it is possible to test the restricted model described by (1.1). If the restricted model does not fit the data, there exist nonzero interaction terms l_{ij}. By testing each individual interaction term for significance, one can investigate for which pair of variables the simple model given by (1.1) does not apply, and for which triple of variables an intransitivity of the location measure may occur.

The above exposition applies to both ordinal categorical and ordinal continuous random variables, although in this book only the categorical case is considered. For categorical data, with X_i and X_j having ordered categories $1, \ldots, I$, and $\pi_l^{X_i}$ denoting the marginal probability that $X_i = l$, we can write

$$L_{ij} = \sum_{l > m} \pi_l^{X_i} \pi_m^{X_j} - \sum_{l < m} \pi_l^{X_i} \pi_m^{X_j}$$

where m and l range over $1, \ldots, I$. With $p_m^{X_i}$ as the sample proportion (being the ML estimate of $\pi_m^{X_i}$), the ML estimate of L_{ij} is

$$\hat{L}_{ij} = \sum_{l > m} p_l^{X_i} p_m^{X_j} - \sum_{l < m} p_l^{X_i} p_m^{X_j}.$$

For the log-multiplicative variant L'_{ij}, similar definitions can be given. Note that ties in the observations do not contribute to the value of L or of $\hat{L}$.

By way of a practical example for the computation of $\hat{L}_{ij}$, let A and B be two categorical variables, each with three ordered categories 1, 2, and 3. The cell frequencies are denoted as n_1^A, n_1^B, n_2^A, etc., with sample size N_A for variable A and N_B for variable B. In this simple example, let $N_A = N_B = N$. The sample proportions are then computed as $p_1^A = n_1^A/N$, etc.

	1	2	3	Total
A	p_1^A	p_2^A	p_3^A	1
B	p_1^B	p_2^B	p_3^B	1

and the ordinal coefficient $\hat{L}_{AB}$ (without the BT restriction discussed above) is estimated as

$$\hat{L}_{AB} = (p_2^A p_1^B + p_3^A p_1^B + p_3^A p_2^B) - (p_1^A p_2^B + p_1^A p_3^B + p_2^A p_3^B).$$

Coefficient $\hat{L}_{ij}$ is related to the Mann-Whitney statistic for testing whether or not two ordinal random variables differ in location, and to ordinal association coefficients like Kendall's tau and Goodman and Kruskal's gamma (see also Section 1.3.3 below).

1.3.2 Measuring Differences in Dispersion

In this section, some standard measures of dispersion for interval type data and for nominal type data will be described; more specifically, the variance, the diversity index and the entropy measure. For ordinal data, a comparative measure will be proposed following much the same logic that underlies the comparative location coefficient L_{ij}.

Interval Data

The best-known measure of the variability of data measured on an interval scale is the variance, defined for a random variable X with mean $\mu(X)$ as

$$\sigma^2(X) = E[X - \mu(X)]^2 .$$

The variance is the average squared deviation from the mean and has sample value

$$\hat{\sigma}^2(X) = \sum_i (x_i - \hat{\mu}(X))^2 p_i^X .$$

where the x_i are the scores assigned to category i of variable X. Note that $\hat{\sigma}^2(X)$ is the biased ML estimator of $\sigma^2(X)$.

A disadvantage of the variance as a measure of dispersion is its sensitivity to outliers. The occurrence of some extremely large or small scores may have a large effect on the value of the variance since the deviations from the mean are squared before they are additively combined. A dispersion measure that is less sensitive to outliers is Gini's mean (absolute) difference g (see David, 1968, 1983).

With X' an independent replication of X, the population value of g is formally defined as

$$g = E|X - X'|,$$

which has the sample value

$$\hat{g} = \sum_i \sum_j |x_i - x_j| p_i^X p_j^X .$$

As with the variance, this is the (biased) ML estimator of g.

Ordinal Data

One often proposed measure for the dispersion of ordinal variables is the interquartile range $Q_3 - Q_1$. The use of this measure here encounters the same problems as the median (Q_2) discussed above, and for the same reasons it will not be dealt with here. Therefore, another dispersion coefficient for ordinal random variables is introduced, developed by the first author. This coefficient W_{ij} can best be seen as the dispersion analogue of the location measure L defined in the previous section. It is a comparative measure that can be used for comparing the dispersions of K ordinal random variables $X_1, \ldots, X_K$. An intuitive approach towards comparing the dispersions of X_i and X_j is to see what happens if we draw independently with replacements from the marginal distribution of X_i and from the marginal distribution of X_j: is there a tendency that two random draws from X_i enclose the values of two random draws from X_j, or are the two values of the X_i in between the two X_j draws? To define the ordinal dispersion coefficient W_{ij}, let X_i and X_i' be two independently drawn observations from the first population i, and in a similar way, let X_j and X_j' be two independently drawn observations from the second population j. Then, the dispersion of the observations in population i can be compared with the dispersion in population j by the ordinal dispersion measure W_{ij}

$$W_{ij} = 4P(X_i < X_j \leq X_j' < X_i') - 4P(X_j < X_i \leq X_i' < X_j') .$$

Factor 4 occurs in order to correct for the fact that X_i and X_i', and X_j and X_j' may be permuted independently, making a total of four permissible permutations. Consequently, W_{ij} is the probability that the two observations from the ith sample enclose the two observations from the jth sample, minus the probability that the opposite happens. Obviously, if the tendency for two X_j values to lie in between two X_i values is larger than the tendency for two X_i values to lie in between two X_j values, the values of X_i are further apart and show more dispersion than the values of X_j. So, $W_{ij} = 0$ indicates similar ordinal dispersion of X_i and X_j, $W_{ij} > 0$ (with maximum 1) indicates X_i has a larger dispersion than X_j, and $W_{ij} < 0$ (with minimum -1) indicates X_i has a smaller dispersion than X_j. Note that $W_{ij} = -W_{ji}$. Given the formal definition of W_{ij} presented above, a number of draws are ignored. In particular, pairs of observations of a particular variable that are not enclosed by the pair of values of the other variable do not contribute to W_{ij}, since this inconclusive outcome pattern does not give clear clues about ordinal dispersion differences between the two variables. Moreover, all tied observations on X_i and/or X_j are ignored, except those in which the tied observations from one population are enclosed by the untied observations from the other population.

When interpreting coefficient W, it should be borne in mind that the dispersion measure W is also partially determined by the difference in location of the two variables, as is often the case with location and dispersion measures (except in symmetric marginal distributions). When one distribution is completely to the right of the other, W is zero. Moreover, W_{ij} will be most sensitive to the difference in dispersion when the two variables have the same location, i.e., when $L_{ij} = 0$.

Unfortunately, as with the L_{ij}'s, when comparing more than two variables, the W_{ij}'s have the unsatisfactory intransitivity property: it is possible that $W_{ij} > 0$, $W_{jk} > 0$ and $W_{ki} > 0$, i.e., X_i is more dispersed than X_j, X_j is more dispersed than X_k, but X_k is more dispersed than X_i. To overcome this, we propose again a modified Bradley-Terry type model. A saturated decomposition for the W_{ij} is

$$W_{ij} = w_i - w_j + w_{ij} ,$$

subject to the effect coding identifying restrictions $\sum_i w_i = 0$ and $\sum_j w_{ij} = \sum_j w_{ji} = 0$ for all i. By setting $w_{ij} = 0$, we obtain the Bradley-Terry type model

$$W_{ij} = w_i - w_j .$$

If this model holds, the intransitivities disappear and $W_{ij} + W_{jk} = W_{ik}$. Again, the w_i are only defined comparatively. This model allows the comparison of dispersion in ordinal samples. How to test this BT variant will be discussed in Chapter 3.

For categorical data, the population value of W_{ij} is defined as

$$W_{ij} = 4 \sum_{l < m \leq q < r} \pi_l^{X_i} \pi_m^{X_j} \pi_q^{X_j} \pi_r^{X_i} - 4 \sum_{l < m \leq q < r} \pi_l^{X_j} \pi_m^{X_i} \pi_q^{X_i} \pi_r^{X_j}$$

where m and l range over $1, \ldots, I$. The sample value of W_{ij} (without the BT restriction) is

$$\hat{W}_{ij} = 4 \sum_{l < m \leq q < r} p_l^{X_i} p_m^{X_j} p_q^{X_j} p_r^{X_i} - 4 \sum_{l < m \leq q < r} p_l^{X_j} p_m^{X_i} p_q^{X_i} p_r^{X_j} .$$

Applied to the same simple A and B table with three categories for each variable as used for illustrating the computation of $\hat{L}_{AB}$, coefficient $\hat{W}_{AB}$ is estimated as

$$\hat{W}_{AB} = 4 p_1^A p_2^B p_2^B p_3^A - 4 p_1^B p_2^A p_2^A p_3^B .$$

As may be clear from this illustration, a lot of possible comparisons are ignored because they do not provide conclusive information about possible dispersion differences between A and B. This makes W not very well-suited as a comparative dispersion measure for variables with just a few categories, as will be confirmed in the analyses in later chapters.

For alternative measures of ordinal variation and dispersion, which also could be handled by the marginal-modeling approach advocated in this book, see Blair and Lacy (2000).

Nominal Data

Several indices have been proposed for measuring the dispersion of the scores on nominal-level variables. A prerequisite of any such index is that it is not based on numerical scores assigned to the categories of the nominal variable.

A first index of this type is the *index of diversity* (also called Gini's measure of concentration or variation), which is defined as

$$D = 1 - \sum_{i=1}^{I} \left(\pi_i^X \right)^2 = \sum_{i=1}^{I} \pi_i^X \left(1 - \pi_i^X \right)$$

with π_i^X the probability of observing category i. This index gives the probability that two randomly selected observations will be in a different category. It has been studied extensively by Agresti and Agresti (1977) who showed that it assumes a value of zero if and only if all the responses fall in the same category. It takes on its maximal value if and only if the responses are uniformly distributed over the categories.

A second well-known measure of dispersion for nominal variables is the information index H (also called the entropy or the uncertainty index),

$$H = - \sum_{i=1}^{I} \pi_i^X \log \pi_i^X$$

with log representing the natural logarithm with base e.

This index originates from mathematical communication theory. See Krippendorff (1986) for applications of information-theoretical measures in the social sciences. Just as index D, the information index H is zero if and only if all responses fall in the same category, and attains its maximal value if and only if the responses are uniformly distributed over the response categories.

The sample values $\hat{D}$ of D and $\hat{H}$ of H are obtained as usual by replacing the population proportions π_i^X by the sample proportions p_i^X.

1.3.3 Measuring Association

When a researcher wants to compare bivariate marginal distributions, particular aspects of the bivariate marginals rather than the whole distribution will almost always be relevant. This is even more true here than for the comparison of one-dimensional marginals, if not for substantive than certainly for practical reasons: higher dimensional tables contain just too many cells to allow a simple comparison and easy interpretation. In this section, several useful coefficients for measuring the association or correlation between two variables will be introduced. First, the simplest case of two dichotomous variables whose joint distribution can be represented in a 2×2 table will be discussed. Next, more general cases are considered. If both variables are measured on an interval scale, the correlation coefficient can be used. For two ordinal variables, coefficients based on concordance and discordance probabilities can be defined. Finally, local and global odds ratios for $I \times J$ tables will be discussed that

Table 1.1. Example of a 2×2 table

	Success	Failure
Men	π_{11}^{SE}	π_{12}^{SE}
Women	π_{21}^{SE}	π_{22}^{SE}

are appropriate from the nominal level on. If one of the variables can be considered as an explanatory variable for the second response variable, asymmetric measures of association can be defined that actually express the strength of the effect of the explanatory on the response variable.

Association in 2×2 Tables

Suppose a sample of men and women have taken an exam, and consider Table 1.1 that describes the joint distribution of the respondent's sex ($S = 1$ for Men and $S = 2$ for Women) and success or failure in the exam ($E = 1$ for Success and $E = 2$ for Failure). Let the probabilities of success and failure be denoted as π_{11}^{SE} and π_{12}^{SE} for men and π_{21}^{SE} and π_{22}^{SE} for women.

The effect of *Sex* on the probability of success can be assessed in various ways. Two intuitively appealing measures are described below: the odds ratio and the difference in proportions.

For men, the odds of having success rather than failure are

$$\pi_{11}^{SE} / \pi_{12}^{SE} \, ,$$

and for women

$$\pi_{21}^{SE} / \pi_{22}^{SE} \, .$$

The odds ratio is then defined as

$$\alpha = \frac{\pi_{11}^{SE} / \pi_{12}^{SE}}{\pi_{21}^{SE} / \pi_{22}^{SE}}$$
$$= \frac{\pi_{11}^{SE} \, \pi_{22}^{SE}}{\pi_{21}^{SE} \, \pi_{12}^{SE}} \, .$$

The odds ratio α measures how much larger the odds of success are for men than for women.

A second commonly used measure for assessing the dependence of success on *Sex* is the difference of the conditional proportions

$$\varepsilon = \pi_{1|1}^{E|S} - \pi_{1|2}^{E|S}$$
$$= \frac{\pi_{11}^{SE}}{\pi_{1+}^{SE}} - \frac{\pi_{21}^{SE}}{\pi_{2+}^{SE}} \, ,$$

which, in this case, is the proportion of successful men minus the proportion of successful women. Coefficient ε is an asymmetric measure of association since its value depends on how the variables are treated as explanatory and response variables. A discussion of these measures is given in Agresti (2002), and see Rudas (1997) for an extensive discussion of the odds ratio. Their sample values, obtained by replacing population by sample proportions, are denoted by $\hat{\alpha}$ and $\hat{\varepsilon}$.

Interval Data

Association between interval-level variables X and Y is the extent to which large values of X tend to go with large values of Y. The most commonly used measure is the correlation coefficient that measures the linear association between the variables

$$\rho = \frac{\text{cov}(X,Y)}{\sqrt{\text{var}(X)\text{var}(Y)}} = \frac{E[X - \mu(X)][Y - \mu(Y)]}{\sqrt{E[X - \mu(X)]^2 E[Y - \mu(Y)]^2}} .$$

The population value and ML estimate of the variance was given above. The population value of the covariance for categorical data is

$$\text{cov}(X,Y) = \sum_{i,j}[x_i - \mu(X)][y_j - \mu(Y)]\pi_{ij}^{XY} .$$

The sample value and maximum likelihood estimate of the covariance is

$$\hat{\text{cov}}(X,Y) = \sum_{i,j}[x_i - \hat{\mu}(X)][y_j - \hat{\mu}(Y)]p_{ij}^{XY} ,$$

and the sample value and ML estimate of the correlation coefficient is

$$r = \frac{\hat{\text{cov}}(X,Y)}{\hat{\sigma}(X)\hat{\sigma}(Y)} .$$

Ordinal Data

For various summary measures of association between two ordinal variables, the notions of concordance and discordance play a central role. Two observations from the bivariate distribution of (X,Y) are concordant if the rank order of the two observations on X agrees with that on Y, i.e., if either

$$X_1 < X_2 \quad \text{and} \quad Y_1 < Y_2$$

or

$$X_1 > X_2 \quad \text{and} \quad Y_1 > Y_2$$

holds. They are discordant if the two rank orders disagree, i.e., if either

$$X_1 < X_2 \quad \text{and} \quad Y_1 > Y_2$$

or

$$X_1 > X_2 \quad \text{and} \quad Y_1 < Y_2$$

holds. They are tied if they are neither concordant nor discordant, i.e., if they have the same X value, or the same Y value, or both.

The probabilities of concordance, discordance and a tie are now given by

$$\Pi^C = 2 \sum_{i<k} \sum_{j<l} \pi_{ij}^{XY} \pi_{kl}^{XY}$$

$$\Pi^D = 2 \sum_{i<k} \sum_{j>l} \pi_{ij}^{XY} \pi_{kl}^{XY}$$

$$\Pi^T = 1 - \Pi^C - \Pi^D .$$

A well-known association measure based on these probabilities is γ, defined as

$$\gamma = \frac{\Pi^C - \Pi^D}{\Pi^C + \Pi^D} = \frac{\Pi^C - \Pi^D}{1 - \Pi^T} .$$

The denominator serves to correct for tied observations. The sample value $\hat{\gamma}$ is obtained by replacing population by sample proportions.

Nominal Data

The association between two nominal variables X and Y with arbitrary numbers of response categories can be described by a set of local odds ratios, whose logarithms are defined for categories i and k of X and j and l of Y as

$$\alpha_{ij,kl} = \log \frac{\pi_{ij}^{XY} \pi_{kl}^{XY}}{\pi_{il}^{XY} \pi_{kj}^{XY}} .$$

For ordinal or interval type variables, we usually assume $i < k$ and $j < l$ so that a positive value of $\alpha_{ij,kl}$ corresponds to a positive local association between X and Y. Note that some of the $\alpha_{ij,kl}$'s are redundant in the sense that they are functions of the other $\alpha_{ij,kl}$'s. In fact, an appropriately selected subset of $(I-1)(J-1)$ log odds ratios is sufficient to describe the loglinear association in table XY completely.

Odds ratios are the building blocks of classical loglinear models and describe the loglinear association for categorical variables (see Chapter 2 and Agresti (2002) for further details). Chapter 2 contains several examples of analyses in which the log odds are further restrained by linear restrictions that may be relevant if the variables are ordinal or categorical interval.

The asymmetric association coefficient ε, defined above in the context of a 2×2 matrix, can also be extended to the more general case of an $R \times C$ table by defining a set of coefficients

$$\varepsilon_{ij,k} = \pi_{k\,i}^{Y|X} - \pi_{k\,j}^{Y|X} ,$$

with i and j two different categories of X, and k an arbitrary category of Y. The complete set of these coefficients contains many linear dependencies and a selection of linearly independent coefficients should be made for interpretative purposes.

When X is treated as an explanatory variable for the response variable Y, several asymmetric measures of association defined as proportional reduction in variation have been defined (Agresti, 1990, p. 23-25). Let $V(Y)$ be a measure of variation for the marginal distribution of Y, and let $V(Y|i)$ be the same measure for the conditional distribution of Y, given $X = i$. A proportional reduction in variation measure is defined as

$$\frac{V(Y) - E[V(Y|X)]}{V(Y)}$$

with $E[V(Y|X)]$ the expected value of the conditional variation measures with respect to the distribution of X. Using the index of diversity D as the (nominal) measure of variation, Goodman and Kruskal's tau (τ) is a proportional reduction in variation measure and can be written as

$$\tau_{Y|X} = \frac{\sum_i \sum_j (\pi_{ij}^{XY})^2 / \pi_{i+}^{XY} - \sum_j (\pi_{+j}^{XY})^2}{1 - \sum_j (\pi_{+j}^{XY})^2}.$$

The uncertainty coefficient U is obtained and can be interpreted as a proportional reduction in variation measure by taking the information measure H as measure of variation:

$$U_{Y|X} = -\frac{\sum_i \sum_j \pi_{ij}^{XY} \log(\pi_{ij}^{XY} / \pi_{i+}^{XY} \pi_{+j}^{XY})}{\sum_j \pi_{+j}^{XY} \log \pi_{+j}^{XY}}.$$

Both $\tau_{Y|X}$ and $U_{Y|X}$ are asymmetric measures of association. When X is taken as the response variable and Y as the explanatory variable, coefficients $\tau_{X|Y}$ and $U_{X|Y}$ can be defined similarly, and their values will in general be different from $\tau_{Y|X}$ and $U_{Y|X}$, respectively. Maximum likelihood estimates $\hat{\tau}_{Y|X}$ and $\hat{U}_{Y|X}$ are obtained by replacing the population probabilities π by the sample proportions p.

1.3.4 Measuring Agreement

A natural coefficient for assessing the agreement of responses in a square table is given by the raw agreement index $\sum_i \pi_{ii}$, which is the sum of the joint probabilities on the main diagonal of the matrix (Suen & Ary, 1989). This index attains its maximal value of 1 when all respondents are on the main diagonal of this matrix. A disadvantage is, however, that it may be spuriously inflated by chance agreement: even when the respondents' responses at the two time points are statistically independent, the raw agreement coefficient can attain high values. Cohen's kappa coefficient is an agreement index that corrects for this chance agreement and is generally preferred to the raw agreement index (Cohen, 1960). Kappa is defined as

$$\kappa = \frac{\pi_o - \pi_e}{1 - \pi_e},$$

with $\pi_o = \sum_i \pi_{ii}$ the actual probability of agreement and π_e the probability of agreement by chance, which is given by

$$\pi_e = \sum_i \pi_{i+}\pi_{+i} \, .$$

The quantity π_e would be the probability of agreement when the respondents' responses were independent.

Kappa is zero when the responses for the two variables in the square table are independent of each other, and in this sense it can also be considered as a measure of association in a square table. In panel data, kappa can also be used to assess the stability of the responses over time. The ML estimate $\hat{\kappa}$ is obtained by replacing π by the corresponding sample values p.

2

Loglinear Marginal Models

Loglinear models provide the most flexible tools for analyzing relationships among categorical variables in complex tables. It will be shown in this chapter how to apply these models in the context of marginal modeling. First, in Section 2.1, the basics of ordinary loglinear modeling will be explained. The main purpose of this section is to introduce terminology and notation and those aspects of loglinear modeling that will be used most in the remainder of this book. It will be assumed that the reader already has some familiarity with loglinear modeling and, therefore, the discussion will be concise. An advanced overview of loglinear models is provided by Agresti (2002); an intermediate one by Hagenaars (1990) and an introduction is given by Knoke and Burke (1980) among many others. In Section 2.2, several motivating examples will be presented showing what types of research questions can be answered by means of loglinear marginal modeling. Finally, in Section 2.3, a general ML estimation procedure will be discussed for testing and estimating loglinear marginal models.

2.1 Ordinary Loglinear Models

2.1.1 Basic Concepts and Notation

The most simple applications of loglinear models are to two-dimensional tables such as Table 2.1, in which the self-reported *Political Orientation* (P) and *Religion* (R) of a sample of 911 U.S. citizens is cross-classified. Table 2.1 contains the raw frequencies as well as the vertical percentages. Variable P has seven categories, ranging from extremely liberal ($P = 1$) to extremely conservative ($P = 7$). *Religion* has three categories: Protestant ($R = 1$), Catholic ($R = 2$) and None ($R = 3$). The joint probability that $P = i$ and $R = j$ is denoted by π_{ij}^{PR}. The number of categories of P is $I = 7$ and of R is $J = 3$. For a first interpretation of the data, the vertical percentages in Table 2.1 are useful for comparing the conditional distributions of *Political Orientation* for the three religious groups. It can easily be seen that the nonreligious people are more liberal than the Protestants or the Catholics, but the differences between the latter two groups are less clear. Even in this simple example, a more formal approach may

W. Bergsma et al., *Marginal Models: For Dependent, Clustered, and Longitudinal Categorical Data*, Statistics for Social and Behavioral Sciences, DOI: 10.1007/978-0-387-09610-0_2, © Springer Science+Business Media, LLC 2009

Table 2.1. *Political Orientation* and *Religion* in the United States in 1993 (Source: General Social Survey 1993)

Political Orientation (P)	Religion (R)			Total
	1. Protestant	2. Catholic	3. None	
1. Extremely liberal	11 (1.8%)	2 (1.0%)	4 (4.4%)	17 (1.0%)
2. Liberal	49 (8.0%)	21 (10.1%)	23 (25.3%)	93 (10.2%)
3. Slightly liberal	79 (12.9%)	23 (11.1%)	19 (20.9%)	121 (13.3%)
4. Moderate	220 (35.8%)	96 (46.4%)	30 (33.0%)	346 (38.0%)
5. Slightly conservative	112 (18.3%)	36 (17.4%)	9 (9.9%)	157 (17.2%)
6. Conservative	119 (19.4%)	27 (13.0%)	4 (4.4%)	150 (16.5%)
7. Extremely conservative	23 (3.8%)	2 (1.0%)	2 (2.2%)	27 (3.0%)
Total	613 (100%)	207 (100%)	91 (100%)	911 (100%)

Note: The small Jewish and Other religious groups are omitted

be needed to separate true population differences from sampling fluctuations and to arrive at a clear and parsimonious description of the data.

Saturated loglinear models decompose the observed logarithms of the cell probabilities in terms of loglinear parameters without imposing any restrictions on the data:

$$\log \pi_{ij}^{PR} = \lambda + \lambda_i^P + \lambda_j^R + \lambda_{ij}^{PR} .$$

The parameter λ is called the overall effect, λ_i^P is the effect of category i of P, λ_j^R is the effect of category j of R, and λ_{ij}^{PR} is the two-variable effect of categories i and j of P and R. Note that the term 'effect' is not intended to have a causal connotation here: it simply refers to a parameter in the loglinear model and the term 'effect' is only used for convenience to avoid complicated and awkward phrases (see also Chapter 5).

The loglinear model can also be represented in its multiplicative form as a direct function of the cell frequencies or probabilities, rather than of the log cell frequencies or log probabilities:

$$\pi_{ij}^{PR} = \tau \, \tau_i^P \, \tau_j^R \, \tau_{ij}^{PR} .$$

The multiplicative parameters, denoted as τ, have nice interpretations in terms of odds and odds ratios. However, formulas and computations are simpler in their loglinear representations, and therefore we will mostly use the additive loglinear form of the model. It is, of course, easy to switch between the two representations by means of the transformation $\tau = e^\lambda$.

For the purposes of this book, a somewhat different notation than this standard notation is often needed, because in marginal analyses it is generally necessary to indicate from which marginal table a particular loglinear parameter is calculated. In this new notation, the superscripts will indicate the relevant marginal table. In the loglinear equation above, all parameters are calculated from table *PR*. Therefore, all parameters will get *PR* as their superscript. To indicate to which effect a particular symbol refers, the pertinent variable(s) will be indexed while the others get an asterisk ($*$) as their subscript. For example, parameter λ_{i*}^{PR} (which is the same as λ_i^P in traditional notation) is the effect of category i of P calculated from table *PR*. Throughout this book, we will generally use this 'marginal' notation, unless its use becomes too cumbersome. In all cases, the meaning of the notation used will be made clear or will be evident from the context.

The equation for the saturated loglinear model above now looks as follows in the marginal notation:

$$\log \pi_{ij}^{PR} = \lambda_{**}^{PR} + \lambda_{i*}^{PR} + \lambda_{*j}^{PR} + \lambda_{ij}^{PR} .$$

Without further restrictions, the λ-parameters are not identified. For example, there are already as many unknown two-variable parameters λ_{ij}^{PR} as there are known cell frequencies. One common identification method is to use *effect coding* (as in traditional ANOVA models), where for all effects the loglinear parameters sum to zero over any subscript. Letting the '+'-sign in a subscript represents summation over that subscript, e.g.

$$\lambda_{+*}^{PR} = \sum_i \lambda_{i*}^{PR} ,$$

the following identifying restrictions are imposed:

$$\lambda_{+*}^{PR} = \lambda_{*+}^{PR} = 0$$

and

$$\lambda_{i+}^{PR} = \lambda_{+j}^{PR} = 0 \quad \text{for all } i,j .$$

Using effect coding, the parameters can be computed as follows:

$$\lambda_{**}^{PR} = \frac{1}{IJ} \sum_{k=1}^{I} \sum_{l=1}^{J} \log \pi_{kl}^{PR} ,$$

$$\lambda_{i*}^{PR} = \frac{1}{IJ} \sum_{k=1}^{I} \sum_{l=1}^{J} \log \frac{\pi_{il}^{PR}}{\pi_{kl}^{PR}} ,$$

$$\lambda_{*j}^{PR} = \frac{1}{IJ} \sum_{k=1}^{I} \sum_{l=1}^{J} \log \frac{\pi_{kj}^{PR}}{\pi_{kl}^{PR}} ,$$

$$\lambda_{ij}^{PR} = \frac{1}{IJ} \sum_{k=1}^{I} \sum_{l=1}^{J} \log \frac{\pi_{ij}^{PR} \pi_{kl}^{PR}}{\pi_{il}^{PR} \pi_{kj}^{PR}} .$$

The overall effect λ_{**}^{PR} is in principle always present in a loglinear model. It is a normalizing constant that guarantees that the estimated probabilities sum to 1 or the estimated cell frequencies to sample size N. The overall effect equals the mean of the log (expected) probabilities in the table, as can be seen from the way it is computed. The one-variable effect λ_{i*}^{PR} is the mean of the log odds (or logits) in the table that have π_{il}^{PR} in the numerator. Roughly speaking, it indicates how much larger the probability is that someone belongs to $P = i$ rather than to any of the other categories of

P, on average among the religious groups. The one-variable effect λ_{*j}^{PR} is the mean of those log odds in the table that have π_{kj}^{PR} in the numerator. It is especially important in the context of this book to realize that generally the one-variable parameters do not reflect the marginal distribution of P or R, but the average conditional distribution of P and R, respectively. Restrictions on the one-variable parameters are therefore not restrictions on the one-variable marginal distributions, but on the average conditional one-variable distributions (and this extends analogously to multiway tables and multiway marginals). The one-variable effects are almost always included in a loglinear model, unless one wants to explicitly test hypotheses about the average conditional distribution of a particular variable, which is rarely the case. Finally, parameter λ_{ij}^{PR} equals the mean of the logs of the odds ratios in the table which have π_{ij}^{PR} in the numerator. The two-variable parameters reflect the sizes of the log cell probability due to the association between P and R, and indicate how much bigger or smaller a particular cell probability is than expected on the basis of the lower-order effects. The variables P and R are statistically independent of each other if and only if $\lambda_{ij}^{PR} = 0$ for all i and j.

A second common method for obtaining an identified model is *dummy coding*, where each parameter that refers to any of the reference categories of the variables is set equal to zero. Using the first category of each variable as the reference category, this method amounts to setting

$$\lambda_{1*}^{PR} = \lambda_{*1}^{PR} = 0$$

and

$$\lambda_{i1}^{PR} = \lambda_{1j}^{PR} = 0 \quad \text{for all } i, j .$$

The choice of the first category of each variable as the reference category is arbitrary, and for each variable any of its categories could, in principle, be used as the reference category.

It is important to note that the values of the parameters will differ from each other depending on the kinds of identifying restrictions chosen: effect coding yields different parameter values from dummy coding and, when using dummy coding, selecting the first category as the reference category leads to different parameter values than using the last category. However, whatever reference category is used, the substantive interpretations in terms of what goes on in the table will remain the same, provided the appropriate interpretation of the parameters is employed, taking the nature of the chosen identifying restrictions into account. The values of the odds and the odds ratios estimated under a particular model will be the same regardless of the particular identification constraints chosen. In this book, we will use effect coding unless stated otherwise. The two-variable parameter estimates for Table 2.1 are given in Table 2.2, using effect coding.

Like the vertical percentages in Table 2.1, the estimates $\hat{\lambda}$ of the saturated loglinear model presented in Table 2.2 clearly indicate that nonreligious people are more liberal than Protestants or Catholics. But it is harder to discover a clear pattern for the differences between the two religious groups, i.e., between Catholics and Protestants.

Table 2.2. *Political Orientation* and *Religion* in the United States in 1993: Estimates $\hat{\lambda}_{ij}^{PR}$ for Table 2.1. Effect coding is used; * significant at .05 level

Political Orientation (P)	Religion (R)		
	1. Protestant	2. Catholic	3. None
1. Extremely liberal	−.20	−.52	.72*
2. Liberal	−.57*	−.04	.61*
3. Slightly liberal	−.22	−.07	.29
4. Moderate	−.17	.39*	−.22
5. Slightly conservative	.11	.36*	−.47*
6. Conservative	.52*	.42*	−.93*
7. Extremely conservative	.52*	−.54	.01

A comparison of particular restricted, nonsaturated models might provide some better insights into what is going on.

Nonsaturated loglinear models are usually tested by means of two well-known test statistics: the likelihood ratio test statistic

$$G^2 = -2N \sum_i p_i \log \frac{\hat{\pi}_i}{p_i}$$

and Pearson's chi-square test statistic

$$X^2 = N \sum_i \frac{(p_i - \hat{\pi}_i)^2}{\hat{\pi}_i} .$$

If the postulated model is true, these test statistics have an asymptotic chi-square distribution. The degrees of freedom (*df*) equal the number of independent restrictions on the nonredundant loglinear parameters (often the number of nonredundant parameters that are set to zero) or, equivalently, to the number of independent constraints on the cell probabilities. In many circumstances, G^2 can be used to obtain a more powerful conditional test by testing a particular model, not (as implied above) against the saturated model, but against an alternative that is more restrictive than the saturated model (but less restrictive than the model to be tested). Given that interest lies in a model M_1 with df_1 degrees of freedom, a conditional test requires that an alternative hypothesis M_2 is considered with df_2 degrees of freedom that contains model M_1 as a special case, i.e., $M_1 \subset M_2$. The conditional test statistic is then defined as

$$G^2(M_1|M_2) = G^2(M_1) - G^2(M_2)$$

and has an asymptotic chi-square distribution with $df = df_1 - df_2$ if M_1 is true. This conditional testing procedure is valid only under the condition that the more general model M_2 is (approximately) valid in the population.

To indicate (non)saturated hierarchical loglinear models, use will be made of the standard short-hand notation. This short-hand notation can be most easily described in terms of the standard, nonmarginal notation for the λ parameters. In this short-hand notation then, a loglinear model is denoted by the superscripts of all its highest order interaction terms. Because of the hierarchical nature of the model, all lower-order effects that can be formed from these superscripts are also included in the loglinear model. For Table 2.1,the saturated model is denoted as model $\{PR\}$ and the independence model with $\lambda_{ij}^{PR} = 0$ as $\{P,R\}$.

The hypothesis of statistical independence between P and R in Table 2.1 is definitely not a viable hypothesis: likelihood ratio chi-square $G^2 = 54.9$, $df = 12$ ($p = .000$; Pearson chi-square $X^2 = 57.4$). However, a partial independence model can be formulated in which the conditional probability distributions of *Political Orientation* are the same for Protestants and Catholics, but different for the nonreligious people. In loglinear terms, this form of partial independence is identical to the restriction that the two-variable λ's are the same for Protestants and Catholics: $\lambda_{i1}^{PR} = \lambda_{j2}^{PR}$ for all $i = j$. Several programs are available for handling such restrictions, e.g., Vermunt's free software LEM (Vermunt, 1997a). The test results are $G^2 = 15.0$, $df = 6$ ($p = .024$; Pearson chi-square $X^2 = 13.8$). This is a somewhat inconclusive result, the interpretation of which strongly depends on the (arbitrarily) chosen significance level of .05 or .01. Assuming that the partial independence model is valid in the population, independence model $\{P,R\}$ can be tested conditionally against this less restricted alternative: $G^2 = 54.9 - 15.0 = 39.9$, $df = 12 - 6 = 6$, $p = .000$. The complete independence model definitely has to be rejected in favor of the partial independence model.

A more powerful investigation of what goes on in the table might be obtained by explicitly taking into account the ordered nature of variable P. There are essentially three partly overlapping ways in which we can deal explicitly with loglinear models for ordered data. If the ordered nature of the data is considered to be the result of strictly ordinal measurement, it makes sense to assume (weakly) monotonically increasing or declining relationships between the variables and impose inequality restrictions on the loglinear association parameters. If the ordered data are considered as interval-level data, fixed numerical scores can be assigned to the interval-level variables and the loglinear parameters may be linearly restricted to obtain linear relationships in the loglinear models. In the third approach, the scores for the variables are not fixed, but linear relationships are assumed to be true and the variable scores are estimated in such a way that the relationships in the loglinear model will be linear. An extensive literature on loglinear modeling of ordinal data exists (see Croon, 1990; Vermunt, 1999; Hagenaars, 2002; Clogg & Shihadeh, 1994).

By way of example, variable P in table PR may be considered as an interval-level variable with fixed scores P_i and R may be treated as nominal-level variable, resulting in an *interval by nominal* loglinear model (also called a column association model). The model has the form

$$\log \pi_{ij}^{PR} = \lambda_{**}^{PR} + \lambda_{i*}^{PR} + \lambda_{*j}^{PR} + P_i \alpha_j^R$$

in which the two-variable effect λ_{ij}^{PR} is replaced by the more parsimonious term $P_i \alpha_j^R$. In terms of ordinary regression analysis, the term α_j^R is similar to a regression coefficient: in this case, one for each category of R, and the scores P_i define the independent interval-level variable X. To maintain the identifying effect coding restrictions for the restricted λ_{ij}^{PR} effects, the scores P_i must sum to 0 (we will use the equal unit distance interval scores $-3, -2, \cdots, 2, 3$) and we need $\sum_j \alpha_j^R = 0$. This linear model implies that the log odds of belonging to religious group j rather than j' increase or decrease linearly with an increasing score on P. Or formulated the other way around, the log odds of belonging to category i of P rather than $i+1$ are systematically larger (or systematically smaller) for $R = j$ than for $R = j'$, where these log odds differences between religious groups j and j' are the same for all values of i:

$$\alpha_j^R - \alpha_{j'}^R = \log \frac{\pi_{i\,j}^{PR} / \pi_{i+1,j}^{P\ \ R}}{\pi_{i,j'}^{PR} / \pi_{i+1,j'}^{P\ \ R}}.$$

As can be seen from this formula, the odds ratio on the right-hand side has the same value for all i. The difference $\alpha_j^R - \alpha_{j'}^R$ indicates how much higher the log odds of scoring one category higher on the political orientation scale is for people of religion j than for people of religion j'. Since there are three religious denominations here, there are three relevant (and two independent) differences of this kind.

For the data in Table 2.1, the ordinal by nominal model fits well: $G^2 = 14.4$, $df = 10$ ($p = .16$, $X^2 = 15.16$). The estimates of the regression coefficients are $\hat{\alpha}_1^R = .226$, $\hat{\alpha}_2^R = .096$, and $\hat{\alpha}_3^R = -.322$, which shows that the Protestants are the most conservative and the nonreligious people are the most liberal, while the Catholics occupy an intermediate position very close to the Protestants. As explained below, for the fitted sample data, the odds of scoring one category higher on the liberal-conservative scale is just 1.14 times higher for Protestants than for Catholics, but 1.73 times higher for Protestants than for nonreligious people, and, finally, 1.52 times higher for Catholics than for nonreligious people. Coefficient 1.14 for the comparison Protestants-Catholics is computed as follows: $1.14 = \exp(.226 - .096)$; using the estimated standard errors of $\hat{\alpha}^R$ (not reported here) its 95% confidence interval (CI) equals $[1.01, 1.29]$; the coefficient for the comparison of Protestants-nonreligious is $1.73 = \exp(.226 + .322)$ and its CI equals $[1.45, 2.06]$; the coefficient for the comparison Catholic-nonreligious is $1.52 = \exp(.10 + .32)$ and its CI equals $[1.25, 1.84]$. The difference in political orientation of Catholics and Protestants is not very large and the reported confidence interval for the pertinent odds ratio almost includes the no difference value of 1. To test whether the (linear) difference between Catholics and Protestants is significant, the same interval by nominal model can be defined, but now with the extra restriction that $\alpha_1^R = \alpha_2^R$. The test outcomes for this model are $G^2 = 18.8$, $df = 11$ ($p = .07$, $X^2 = 18.84$). In this restricted interval by nominal model, $\hat{\alpha}_1^R = \hat{\alpha}_2^R = .17$ and $\hat{\alpha}_3^R = -.34$. On the basis of the unconditional test outcome against the alternative hypothesis that the saturated model is true ($p = .07$), one might decide to accept the restricted model and conclude that Protestants and Catholics have the same political orientation. The more powerful conditional test for the thus restricted model against the alternative hypothesis that the original interval by nominal model holds yields $G^2 = 18.8 - 14.4 = 4.4$, $df = 11 - 10 = 1$ ($p = .04$).

Table 2.3. *Political Orientation, Religion,* and *Opinion on teenage birth control* in the United States in 1993 (Source: General Social Survey 1993)

		Opinion on Teenage Birth Control (B)											
		1. Strongly agree			2. Agree			3. Disagree			4. Strongly disagree		
	Religion (R)	1	2	3	1	2	3	1	2	3	1	2	3
Pol. or. (P) 1		5	1	3	4	0	0	0	0	1	2	1	0
2		18	6	10	15	6	10	9	6	3	7	3	0
3		24	7	7	29	11	7	18	5	4	8	0	1
4		61	31	13	69	30	7	54	20	4	36	15	6
5		19	11	5	32	11	3	37	8	0	24	6	1
6		13	6	2	31	8	1	32	6	0	43	7	1
7		5	0	1	5	1	0	4	0	1	9	1	0

At the 5% significance level, the unrestricted interval by nominal model is accepted but its restricted version is rejected.

From all these test outcomes, it can be clearly concluded that first, there is no reason to reject the linear nature of the relationships in the interval by nominal model; second, that nonreligious people are definitely more liberal than Catholics or Protestants, and third, that the differences in political orientation between Catholics and Protestants are small. For the time being, it may be accepted that Catholics are slightly more liberal than Protestants but new data are needed to conform this outcome. Suspension of judgement is the best option here (Hays, 1994, p. 281).

2.1.2 Modeling Association Among Three Variables

Basic loglinear modeling for two-way tables can easily be extended to tables of much higher dimensions. As a simple example, the data in three-way Table 2.3 will be used to investigate how the two variables dealt with so far, *Religion* (R) and *Political Orientation* (P), affect opinion on teenage *Birth Control* (B). Variable B has $K = 4$ categories, ranging from strongly agree to strongly disagree.

For the three-dimensional table *PRB*, saturated loglinear model {*PRB*} decomposes the log probabilities as follows:

$$\log \pi_{i\,jk}^{PRB} = \lambda_{***}^{PRB} + \lambda_{i**}^{PRB} + \lambda_{*j*}^{PRB} + \lambda_{**k}^{PRB} + \lambda_{i\,j*}^{PRB} + \lambda_{i*k}^{PRB} + \lambda_{*jk}^{PRB} + \lambda_{i\,jk}^{PRB}.$$

Note that here the superscripts of the λ parameters are all *PRB*, indicating that the parameters refer to table *PRB* rather than to table *PR* from the previous subsection. Where the highest order effect parameter in the previous subsection was a two-variable effect, now we have also a three-variable parameter $\lambda_{i\,j\,k}^{PRB}$ that indicates to what extent the conditional associations between any of the two variables vary among the categories of the third variable.

The effect coding identifying restrictions are

$$\lambda_{+**}^{PRB} = \lambda_{*+*}^{PRB} = \lambda_{**+}^{PRB} = 0$$
$$\lambda_{i+*}^{PRB} = \lambda_{+j*}^{PRB} = \lambda_{i*+}^{PRB} = \lambda_{+*k}^{PRB} = \lambda_{*j+}^{PRB} = \lambda_{*+k}^{PRB} = 0$$
$$\lambda_{ij+}^{PRB} = \lambda_{i+k}^{PRB} = \lambda_{+jk}^{PRB} = 0$$

for all i, j, and k. Because there are sampling zeroes in the observed table PRB, the sample values of the loglinear parameters of the saturated model are either plus or minus infinity or undefined.

In general, as indicated above, the loglinear effects pertaining to the same variables are different when calculated in different (marginal) tables; even their signs may be different. For the data in Table 2.3, we have $\hat{\lambda}_1^P = -1.65$, $\hat{\lambda}_{1*}^{PR} = -1.50$, and $\hat{\lambda}_{1**}^{PRB} = -\infty$ (minus infinity). All three parameters pertain to the distribution of P and represent the effect of the first category of P, but are calculated in the marginal tables P, PR, and the full table PRB, respectively (and assuming saturated models for the pertinent tables). Parameter λ_1^P reflects the cell size of $P = 1$ in the univariate marginal distribution of P; parameter λ_{1*}^{PR} indicates the cell size of $P = 1$ on average in the J conditional distributions of P in table PR; and parameter λ_{1**}^{PRB} mirrors the average cell size $P = 1$ in the $J \times K$ conditional distributions of P in table PRB. The two-variable parameters for table PRB are now partial coefficients indicating the direct relationship between two variables on average within the categories of the third variable, in this way controlling for the third variable.

Table 2.4 contains the test outcomes of a few relevant hierarchical loglinear models for Table 2.3 concerning the influence of P and R on B. The models in Table 2.3 can also be seen as logit models for the effects of P and R on B (Agresti, 2002, Section 8.5). The models are again represented in the usual short-hand notation by means of which hierarchical models are indicated by their highest order interactions, implying the presence of all pertinent lower order effects. The second column in Table 2.3 gives an interpretation of the model in terms of (conditional) independence relations ($\perp\!\!\!\perp$) or the absence of interaction terms. The final four columns summarize the results of the testing procedures.

In the last row of Table 2.4, the results of the no three-factor interaction model are given. The no three-factor interaction model has the form

$$\log \pi_{ijk}^{PRB} = \lambda_{***}^{PRB} + \lambda_{i**}^{PRB} + \lambda_{*j*}^{PRB} + \lambda_{**k}^{PRB} + \lambda_{ij+}^{PRB} + \lambda_{i*k}^{PRB} + \lambda_{*jk}^{PRB} .$$

As can be seen in Table 2.4, this model $\{PR, PB, RB\}$ fits the data well. However, the estimated (and observed) table is sparse, which may invalidate the approximation of the chi-square distribution. It is not certain whether the reported p-value for the model is correct. One may become more confident that the model can be accepted by observing that the value of Pearson's chi-square statistic is 35.2, which is not too different from G^2. The no three-factor interaction model will be accepted here and used as an alternative hypothesis for testing more parsimonious models: especially with sparse tables, conditional tests more readily approximate the theoretical chi-square distribution and are in many circumstances more powerful than unconditional tests.

Table 2.4. Goodness of fit of various hierarchical loglinear models for Table 2.3

Model	Interpretation	G^2	df	p-value	X^2
1. $\{PR, B\}$	$PR \perp B$	120.2	60	.000	105.3
2. $\{PR, RB\}$	$P \perp B\|R$	95.6	54	.000	84.7
3. $\{PR, PB\}$	$R \perp B\|P$	53.7	42	.107	46.8
4. $\{PR, PB, RB\}$	No 3-factor interaction	39.2	36	.328	35.2

The model in which neither P nor R have an effect on B does not fit the data (model 1 in Table 2.3). The same is true for conditional independence model 2 in which P has no direct influence on B, but R has. At first sight, conditional independence model 3, in which *Religion* has no direct effect on opinion of teenage *Birth Control*, provides an acceptable fit to the data ($p = .107$). However, testing this model against the no three-factor interaction model yields $G^2 = 14.7$ with $df = 6$ ($p = .025$). This conditional test has more power to detect the (possibly small) effects of R on B in the population than the corresponding unconditional test. Although the conclusion regarding $p = .025$ again strongly depends on the chosen significance level .01 or .05, we will proceed cautiously and at least for the time being accept the possibility of (small) effects of R on B in the population. Model 3 will be rejected in favor of the no three-factor interaction model 4.

In the no three-factor interaction model 4, the conditional association between P and B given R is described by the 28 parameters λ_{i*k}^{PRB} and the conditional association between R and B given P is described by the 12 parameters λ_{*jk}^{PRB}. A simpler description of the models might be obtained by taking the ordered character of variables P and B into account. More precisely, variable B will be treated as an interval-level variable with scores $B_k = -1.5, -.5, +.5, +1.5$ and also variable P will be considered (as before) as an interval-level variable with scores $P_i = -3, -2, \cdots, +2, +3$. Further, the following restrictions will be applied:

$$\lambda_{i*k}^{PRB} = \vartheta P_i B_k$$
$$\lambda_{*jk}^{PRB} = \rho \gamma_j^R B_k .$$

These restrictions are similar to the ones used above in the interval by nominal model for Table *PR*, but a slightly different notation than above is employed to indicate somewhat different aspects of models for ordered data. The resulting loglinear model has the form

$$\log \pi_{ijk}^{PRB} = \lambda_{***}^{PRB} + \lambda_{i**}^{PRB} + \lambda_{*j*}^{PRB} + \lambda_{**k}^{PRB} + \lambda_{ij*}^{PRB} + \vartheta P_i B_k + \rho \gamma_i^R B_k . \quad (2.1)$$

The parameter ϑ is a kind of regression coefficient for the linear effect of P on B, and $\rho \gamma_j^R$ is the regression coefficient for the linear relationship between R and B, one for each category of R. One might also say that ρ is the regression coefficient and consider γ_j^R as scores to be estimated for R, given a linear relationship between R and B. In order to guarantee model identification, and more specifically to guarantee identification of the product $\rho \gamma_j^R$, the additional constraint $\sum_j \gamma_j^R = 0$ is imposed.

If ρ and γ_j^R have to be identified separately, which is not necessary here for our purposes, the variance of the estimated scores γ_j^R has to be fixed, e.g., by means of the restriction $\sum_j (\gamma_j^R)^2 = 1$.

According to this model, the direct relationship between P and B is linear in the sense that the log odds of choosing category k of B rather than k' increase (or decrease) linearly with increasing values for P, or vice versa, but less appropriate here given the assumed 'causal' order of the variables: the log odds of choosing category i of P rather than i' increase (or decrease) linearly with increasing values for B. The relation between R and B is linearly restricted in the sense of the interval by nominal model discussed in the previous section: for two religions j and j', there is a linear relationship between *Religion* and the opinion about teenage *Birth Control*. One way to clarify the meanings of the effects of P and R on B, i.e., of the consequences of having different scores on P or R for the scores on B is the following:

$$\vartheta(P_i - P_{i'}) = \log \frac{\pi_{i\,j\,k}^{PRB} / \pi_{i\,j\,k+1}^{PRB}}{\pi_{i'\,j\,k}^{PRB} / \pi_{i'\,j\,k+1}^{PRB}} \tag{2.2}$$

$$\rho(\gamma_j^R - \gamma_{j'}^R) = \log \frac{\pi_{i\,j\,k}^{PRB} / \pi_{i\,j\,k+1}^{PRB}}{\pi_{i\,j'k}^{PRB} / \pi_{i\,j'k+1}^{PRB}} \;. \tag{2.3}$$

The odds ratio on the right-hand side of (2.2) is the conditional odds ratio indicating the direct relationship between P and B for $R = j$. It turns out to be the same for all values of j, a necessary consequence of the no three-variable-interaction model in (2.1). Further, the conditional odds ratio on the right-hand side of (2.3) indicating the direct relationship between R and B for $P = i$ is the same for all values of i. The left-hand side element $\rho(\gamma_j^R - \gamma_{j'}^R)$ shows how much higher the log odds of scoring one category higher on B is for people in category j of R than for people in category j' of R, conditionally on P. The left-hand side element $\vartheta(P_i - P_{i'})$ indicates how much higher the log odds of scoring one category higher on B is for people in category i of P than for people in category i' of P, conditionally on R. The linearly restricted model for the direct relation between P and B is called an 'interval by interval' or a 'linear by linear' model, and also a 'uniform association model', because all local partial odds ratios for the direct relation between P and B are the same, their logarithm being ϑ. In terms of the original lambda parameters,

$$\vartheta = \lambda_{i\,*k}^{PRB} - \lambda_{i+1\,*\,k}^{P\,R\,B} - \lambda_{i\,*\,k+1}^{PRB} + \lambda_{i+1\,*\,k+1}^{P\,R\,B},$$

for all values of i and k.

Testing model (2.1) yields $G^2 = 57.7$ with $df = 57$ ($p = .449$, $X^2 = 52.7$); testing it against the no three-factor interaction model, it is found $G^2 = 57.7 - 39.2 = 18.5$ with $df = 57 - 42 = 21$ ($p = .617$). There is no reason to reject this very parsimonious model for describing the association structure between the three variables. The relevant estimated effects are

$$\hat{\vartheta} = .155,$$

$$\widehat{\rho\gamma_1^R} = .182,$$

$$\widehat{\rho\gamma_2^R} = .026,$$
$$\widehat{\rho\gamma_3^R} = -.208.$$

The estimated direct linear effect of *Political Orientation* on the opinion on teenage *Birth Control* is significant (estimated standard errors not reported here) and in the expected direction: the more conservative one is, the more one is opposed to teenage birth control. The (significant) direct effects of *Religion* indicate that nonreligious people are less opposed to birth control than the Protestants with the Catholics in an intermediate position.

The effects of *Political Orientation* on the opinion on teenage *Birth Control* are much stronger than the effects of *Religion*. One way to see this clearly is to estimate the maximum effects for the variables, that is, the largest (log) odds ratios that can be obtained in the pertinent tables. Because of the linear relationship and the number of categories of the variables, the log of the maximum odds ratio for the effect of P on B turns out to be $6 \times 3 \times \hat{\vartheta} = \hat{\lambda}_{1*1}^{PRB} - \hat{\lambda}_{1*4}^{PRB} - \hat{\lambda}_{7*1}^{PRB} + \hat{\lambda}_{7*4}^{PRB} = 6 \times 3 \times .1548 = 2.786$. The corresponding maximum odds ratio equals $\exp(2.786) = 16.22$. Similar computations lead to a maximum (log) effect of R on B of $1.169 (= 3 \times (.182 - (-.208))$ and a corresponding odds ratio of 3.220: the effect of P on B is five times stronger than the effect of R on B.

2.2 Applications of Loglinear Marginal Models

The loglinear models applied in the previous section to analyze a joint probability distribution can also be employed for jointly analyzing two or more marginal distributions. In this section, several concrete research problems and designs will be discussed that require marginal-modeling methods, and for which loglinear marginal models are very useful to answer the pertinent research questions. Real-world examples and data will be used to illustrate these kinds of research questions and the ways they can be translated into the language of loglinear modeling. Maximum likelihood estimates of the parameters and significance tests for these examples will be given, along with their substantive explanation. In the last section, a general algorithm will be presented to obtain maximum likelihood estimates for loglinear marginal models. The contents of some parts of this section will be more demanding from a statistical point of view and are indicated by ***.

2.2.1 Research Questions and Designs Requiring Marginal Models

As discussed before, marginal modeling is about the simultaneous analyses of marginal distributions where the different marginal distributions involve dependent observations, but where the researcher is in principle not interested in the nature of the dependencies. As the remainder of this book will show, this is a research situation that actually occurs a lot in practice. For a simple concrete example, let us turn to the analysis of family data. Family data are of interest for social scientists studying such diverse topics as social mobility, political change, changing family relations or the societal role of generational differences. In this respect, social scientists want to compare family members, wives and husbands, children and parents, and sisters and brothers regarding their political preferences, social and occupational status, education, religious beliefs, etc. These comparisons usually involve comparing dependent marginal tables, not only one-way, but also higher-way marginal tables that involve dependent observations. For example, clustered family data are needed to answer concrete research questions such as

- Are the relative direct influences of religion and social class on political preference the same for the children and their parents?
- Is the agreement in attitudes between fathers and sons of the same size and nature as between mothers and their daughters, and is the agreement less in pairs of opposite sex, i.e., between fathers and their daughters or between mothers and their sons?
- Are sisters more like each other than sisters and brothers?

Standard analysis techniques that ignore the dependencies in the data, i.e., ignore the hierarchical or clustered nature of the data are not appropriate here. Especially if an investigator wants answers to these research questions without at the same time wanting to make assumptions about the nature of the dependencies in the data, marginal-modeling methods provide an excellent way to analyze the family data reckoning with the fact that the observations are dependent.

Marginal modeling is also needed to answer particular kinds of research questions that make use of data that are seemingly not clustered. This happens, for instance, when a political scientist has conducted a one-shot cross-sectional survey based on simple random sampling, in which respondents are asked to state their degree of sympathy for different political parties on a five point scale. To answer the question of whether the distributions of the sympathy scores are the same for all political parties, standard chi-square tests cannot be used because the comparisons of the several one-variable distributions pertain to the same respondents. The data are actually clustered within individuals given the research question of interest, despite the cross-sectional design and the simple random sampling scheme.

Many other research questions in similar situations require marginal modeling. The same political scientist may also want to measure political interest by means of several items in the form of seven-point rating scales. These items are supposed to form a summated (Likert) scale. In its strictest form, it is assumed in Likert scaling that the items are parallel measurements of the same underlying construct, having independently distributed error terms with identical error variances (for a more precise technical definition of parallel measurements, see Lord & Novick, 1968). This strict measurement model implies that all marginal distributions of all items are the same, as are all pairwise associations. Again, marginal modeling is needed to test such implications.

In the following two subsections of this chapter, empirical illustrations will be provided for exactly the above kinds of research topics, showing how to translate these questions into the language of loglinear modeling. In the next chapter, the same data and general research questions will be used but then formulated in terms of nonloglinear marginal models.

2.2.2 Comparing One Variable Distributions

Comparing One Variable Distributions in the Whole Population

To gain a practical insight into the nature of marginal modeling, the best starting point is the comparison of a number of simple one-way marginals. Our example concerns a study into the way people perceive their body. A group of 301 university students (204 women and 97 men) answered questions about their degrees of satisfaction with different parts or aspects of their body by completing the *Body Esteem Scale* (Franzoi & Shields, 1984; Bekker, Croon, & Vermaas, 2002). This scale consisted of 22 items (not counting the items concerning gender-specific body parts), seven of which will be considered here. These seven items loaded highest on the first unrotated principal component, with loadings higher than .70. Principal component analysis was used to discover whether the separate expressions of satisfaction with the different body aspects can be seen as just an expression of the general underlying satisfaction with the body as a whole or whether more underlying dimensions are needed (for the interested reader: two rotated factors were needed to explain the correlations among all the 22 items, one having to do with the general appearance of the body and the other with the satisfaction with the parts of one's face; the items chosen here all belong to the first factor). Such dimensional analyses tell the researcher how strongly satisfaction on one particular item goes together with satisfaction with other parts or aspects of the body. However, even if a correlation between two particular items is positive (and it turned out that the correlations among all 22 items had positive values), it does not follow automatically that the respondents react in the same way to these items in all respects. Despite the positive correlation, people may be much more satisfied on average with one part of their body than with another, or the disagreement among the respondents regarding the satisfaction with one bodily aspect may be much larger than the disagreement with another one. Such differences may be important and may reveal, in addition to correlational analyses, relevant details about how the body is perceived. To investigate these kinds of differences, one must compare the overall reactions of the respondents to the different body parts, in other words, compare the marginal distributions. For the selected seven items, the marginal distributions are shown in Table 2.5 (the complete seven-way table can be found on the website mentioned in the last chapter). The response categories are 1 = very dissatisfied; 2 = moderately dissatisfied; 3 = slightly satisfied; 4 = moderately satisfied; and 5 = very satisfied.

The items in Table 2.5 will be denoted by I_i: I_1 refers to item Thighs, I_2 to Build, etc. To see whether the seven response distributions differ significantly from each other, one could start from the marginal homogeneity model (MH) that states that the marginal distributions of the variables I_i are the same for all i, i.e.,

$$\pi_i^{I_1 I_2 I_3 I_4 I_5 I_6 I_7} {}_{++++++} = \pi_{+i}^{I_1 I_2 I_3 I_4 I_5 I_6 I_7} {}_{+++++} = \cdots = \pi_{+++++i\,+}^{I_1 I_2 I_3 I_4 I_5 I_6 I_7} = \pi_{++++++i}^{I_1 I_2 I_3 I_4 I_5 I_6 I_7}$$

Table 2.5. *Body Esteem* Scales

	1. Thighs	2. Body Build	3. Buttocks	4. Hips	5. Legs	6. Figure	7. Weight	MH
1	22	10	22	22	18	15	20	21.12
2	67	45	59	51	57	45	48	52.08
3	79	78	93	88	79	70	74	76.32
4	105	127	95	111	110	140	104	110.56
5	28	41	32	29	37	31	55	40.91
Total	301	301	301	301	301	301	301	301.00
Mean	3.17	3.48	3.19	3.25	3.30	3.42	3.42	3.33
SD	1.099	1.010	1.093	1.075	1.093	1.024	1.152	1.124

Notes: See text for explanation. Source: Franzoi and Shields, 1984

for all i. When it does not cause any misunderstandings, a shorter and simpler notation will be used, omitting the variables over which the multidimensional table is marginalized:

$$\pi_i^{I_1} = \pi_i^{I_2} = \pi_i^{I_3} = \pi_i^{I_4} = \pi_i^{I_5} = \pi_i^{I_6} = \pi_i^{I_7}.$$

To test this MH model, application of the standard chi-square test to Table 2.5 is not appropriate since it is not an ordinary two-way contingency table. The columns of table *SI* (*Satisfaction* × *Item*) contain the marginal score distributions for the seven items that are all based on the same sample of respondents. Nevertheless, we will often refer to such a table as an *SI* table for the sake of obtaining a simpler notation and a much simpler indication of relevant models. The MH model for Table 2.5 must be fitted using the marginal-modeling methods that are explained in the following section (and implemented in the programs described in the last chapter). The MH model turns out to fit badly ($G^2 = 55.76$, $df = 24$, $p = .000$, $X^2 = 45.42$) and it must be concluded that the seven marginal item distributions are not identical.

To investigate the differences among the marginal distributions, several paths may be followed. First, adjusted or standardized residual frequencies can be calculated comparing the observed and expected frequencies. The expected frequency distribution for all items under marginal homogeneity is also reported in Table 2.5 in the last column labeled MH. Note that the estimated frequencies under MH are not simply the average of the seven column frequencies. Among other things, the adjusted residual frequencies (not reported here) clearly indicate that more people are dissatisfied with their Thighs and Buttocks than estimated under the MH model, while more people are satisfied with their general Build than expected under MH. None of the residuals for Legs were significant.

Another way of investigating the differences among the marginal item distributions is to compare certain aspects of the marginal item distributions, e.g., the means or standard deviations. These two characteristics are reported in the last two rows

Table 2.6. *Body Esteem* Scales. Loglinear parameters $\hat{\lambda}_{ij}^{SI}$ and their standard errors for data in Table 2.5

	1. Thighs	2. Build	3. Buttocks	4. Hips	5. Legs	6. Figure	7. Weight
1	.181 (.123)	-.482 (.167)	.167 (.109)	.196 (.102)	-.013 (.121)	-.010 (.120)	.051 (.141)
2	.213 (.090)	-.060 (.109)	.072 (.095)	-.045 (.092)	.059 (.096)	-.083 (.102)	-.155 (.107)
3	-.039 (.085)	.074 (.090)	.110 (.085)	.084 (.078)	-.032 (.086)	-.058 (.088)	-.139 (.099)
4	-.096 (.069)	.220 (.070)	-.210 (.073)	-.026 (.061)	-.042 (.070)	.294 (.067)	-.140 (.079)
5	-.258 (.100)	.249 (.092)	-.139 (.098)	-.208 (.096)	.028 (.093)	-.054 (.100)	.383 (.098)

of Table 2.5. However, their comparisons involve nonloglinear marginal models and will be dealt with in the next chapter.

Finally, within the loglinear context, the differences among the marginal item distributions can be described by means of the loglinear parameters. Essentially, the two-variable parameters and their standard errors in saturated model $\{SI\}$ applied to Table 2.5 are estimated, but then in the correct way by taking the dependencies among the observations into account by means of marginal model estimation procedures (see also below). The results in terms of the two-variable parameters $\hat{\lambda}_{ij}^{SI}$ are reported in Table 2.6, with their respective standard errors between parentheses.

The $\hat{\lambda}_{ij}^{SI}$ estimates provide a detailed description of the differences among the marginal distributions. In agreement with the adjusted residuals, they show that there is a relative overrepresentation of dissatisfied people for Thighs and Buttocks (and a corresponding underrepresentation of satisfied people); the opposite tendency is noted for Build. Satisfaction with Legs resembles the average (log)distribution the most. Relative outlying cell frequencies can be seen for Figure and Weight, where the categories *moderately satisfied* and *very satisfied*, respectively, are comparatively strongly overrepresented.

To gain further insight into the properties of marginal modeling, it is useful to compare these results with analyses directly applied to the data in Table 2.5, incorrectly ignoring the dependencies among the column distributions. Table 2.5 is then treated as if it were a normal SI two-way table and the equality of the item distributions is tested by applying (inappropriately) the standard chi-square test for independence (model $\{S,I\}$) directly to this table. The test results are $G^2 = 45.50$ for $df = 24$ ($p = .005, X^2 = 46.41$). The test statistic G^2 for the independence hypothesis is smaller than G^2 for MH obtained above which had the value 55.77 and the same degrees of freedom. This was to be expected. The item distributions in Table 2.5 are the result of repeated measurements of the same 301 respondents. As is well-known for repeated measurements, if the dependencies among the repeated measurements are positive, then in general, the standard errors of the estimates will be smaller than for independent observations and consequently test statistics will be larger (Hagenaars, 1990, p. 205-210). For negative dependencies, the situation will be reversed: larger standard errors and smaller test statistics. Because the body items here all correlate positively, the MH test will have more power to detect the differences in the population than the (inappropriately applied) independence test.

As implicitly indicated above, it is important to note explicitly that the two sets of expected frequencies estimated using maximum likelihood methods under the independence hypothesis (as defined above, treating the data incorrectly as coming from independent observations) and under MH will generally not be the same. The estimated frequencies for each item distribution under MH in Table 2.5, Column MH result in the following proportions for the satisfaction categories 1 through 5: .070, .173, .254, .367, and .136. However, the corresponding estimated proportions stemming from the (inappropriate) application of model $\{S, I\}$ directly to Table 2.5 are: .061, .177, .266, .376, and .120. The latter distribution simply results from adding all cell frequencies in a particular row of Table 2.5 and dividing it by the total number of observations (here: 7×301). In other words, the inappropriate independence model $\{S, I\}$ reproduces the observed marginal distribution of S since the marginal distribution of S is a sufficient statistic for model $\{S, I\}$ when the observations are independent. However, this is not true when independence in Table 2.5 is tested in the correct way by means of the model. Note that the MH model has no simple sufficient statistics like the ordinary nonmarginal independence model has. Ignoring the dependencies between the observations for the different items may not only distort the chi-square values, but also the estimates of the item distributions. Finally, regarding the application of saturated model $\{SI\}$ directly to Table 2.5, it does not matter whether we compute the $\hat{\lambda}_{ij}^{SI}$-estimates just from Table 2.5 or use marginal-modeling procedures. The two methods yield the same values for the $\hat{\lambda}_{ij}^{SI}$ estimates, that describe the observed differences between the item distributions. However, the estimated standard errors of the $\hat{\lambda}_{ij}^{SI}$-estimates are different for the two procedures. As expected, the inappropriate estimates, obtained directly from Table 2.5 assuming independent observations, are all larger than the ones obtained from the correct marginal-modeling procedures. Most differences are within the range .02–.04, which makes a difference for the significance level of several $\hat{\lambda}_{ij}^{SI}$ estimates.

Subgroup Comparisons of One Variable Distributions

The satisfaction with body parts is not only known for the whole sample, but also separately for men and women. As conventional (and scientifically based) wisdom holds, women perceive their body in ways different from men. The observed marginal distributions of satisfaction with the body parts and aspects are shown separately for men and women in Table 2.7. What strikes one immediately in Table 2.7 is that men seem to be more satisfied with their body than women. These and other differences between men and women will be investigated using loglinear marginal models. Most of these models will be indicated by referring to Table 2.7 as if it were a normal table *GSI* (*Gender* × *Satisfaction* × *Item*) without repeating every time that this table is not a normal table and that the models must be tested and its parameters estimated by means of the appropriate marginal-modeling procedures.

As seen above, the MH model had to be rejected for the total group. But perhaps this is because one subgroup (maybe women) expresses different degrees of satisfaction with different body parts, while the other group (maybe men) is equally (dis)satisfied with all body parts. Testing the MH hypothesis among the men yields:

Table 2.7. *Body Esteem* Scales for Men and Women. See also Table 2.5

Men

	1. Thighs	2. Body Build	3. Buttocks	4. Hips	5. Legs	6. Figure	7. Weight
1	1	3	2	2	1	3	4
2	8	9	9	8	6	11	13
3	20	26	25	24	18	18	15
4	47	40	42	48	49	50	35
5	21	19	19	15	23	15	30
Total	97	97	97	97	97	97	97
Mean	3.81	3.65	3.69	3.68	3.90	3.65	3.76
SD	.901	.995	.956	.903	.867	.974	1.064

Women

	1. Thighs	2. Body Build	3. Buttocks	4. Hips	5. Legs	6. Figure	7. Weight
1	21	7	20	20	17	12	16
2	59	36	50	43	51	34	35
3	59	52	68	64	61	52	59
4	58	87	53	63	61	90	69
5	7	22	13	14	14	16	25
Total	204	204	204	204	204	204	204
Mean	2.86	3.40	2.95	3.04	3.02	3.31	3.25
SD	1.050	1.007	1.072	1.088	1.075	1.029	1.117

$G^2 = 29.41$, $df = 24$ ($p = .205$, $X^2 = 20.66$); for women, the results are $G^2 = 59.95$, $df = 24$ ($p = .000$, $X^2 = 44.33$). Testing the overall hypothesis that there is MH in both (independently observed) subgroups is possible by simply summing the test statistics: $G^2 = 29.41 + 59.95 = 89.35$, $df = 24 + 24 = 48$ ($p = .000$). This overall hypothesis can also be indicated as conditional independence between S and I in table GSI, i.e., model $\{GS, GI\}$ for table GSI. The test results point out that it is clearly not true that there is MH in both subgroups. Looking at the separate subgroup tests, one might be inclined to conclude that there is definitely no MH among women, but that MH might be accepted for men. Note, however, that there are 204 women but only 97 men. If we had observed the proportional data in a sample of 204 men instead of only 97, given these results the expected value of G^2 would have been $29.41 \times \frac{204}{97} = 61.85$, which is about the same value as obtained for the women. In other words, the number of men is too small and the test does not have enough power to definitely draw the conclusion that MH is true for men but not for women.

In light of this, the observed frequencies in Table 2.7 might be taken for granted as the best guesses of the population values, and they might be parameterized by means of the parameters of saturated model $\{GSI\}$ for table GSI. These parameters can be used to describe how much more men are (dis)satisfied with their body (parts) than women and how this difference between men and women varies among the items. It can also be formulated the other way around: how much more satisfied the respondents are with particular body parts than with others and how these item differences vary between men and women.

But before carrying out such detailed descriptions, other models might be considered that are more parsimonious than the saturated model, but less parsimonious than MH for both subgroups. The no three-variable interaction model $\{GS, GI, SI\}$ for table GSI is a first interesting candidate. According to this model the items may have different satisfaction distributions, and men may be more or less satisfied than women, but the item differences are the same for men and women and the gender differences are the same for all items. Model $\{GS, GI, SI\}$ then implies that the odds ratios in subtable SI for men are equal to the corresponding odds ratios in subtable SI for women. In other words, the corresponding conditional loglinear parameters for the association between S and I in the subgroups Men and Women are all equal: $\lambda_{ijm}^{SI|G} = \lambda_{ijw}^{SI|G}$ in which the conditional loglinear parameters (Hagenaars, 1990, p. 43-44) are defined as

$$\lambda_{ijk}^{SI|G} = \lambda_{ij*}^{SIG} + \lambda_{ijk}^{SIG} .$$

Formulated from the viewpoint of the relationship between G and S, the no three-variable interaction model similarly implies (with obvious notation): $\lambda_{kij}^{GS|I} = \lambda_{kij'}^{GS|I}$.

Model $\{GS, GI, SI\}$ will be treated here as a logit model for the investigation of the effects of G and I on S, conditioning on the observed distribution of GI. The test results for model $\{GS, GI, SI\}$ (using the correct marginal-modeling estimation procedures) are: $G^2 = 40.20$, $df = 24$ ($p = .020$, $X^2 = 28.66$). Where the previous analyses for testing MH among men and women may have led to the conclusion that men are equally (dis)satisfied with their different body parts, while women react differently to different bodily aspects, acceptance of model $\{GS, GI, SI\}$ would imply that both men and women are differently satisfied with different body parts but that these item differences are the same for both subgroups. It is, however, not clear what to do: reject or accept model $\{GS, GI, SI\}$ given the p-value and the rather large discrepancy between G^2 and X^2. The parameter estimates in model $\{GS, GI, SI\}$ (not reported here) suggest a linear relationship between Gender and Satisfaction and an *interval by nominal* association between Satisfaction and Item. Degrees of freedom might be gained from imposing these linear restrictions obtaining a more parsimonious no three-variable interaction model. The test outcomes for such a linearly restricted model are $G^2 = 59.88, df = 45$ ($p = .068$, $X^2 = 43.50$). However, one should be careful applying such data dredging procedures, certainly in the light of the p-values obtained and the discrepancies between test statistics G^2 and X^2. Actually, more data are needed to arrive at firm conclusions.

The adjusted residuals of model $\{GS, GI, SI\}$ or, for that matter, the parameter estimates of the saturated model $\{GSI\}$, give interesting clues as to why model $\{GS, GI, SI\}$ fails to fit unequivocally. They show that the biggest discrepancies regarding men's and women's satisfaction are for items Thighs and Build. On average, men are much more satisfied with all their body parts than women. The parameter estimates for the average two-variable relation GS in model $\{SGI\}$ for all items go from $\hat{\lambda}_{m1*}^{GSI} = -.529$ for men in the very dissatisfied category almost linearly to $\hat{\lambda}_{m5*}^{GSI} = .623$

for men in the very satisfied category; the corresponding estimates for women are .529, and $-.623$ (remembering effect coding is being used). From this, the (extreme) odds ratio for the average relation between G and S among the body items can be computed: $\exp(-.529-.623-.623-.529) = \exp(-2.304) = .100$. The odds that a man is very satisfied with a body part rather than very dissatisfied is ten times $(1/.100)$ higher than the corresponding odds for a woman. The three-variable interaction parameter estimates indicate that this average difference between men and women is even very much (and statistically significantly) stronger for Thighs (inverse extreme odds ratio $1/.0159 = 63.02$), while it is very much (and statistically significantly) weaker for Build (inverse extreme odds ratio $1/.532 = 1.88$). Regarding these men-women differences, Buttocks and Hips follow Thighs, although the men-women differences are much smaller for Buttocks and Hips than for Thighs, while Legs, Figure and Weight follow Build, but also with smaller differences.

Finally, as in the analyses for the whole group, it is seen here again that the parameter estimates for the nonsaturated loglinear models when (inappropriately) estimated directly from Table 2.7 differ from the correct estimates, taking the dependencies of the observations into account. Further, the standard errors of the estimates of the loglinear parameters are all smaller when computed in the correct way and, consequently, the test statistics larger. Finally, the observed marginal proportions that are exactly reproduced in the inappropriate analyses assuming independence of the observations are not exactly reproduced by the marginal-modeling procedures.

These consequences and tendencies have been found many times in later analyses and will therefore not always be reported anymore; attention will mainly be paid to exceptions or special cases. One must keep in mind that in the relative simple cases such as discussed here, standard errors will be larger and test statistics smaller, when the correlations between the dependent observations are negative or the data in the joint table lie mainly outside the main diagonal (Hagenaars, 1990, p. 208-209). However, in situations with more complex dependencies patterns, the consequences of incorrectly assuming independent observations may not be this simple (see also Berger, 1985; Verbeek, 1991).

2.2.3 More Complex Designs and Research Questions

Complex Dependency Patterns

The analyses of the previous subsection dealt with data coming from a simple random sampling design and pertained to substantive research questions regarding the comparison of simple (conditional) one-way marginal tables. How to handle more complicated research designs and more complex marginal tables will be illustrated in this subsection. The data that will be used as an example come from the Netherlands Kinship Panel Study (NKPS), a unique in-depth large-scale study into (changing) kinship relationships covering a large number of life domains (Dykstra et al., 2004). NKPS contains several different modules with different modes of data collection. The data used here come from a module in which essentially a random sample from the Dutch population above 18 years old has been interviewed, as well as one randomly chosen parent of each respondent. Because of selective nonresponse, there are many more women than men in the sample. This will be ignored here, but a possible approach for dealing with nonresponse in marginal models will be briefly discussed in the last chapter. We will also ignore the fact that within the module we use, just one child and one parent is selected from each family, regardless of the size of the (nuclear) family. To get a representative sample of individual family members, weights must be applied to correct for the smaller chances of children from large families to get included in the sample. We will ignore this issue in order to not complicate things further, but discuss it very briefly in the last chapter. Further details on the study's design, fieldwork and nonresponse are provided on the NKPS website, *www.nkps.nl*.

Among many other things, family members were questioned about their traditional sex role attitude. A scale was constructed from these questions with (mean) scores running from 0 to 4. Here, an index will be used with three categories:

- 1 = less traditional attitude (more in favor of an egalitarian division of tasks between men and women), corresponding with original scale scores between 0 and 1
- 2 = moderately traditional, corresponding to original scale scores between 1 and 2
- 3 = traditional attitude (in favor of a traditional division of labor, such as women taking care of the house and the children, but men working outside the house earning the household's income, etc.), corresponding with original mean scale scores between 2 and 4.

The data for 1,884 families (parent-child pairs) are shown in Table 2.8. There are four variables in this table: variable P represents *Sex of the parent* (1 = father, 2 = mother), variable C *Sex of the child* (1 = son, 2 = daughter). Variable A is the *Sex role attitude of the parent*, whereas variable B is the *Sex role attitude of the child*. This table is the joint (or fully) observed table since its entries are based on independently sampled households. Therefore, for many research questions regarding the relationships among the four variables, no special marginal-modeling techniques are required. However, this is certainly not true for many other questions.

A relevant question about these data might be whether there are any overall differences between parents and children with respect to their sex role attitudes. The relevant marginals are shown in the first two columns of Table 2.9. These marginals are obtained by summing the row totals (for the parents' attitude A) and summing the column totals (for the children's attitude B) in the four subtables of Table 2.8. The first two columns of Table 2.9 will now be denoted as table TG, where T stands for *Traditionalism* (with three categories) and G for *Generation* (parent/child). Given the sampling design of the NKPS data in the module used here, the data in this table TG are not independently observed as each parent is coupled with one child (from the same family). Marginal modeling methods must be used to take this dependency (clustering, matching) into account when comparing the overall parent and child distributions.

Complete homogeneity of parents' and children's distributions clearly has to be rejected: $G^2 = 343.11$, $df = 2$ ($p = .000$, $X^2 = 297.98$). Ignoring the partial

Table 2.8. *Traditional sex role attitudes*; source NKPS, see text

$P = 1, C = 1$: Father–Son

	B. Child's Attitude			
A. Parent's Attitude	1. Nontrad.	2. Mod. trad.	3. Trad.	Total
1. Nontraditional	37	26	3	66
2. Moderately traditional	60	62	13	135
3. Traditional	19	41	11	71
Total	116	129	27	272

$P = 1, C = 2$: Father–Daughter

	B. Child's Attitude			
A. Parent's Attitude	1. Nontrad.	2. Mod. trad.	3. Trad.	Total
1. Nontraditional	101	25	3	129
2. Moderately traditional	108	62	2	172
3. Traditional	26	37	5	68
Total	235	124	10	369

$P = 2, C = 1$: Mother–Son

	B. Child's Attitude			
A. Parent's Attitude	1. Nontrad.	2. Mod. trad.	3. Trad.	Total
1. Nontraditional	92	55	5	152
2. Moderately traditional	91	123	18	232
3. Traditional	30	49	23	102
Total	213	227	46	486

$P = 2, C = 2$: Mother–Daughter

	B. Child's Attitude			
A. Parent's Attitude	1. Nontrad.	2. Mod. trad.	3. Trad.	Total
1. Nontraditional	204	65	6	275
2. Moderately traditional	222	114	11	347
3. Traditional	63	63	9	135
Total	489	242	26	757

Table 2.9. *Traditional sex role attitudes*; marginal distributions; source NKPS, see text

Sex Role Attitude	1. Parent	2. Child	1. Men	2. Women
1. Nontraditional	622	1053	524	1151
2. Moderately traditional	886	722	663	945
3. Traditional	376	109	212	273
Total	1884	1884	1399	2369

dependency and inappropriately applying the standard independence model $\{T, G\}$ directly to the first two columns of Table 2.9 yields different values for the test statistics: $G^2 = 284.41$, $df = 2$, $p = .000$, $X^2 = 274.62$. All adjusted residuals under the MH model (not reported here) are statistically significant. For parents' attitude, the vector with observed marginal proportions in Table 2.9 is $(.330, .470, .200)$; for children's attitude, it is $(.559, .383, .058)$. Looking at these observed marginal proportional distributions of traditionalism, it is evident that children have much less traditional views on sex roles than their parents. The differences in sex roles attitude between parents and children can be expressed in terms of the loglinear parameters in saturated model $\{TG\}$ applied to table TG using marginal-modeling methods. All two-variable parameters are statistically significant and show an almost perfectly linear relationship between T and G: $\hat{\lambda}_{11}^{GT} = -.416$, $\hat{\lambda}_{12}^{GT} = -.050$, and $\hat{\lambda}_{13}^{GT} = .466$. From these estimates, it can be computed that the extreme odds ratio equals $\exp(-.416 - .466 - .416 - .466) = .171$ (and its inverse $1/.171 = 5.836$): the odds of being traditional $(T = 1)$ rather than nontraditional $(T = 3)$ are almost six times larger for parents than for children.

Another possibly relevant question that gives rise to a more complex dependency pattern is about the overall differences between men and women. To answer this question, the overall marginal distributions of *Traditionalism* for men and women have to be obtained by summing appropriate row totals and column totals of the subtables in Table 2.8. For example, to obtain the marginal distribution for men, one takes the sum of the row and column totals of subtable 1, the row totals of subtable 2 and the column totals of subtable 3. The resulting marginal distributions are reported in the last two columns of Table 2.9, denoted as table TS, where T has three categories as before and S has two (men, women). Note that now the totals of these two columns are no longer equal to each other and not equal to the 1,884 parent-child pairs. In total, there are $2 \times 1884 = 3,768$ responses to the questions about sex roles, 1,399 of which were given by men (fathers and sons) and 2,369 by women (mothers and daughters). These answers come partly from matched observations, viz. when originating from the same subtable in Table 2.8, i.e., from father-son or mother-daughter pairs. For the other part, they are independent observations. Again, marginal-modeling methods have to be used to take the (partial) dependency into account.

Marginal homogeneity for men and women in table TS must now explicitly refer to the equality of the probability distributions because of the different column totals. The test results are $G^2 = 46.91$, $df = 2$ ($p = .000$, $X^2 = 46.33$). Inappropriately assuming completely independent observations and applying model $\{T, S\}$ directly to table TS yields $G^2 = 45.38$, $df = 2$, $X^2 = 45.11$. Men appear to have different opinions than women regarding the roles of the sexes. The observed distribution for men is given by vector $(.375, .474, .152)$, for women by $(.486, .399, .115)$. So, men are more traditional regarding sex roles than women. This is confirmed by the (statistically significant) loglinear parameters of saturated model $\{TS\}$ for table TS. The relationship between S and T is approximately linear: $\hat{\lambda}_{11}^{ST} = -.161$, $\hat{\lambda}_{12}^{ST} = .055$, and $\hat{\lambda}_{13}^{ST} = .106$. The overall differences in traditionalism between parents and children as found above in table TG are definitely larger than the overall differences found here between men and women in table TS. The extreme odds ratio for the relationship between S and T in table TS, computed analogously as above for the relation $G - T$ equals $.534$ and its inverse $1/.534 = 1.873$: the odds that a man gives a traditional answer $(T = 3)$ rather than a nontraditional one $(T = 1)$ are almost two times

Table 2.10. *Traditional sex role attitudes*; men–women and parents–children marginal distributions; source NKPS, see text

P. Parent's sex		Male			
C. Child's sex		Male		Female	
R. Respondent's status		Parent	Child	Parent	Child
	1. Nontraditional	66	116	129	235
T. Sex role attitude	2. Moderately traditional	135	129	172	124
	3. Traditional	71	27	68	369
	Total	272	272	369	369

P. Parent's sex		Female			
C. Child's sex		Male		Female	
R. Respondent's status		Parent	Child	Parent	Child
	1. Nontraditional	152	213	275	489
T. Sex role attitude	2. Moderately traditional	232	227	347	242
	3. Traditional	102	46	135	26
	Total	486	486	757	757

larger than for a woman. It is also possible to test whether the difference in strengths of the marginal association between $S - T$ and $G - T$ is statistically significant using marginal-modeling methods; this will not be done here, but a similar research question will be illustrated below.

Given the overall differences in traditionalism between generations, and between men and women, a next logical question is to ask whether or not the generational differences in traditionalism are larger among men than among women; or, formulated the other way around, are the sex differences in traditionalism larger between fathers and mothers than between sons and daughters. To answer this question, all row and column totals of the subtables in Table 2.8 have to be considered. They have been put together in Table 2.10 in the form of a *PCRT* table (where the symbols P, C, R, and T are explained in the table). Here the observations are again partially dependent. The loglinear models of interest can most easily be formulated in terms of the variables in Table 2.10, using the short-hand notation for hierarchical models. All models will be logit models for the investigation of the effects on T, conditioning on the distribution of *PCR*.

It was found above that the overall differences between parents and children (now variable R) concerning their sex roles attitudes (variable T) were very large. But how large are these differences when we control for the sex of parents and children using table *PCRT*? The most parsimonious model in this respect is the model in which there are no differences left, that is, model $R \perp\!\!\!\perp T|PC$, which states that R and T are conditionally independent given P and C. This model is equivalent with loglinear model $\{PCR, PCT\}$ for Table 2.10, and identical to the hypothesis of simultaneous MH in each of the subtables of Table 2.8. For each of the subtables of Table 2.8, the hypothesis of marginal homogeneity has to be rejected (all $p = .000$), as well as the MH hypothesis for all four subgroups simultaneously: $G^2 = 354.89$, $df = 8$ ($p = .000$, $X^2 = 303.56$). The next logical step is then to ask whether the apparently existing marginal differences between parents and children are the same in all four subtables in Table 2.8. In loglinear terms, is model $\{PCR, PCT, RT\}$ for Table 2.10 true in the population? However, this model has to be rejected too: $G^2 = 30.00$, $df = 6$ ($p = .000$, $X^2 = 29.44$).

In order to find out why model $\{PCR, PCT, RT\}$ did not fit the data, the parameters for the effects on T in saturated model $\{PCRT\}$ for Table 2.10 were estimated. The first striking finding was that all loglinear parameter estimates $\hat{\lambda}$ that had P and T among their superscripts ($PT, PCT, PRT, PCRT$) had insignificant and very small values: all absolute $\hat{\lambda}$-values were smaller than .05. This means that there are no effects at all of P on T and it must be concluded, somewhat unexpectedly, that fathers and mothers do not differ in their attitudes on sex roles in model $\{PCRT\}$ for Table 2.10, despite the overall differences between men and women found above. Apparently, these sex differences only apply to the children, not to the parents. Second, there was a substantial main effect of R on T in model $\{PCRT\}$. It is approximately a linear effect (as, by the way, all direct effects on T in model $\{PCRT\}$ are). The extreme odds ratio for the relationship $R - T$ in model $\{PCRT\}$, i.e., for the differences between parents and children regarding the odds nontraditional ($T = 1$) versus traditional ($T = 3$) is .157 and its inverse is 6.366. This effect is just a little bit stronger than the corresponding overall effect $G - T$ discussed above (that was .171 with its inverse 5.836). But there is also a significant and non-negligible three-variable interaction term CRT. Because of this interaction term, when sons ($C = 1$) are being interviewed, the extreme conditional odds ratio for the differences in traditionalism between parents and children becomes .264 and its inverse is 3.790, while for daughters ($C = 2$), the corresponding extreme conditional odds ratio equals .094 and its inverse is 10.693. In sum, parents are generally more traditional than their children, but sons depart substantially less from their parents than daughters do.

Summing up, there are no differences whatsoever with regard to traditional attitudes towards sex roles between fathers and mothers. Boys and girls, on the other hand differ; boys being more traditional than girls. The largest attitude differences were found for *Generation*: children are much less traditional than their parents, and this is especially true for daughters, but less so for sons.

So far, all marginal models for the NKPS data concerned the comparison of the distributions of the one characteristic attitude towards sex role. However, in NKPS, the respondents were not only asked about their attitudes regarding sex roles but also regarding marriage: to what extent did the respondents feel that marriage is a sacred institute? This attitude was measured by items such as 'having sex before marriage is forbidden', and 'marriage among homosexuals is not allowed'. Traditionalism concerning marriage, which was also originally expressed in terms of mean scale scores,

Table 2.11. *Traditional sex role* and *Marriage attitudes*; marginal distributions for Parents and Children; source NKPS, see text

| I. Item | 1. Sex Role | | 2. Marriage | |
R. Respondent's Status	1. Parent	2. Child	1. Parent	2. Child
T. *Traditional attitude*				
1. Nontraditional	622	1053	783	1310
2. Moderately traditional	886	722	896	501
3. Traditional	376	109	205	73
Total	1884	1884	1884	1884

was coded in three categories in the same way as the categorization of the variable sex role attitude.

A relevant research question then might be whether or not parents and children differ in the same way regarding both characteristics, viz. traditionalism regarding sex roles and marriage. The marginal one-variable distributions of both characteristics for parents and children were formed from the full table (not presented here, but see our website) and shown in Table 2.11.

The observations in the different columns in Table 2.11 are not only dependent because of the partial matching between particular parents and children, but also because the data for marriage traditionalism have been obtained from exactly the same respondents as the data on sex role traditionalism. Estimation and testing procedures have to take these complex patterns of dependencies into account. Table 2.11 will be treated as an *IRT* table with variable T now representing *Traditionalism* (regarding sex roles and marriage). The loglinear models considered for the data in Table 2.11 will be logit models for the effects on T, conditioning on the observed frequencies for marginal table *IR*.

The hypothesis that there are no parent-child and item differences, in other words, that all column distributions in Table 2.11 are homogeneous is represented by model $\{IR,T\}$ for Table 2.11. In the less restrictive model $\{IR,IT\}$, the distributions of T are allowed to be different for the two items, but not between parents and children. Assuming homogeneous distributions for the two items but different distributions for parents and children leads to model $\{IR,RT\}$. Finally, the hypothesis that there are both item and parent-child differences regarding T, but no special three-variable interactions, is represented by model $\{IR,IT,RT\}$.

Note that these models bear strong resemblances to traditional MANOVA models or ANOVA models for repeated measures, as were the models for the body items data in Table 2.7; see Chapters 3 and 5 for more MANOVA-like analyses.

However, all these models for Table 2.11 have to be rejected; all p-values are $p = .000$. That leaves us with the saturated model $\{IRT\}$ for Table 2.11. However, in this saturated model, the three variable interaction effect is very small (although statistically significant) and does not lead to really different conclusions from the no three-variable interaction model $\{IR,IT,RT\}$. The outcomes in model $\{IR,IT,RT\}$

(not presented here) indicate that parents are substantially more traditional than their children regarding both characteristics, and that both parents and children have a bit more traditional views on sex roles than on marriage.

The complex patterns of dependencies in the observations in Table 2.11 as a consequence of the matched parent-child relation, and of the repeated items, led to somewhat unexpected outcomes for standard errors and test statistics when the above correct marginal procedures are compared with the inappropriate analyses ignoring the dependencies. For example, the G^2 statistics for models $\{IR, T\}$ and $\{IR, IT\}$ in Table 2.11 are smaller when calculated in the correct manner than when inappropriately applied assuming independent observations. For models $\{IR, RT\}$ and $\{IR, IT, RT\}$ the opposite is true. In the same vein, the standard errors for some of the parameter estimates in the saturated model are smaller but also larger for some parameters when estimated in the correct way than when estimated incorrectly assuming independent observations. This is different from the patterns found so far. It turns out that predictions about the consequences of ignoring the (complex) dependencies between the observations may not be of a simple nature: it is not guaranteed at all that one gains statistical power by taking the dependencies into account; one might as well lose power. As was seen in many complex analyses, even when expecting 'positive' dependencies, ignoring the dependencies in the observations may lead to smaller standard errors and larger test statistics, but also to the opposite situation: larger standard errors and smaller test statistics.

Associations Among Variables

So far, only one-way marginals have been considered, either for the sample as a whole or for subgroups, and either for one or even more characteristics. But multiway marginals may be at least as interesting for researchers. Comparisons of more complex multiway marginals will be discussed in Chapters 4, 5, and 6. Now, the basic approach will be outlined by means of a simple example showing how to conduct analyses of associations using two-way marginals. As a first example, the data in Table 2.8 will be used in which the relationship was shown between Parents' attitudes (A) and Children's attitudes towards sex roles (B) for different subgroups. One may wonder whether the associations AB for homogeneous pairs (father-son and mother-daughter) are stronger than for nonhomogeneous pairs (father-daughter and mother-son). If the data in Table 2.8 had come from a design different from NKPS, in which the information for the particular combinations of the four relevant respondents, viz. father, mother, son, and daughter had all been collected within the same family, marginal-modeling methods would have been needed to test such a hypothesis. However, given the module of NKPS that is used here, in which one respondent (child) is coupled with one parent, the four subgroups in Table 2.8 have been independently observed and regular loglinear modeling can be used. The hypothesis that the relationship AB expressed in terms of odds ratios is the same in all four subgroups is identical to applying model $\{PCA, PCB, AB\}$ directly and in the standard way to Table 2.8. This hypothesis need not be rejected: $G^2 = 12.26$, $df = 12$ ($p = .414$, $X^2 = 12.26$). Consequently, the idea that the association AB might be stronger in

Table 2.12. Association between *Sex role* and *Marriage attitudes* for Parents and Children; source NKPS, see text. Response categories for Marriage and Sex Role: 1 = nontraditional, 2 = moderately traditional, 3 = traditional

R. *Respondent's Status*		Parent			Child		
B. *Marriage*		1	2	3	1	2	3
A. Sex Role	1	459	152	11	923	120	10
	2	251	542	93	345	339	38
	3	73	202	101	42	42	25

the homogeneous subgroups (mother-daughter and father-son combinations) than in the nonhomogeneous subgroups (father-daughter and mother-son combinations) is not accepted. There is the same strong, positive, and statistically significant relation between A and B within each subtable; the relationship is also approximately linear and the deviations from linearity are not significant (precise results not reported here). The test results for a conditional test of the linear restrictions for AB, assuming a linear $\times$ linear (or uniform) association against model $\{PCA, PCB, AB\}$ without the linear restrictions for AB are $G^2 = 1.85$, $df = 3$, ($p = .605$).

A research question that does require marginal modeling with the NKPS data would be whether or not the association between the sex role and the marriage attitudes are different for parents and children. One may postulate, for example, that different specific attitudes are more crystalized into a consistent attitude system among parents than among children. The full data set is presented on the book's webpage; the relevant marginal data are displayed in Table 2.12, which will be treated as an *RBA* table.

The null hypothesis that the association between A and B is the same for parents and children corresponds to model $\{RB, RA, AB\}$ for Table 2.12. This model fits the data excellently: $G^2 = 3.791$, $df = 4$ ($p = .435$, $X^2 = 3.802$). Assuming independent observations and applying this model directly to the data in Table 2.12 yields $G^2 = 3.891$, $df = 4$, $p = .421$, $X^2 = 3.904$. There is no reason to assume that the relationship between the two attitudes is stronger for the parents than for the children. The common relationship between sex role and marriage attitudes is very strong and monotonically increasing. All local log odds ratios in the common 3×3 table AB are positive. The extreme odds ratio involving cells 11, 13, 31, and 33 equals 52.47. The relationship is not strictly linear since the linear $\times$ linear or uniform association model has to be rejected.

The above illustrations should have made clear when and how substantive research questions involve the comparison of marginal tables and how they can be translated into the language of loglinear models for marginal tables. Later chapters will exemplify still more and more complicated research questions. But first, attention must be paid to the central question of how the maximum likelihood estimates for loglinear marginal models can be obtained.

2.3 Maximum Likelihood Inference for Loglinear Marginal Models

Maximum likelihood inference for marginal models requires that the cell probabilities of the joint table are estimated under the constraints imposed by the marginal model. The main difficulty in fitting and testing marginal models, compared to ordinary loglinear models, is the dependency of the observations. The estimation procedure discussed here is Bergsma's (1997) modification of the method of Lang and Agresti (1994). Fitting marginal models generally requires numerical procedures since closed form expressions are usually not available. Before the details of the proposed estimation procedure are given, several sampling methods that are often used and that are appropriate for maximum likelihood inference are discussed.

2.3.1 Sampling Methods

Let π_i represent the probability of observing response pattern i pertaining to a set of categorical variables for a randomly selected respondent from the population. Each response pattern i defines a cell in the multidimensional contingency table that contains the observed frequencies n_i. Let N be the total number of respondents and I the total number of cells in the table. It is commonly assumed in social and behavioral science research that the observed frequency distribution is given by the multinomial distribution

$$\Pr(n_1, \cdots, n_i, \cdots, n_I) = \prod_i^I \binom{N}{n_i} \pi_i^{n_i} .$$

Most researchers take it for granted that the multinomial distribution is the appropriate sampling distribution of the frequencies in a contingency table. However, one should keep in mind that this is only true if several, not always trivial, conditions are satisfied. For example, the sample size N should be fixed in advance, and should not depend on other aspects of the sampling process. Further, the respondents are supposed to be sampled independently from the population with replacement implying that the theoretical cell probabilities π_i remain constant during the sampling process. If these conditions are not satisfied, the multinomial assumption is not valid and other sampling distributions apply. If sampling is without replacement, the observed frequencies follow a hypergeometric distribution. If the sample size is not fixed in advance, but depends on the number of times a certain event (called a 'success') occurs, the negative binomial distribution is more appropriate. The full information ML estimation procedure for fitting marginal models and the associated statistical procedures discussed in this book are appropriate for multinomially distributed frequencies. They are, in general, not appropriate for the hypergeometric and negative binomial distributions. These less well-known theoretical probability distributions for observed frequencies will not be discussed further in this book. However, unless stated otherwise, the estimation and test procedures developed here remain valid for two other sampling distributions: the product multinomial and the Poisson distribution.

Sometimes the entire population is stratified before sampling, using a stratifying variable S. If parameters of different subgroups are of interest, the researcher may consider taking a fixed number of subjects from each subgroup. The numbers N_k of observations in each stratum are fixed in advance. Let n_{ik} be the number of observations from stratum k in cell i, assuming K strata. Then, the joint distribution of all cell frequencies is often assumed to follow the product multinomial distribution

$$\Pr(n_{11}, \cdots, n_{ik}, \cdots, n_{IK}) = \prod_{k=1}^{K} \prod_{i=1}^{I} \binom{N_k}{n_{ik}} \pi_{ik}^{n_{ik}} .$$

The product multinomial distribution is applicable when the sampling within each stratum satisfies the conditions stated above for the multinomial distribution, and additionally when the sampling from different strata occurs independently.

In other applications, it is not the number of observations that is fixed in advance but some other aspect of the sampling process such as the total observation time. In that case, the observed frequencies may follow a Poisson distribution

$$\Pr(n_1, \cdots, n_i, \cdots, n_I) = \prod_{i=1}^{I} \frac{\mu_i^{n_i} e^{-\mu_i}}{n_i!}$$

with expected frequencies μ_i for $i = 1, \cdots, I$. The Poisson distribution can be used when the events that are counted occur randomly over time or space with outcomes in disjoint periods or regions independent of each other. If the rate of occurrence of an event is the parameter of interest, such as the number of pedestrians passing a shopping street per hour, of the above schemes only Poisson sampling would be appropriate.

2.3.2 Specifying Loglinear Marginal Models by Constraining the Cell Probabilities

In order to describe the ML estimation procedure, it is useful to specify marginal models in matrix notation and define the loglinear models in the form of restrictions on the cell probabilities rather than in terms of loglinear parameters. It will be shown below why this is useful and how this can be done. The notation that will be used is an adaptation of the notation proposed by Grizzle et al. (1969).

A vector of loglinear marginal parameters ϕ can generally be written in the form

$$\phi(\pi) = C' \log A' \pi ,$$

where π is a vector of cell probabilities, and A and C are matrices of constants. The basic principle of this representation will be illustrated by means of a few simple examples. But first, it will be made clear what it means when a scalar function $f(x)$ is applied to a vector of values. Let $f(x)$ be a function of a scalar variable x such as, for example, $f(x) = \exp(x)$ or $f(x) = \log(x)$. If this function is applied to a vector of values like

$$x = \begin{pmatrix} x_1 \\ x_2 \\ x_3 \end{pmatrix},$$

the result is another vector with the values of $f(x_i)$ as its elements:

$$f(x) = \begin{pmatrix} f(x_1) \\ f(x_2) \\ f(x_3) \end{pmatrix}.$$

This definition can easily be extended to vectors containing arbitrary numbers of elements.

Example 1: Marginal Homogeneity in a 3×3 *Table*

Suppose a categorical variable with three categories is observed at two time points. The theoretical cell probabilities for the data from this simple panel study can be written in a 3×3 matrix

$$\Pi = \begin{pmatrix} \pi_{11} & \pi_{12} & \pi_{13} \\ \pi_{21} & \pi_{22} & \pi_{23} \\ \pi_{31} & \pi_{32} & \pi_{33} \end{pmatrix}.$$

The rows of this matrix correspond to the first measurement and its columns correspond to the second one. For further use, this matrix will have to be written as vector. In this book we decided to vectorize a matrix row-wise so that when the elements of a matrix are written in vector form, the last index changes the fastest. Hence, for matrix Π above,

$$\pi = \mathrm{vec}(\Pi) = \begin{pmatrix} \pi_{11} \\ \pi_{12} \\ \pi_{13} \\ \pi_{21} \\ \pi_{22} \\ \pi_{23} \\ \pi_{31} \\ \pi_{32} \\ \pi_{33} \end{pmatrix}.$$

Note that our definition of the vectorization operations differs from the vec operation defined in textbooks on linear algebra as Searle (2006) and Schott (1997) where vectorization is carried out column-wise. This vectorization operation can also be applied to general multidimensional arrays, not just to two-dimensional matrices. Whenever such a multidimensional array is vectorized, its last dimension changes the fastest and its first dimension changes the slowest. So, for a $2 \times 2 \times 2$ array F with entries f_{ijk}, one has

$$\text{vec}(F) = \begin{pmatrix} f_{111} \\ f_{112} \\ f_{121} \\ f_{122} \\ f_{211} \\ f_{212} \\ f_{221} \\ f_{222} \end{pmatrix}.$$

Going back to our example, the univariate marginals of matrix Π are

$$\pi_{i+} = \sum_j \pi_{ij}$$

and

$$\pi_{+j} = \sum_i \pi_{ij}.$$

These marginals can be written in the vector

$$\begin{pmatrix} \pi_{1+} \\ \pi_{2+} \\ \pi_{3+} \\ \pi_{+1} \\ \pi_{+2} \\ \pi_{+3} \end{pmatrix}.$$

Now, consider the following 6×9 matrix:

$$A' = \begin{pmatrix} 1\,1\,1\,0\,0\,0\,0\,0\,0 \\ 0\,0\,0\,1\,1\,1\,0\,0\,0 \\ 0\,0\,0\,0\,0\,0\,1\,1\,1 \\ 1\,0\,0\,1\,0\,0\,1\,0\,0 \\ 0\,1\,0\,0\,1\,0\,0\,1\,0 \\ 0\,0\,1\,0\,0\,1\,0\,0\,1 \end{pmatrix}.$$

Then it is easy to see that

$$\begin{pmatrix} \pi_{1+} \\ \pi_{2+} \\ \pi_{3+} \\ \pi_{+1} \\ \pi_{+2} \\ \pi_{+3} \end{pmatrix} = \begin{pmatrix} 1\,1\,1\,0\,0\,0\,0\,0\,0 \\ 0\,0\,0\,1\,1\,1\,0\,0\,0 \\ 0\,0\,0\,0\,0\,0\,1\,1\,1 \\ 1\,0\,0\,1\,0\,0\,1\,0\,0 \\ 0\,1\,0\,0\,1\,0\,0\,1\,0 \\ 0\,0\,1\,0\,0\,1\,0\,0\,1 \end{pmatrix} \cdot \begin{pmatrix} \pi_{11} \\ \pi_{12} \\ \pi_{13} \\ \pi_{21} \\ \pi_{22} \\ \pi_{23} \\ \pi_{31} \\ \pi_{32} \\ \pi_{33} \end{pmatrix} = A'\pi.$$

Premultiplication of π by matrix A' yields the appropriate marginal distributions, and $\log(A'\pi)$ is then the vector containing the logarithms of those marginal probabilities:

$$\log(A'\pi) = \begin{pmatrix} \log(\pi_{1+}) \\ \log(\pi_{2+}) \\ \log(\pi_{3+}) \\ \log(\pi_{+1}) \\ \log(\pi_{+2}) \\ \log(\pi_{+3}) \end{pmatrix}.$$

Under the marginal homogeneity hypothesis, it is assumed that $\pi_{i+} = \pi_{+i}$ for all categories of the response variable. It then also follows that $\log(\pi_{i+}) = \log(\pi_{+i})$, implying that the six entries of the vector $\log(A'\pi)$ are functions of only three parameters (or: two independent ones, see below). Then, the matrix X is defined as

$$X = \begin{pmatrix} 1 & 0 & 0 \\ 0 & 1 & 0 \\ 0 & 0 & 1 \\ 1 & 0 & 0 \\ 0 & 1 & 0 \\ 0 & 0 & 1 \end{pmatrix},$$

where the design matrix and the effects are arbitrarily expressed in terms of a dummy-variable-like notation rather than effect coding. The hypothesis of marginal homogeneity can now be represented as

$$\begin{pmatrix} \log(\pi_{1+}) \\ \log(\pi_{2+}) \\ \log(\pi_{3+}) \\ \log(\pi_{+1}) \\ \log(\pi_{+2}) \\ \log(\pi_{+3}) \end{pmatrix} = \begin{pmatrix} 1 & 0 & 0 \\ 0 & 1 & 0 \\ 0 & 0 & 1 \\ 1 & 0 & 0 \\ 0 & 1 & 0 \\ 0 & 0 & 1 \end{pmatrix} \cdot \begin{pmatrix} \beta_1 \\ \beta_2 \\ \beta_3 \end{pmatrix} = \begin{pmatrix} \beta_1 \\ \beta_2 \\ \beta_3 \\ \beta_1 \\ \beta_2 \\ \beta_3 \end{pmatrix},$$

or more concisely as

$$\log(A'\pi) = X\beta,$$

with the vector β containing the three unknown parameters β_j for $j = 1,2,3$. This equation provides a parametric representation of the marginal homogeneity hypothesis, with β_j being the (unknown) logarithm of the sum of the cell probabilities in the j-th row and in the j-th column of matrix Π.

A parameter-free representation of the marginal homogeneity model is obtained by noting that it implies the following constraints on the marginal probabilities

$$\log(\pi_{1+}) = \log(\pi_{+1})$$
$$\log(\pi_{2+}) = \log(\pi_{+2})$$
$$\log(\pi_{3+}) = \log(\pi_{+3}).$$

Since the cell probabilities in Π sum to 1, only two (e.g., the first two) constraints of the three given here have to be considered. Now, define the 2×6 matrix B' as

$$B' = \begin{pmatrix} 1\ 0\ 0 -1\ \ 0\ 0 \\ 0\ 1\ 0\ \ \ 0 -1\ 0 \end{pmatrix} ,$$

then

$$\begin{pmatrix} \log(\pi_{1+}) - \log(\pi_{+1}) \\ \log(\pi_{2+}) - \log(\pi_{+2}) \end{pmatrix} = B' \log(A'\pi) .$$

Marginal homogeneity is now equivalent to $B' \log(A'\pi) = 0$. Here we have a representation of the marginal homogeneity model that does not contain any parameter, but is completely formulated in terms of constraints on the cell probabilities in matrix Π. The hypothesis of marginal homogeneity can now be tested by first determining the maximum likelihood estimates of the cell probabilities under the constraints implied by the model, and then testing whether or not this restricted model provides a significantly worse fit than the unconstrained model. In this testing procedure, no unknown parameters (apart from the cell probabilities) will be estimated.

Example 2: Independence in a 3 × 3 Table

In this second example, it will be illustrated how the simple independence model in a two-dimensional table can also be cast in the form of a parameter-free model (switching to a nonmarginal model for our explanations). We return to the 3×3 table Π introduced in the previous example. The loglinear model representing independence of the row and column variable is given by

$$\log(\pi_{ij}) = \lambda_{**}^{RC} + \lambda_{i*}^{RC} + \lambda_{*j}^{RC} .$$

Now, take matrix C as the 9×9 identity matrix. Further, define

$$X = \begin{pmatrix} 1\ 1\ 0\ 0\ 1\ 0\ 0 \\ 1\ 1\ 0\ 0\ 0\ 1\ 0 \\ 1\ 1\ 0\ 0\ 0\ 0\ 1 \\ 1\ 0\ 1\ 0\ 1\ 0\ 0 \\ 1\ 0\ 1\ 0\ 0\ 1\ 0 \\ 1\ 0\ 1\ 0\ 0\ 0\ 1 \\ 1\ 0\ 0\ 1\ 1\ 0\ 0 \\ 1\ 0\ 0\ 1\ 0\ 1\ 0 \\ 1\ 0\ 0\ 1\ 0\ 0\ 1 \end{pmatrix} ,$$

which is the design matrix of the model (arbitrarily using dummy rather than effect coding). Then, our loglinear model can be written as

$$C' \log(\pi) = \log(\pi) = X\lambda ,$$

with vector λ containing the loglinear parameters. As is customary in the discussion of loglinear models, the symbol λ is used (rather than β) to represent the unknown loglinear parameters.

In order to derive a parameter-free representation of the same model, the concept of the null space of a matrix has to be introduced. Take the matrix X as defined above. This matrix is of the order 9×7 and its columns define a vector space whose elements are the linear combinations of the columns of X

$$\mathcal{V} = \{y : y = Xw\}$$

for all weight vectors w consisting here of seven arbitrary weights w_k. The vectors y in $\mathcal{V}$ contain nine elements. Since the vector space $\mathcal{V}$ is generated by the columns of matrix X, it is often called the column space of X.

Do we really need all columns of X to generate this vector space? The answer to this question depends on the column rank of X. The matrix X is of full column rank if the zero weight vector $w = 0$ is the only weight vector for which $Xw = 0$. When we can find a nonzero vector w, i.e., a vector with a least one element different from zero, for which $Xw = 0$, the matrix is of deficient column rank. In that case, there exists a linear dependency among the columns of matrix X. In the present example, twice the first column of X is the sum of the other six columns, implying that for the nonzero weight vector $w'_1 = (-2, 1, 1, 1, 1, 1, 1)$, we have $Xw_1 = 0$. Moreover, the sum of columns 2 to 4 is equal to the sum of columns 5 to 7. Hence, we have also $Xw_2 = 0$ for $w'_2 = (0, 1, 1, 1, -1, -1, -1)$. For the present matrix X, one can prove that only two different linear dependencies exist among its columns. These linear dependencies allow us to express two columns of X as linear combinations of the other columns. For example, for the first (x_1) and second column (x_2), we can write

$$x_1 = x_5 + x_6 + x_7$$

and

$$x_2 = -x_3 - x_4 + x_5 + x_6 + x_7 \, ,$$

showing that arbitrary elements of $\mathcal{V}$ can be generated by a particular selection of five columns of X. A set of linearly independent vectors that generate a vector space is called a 'basis' of the vector space. The vectors in a vector space can be written as linear combinations of the elements in its basis. Bases of vector spaces are not uniquely defined, but the number of generating vectors in them is: the number of vectors in a basis is called the dimension of the vector space. The vectors defining a basis for a vector space will be written column-wise in the matrix X_B. Removing any generating vector from a basis transforms the given vector space into a different one of lower dimensions. If matrix X is of full column rank, there exist no linear dependencies among its columns and we need all its columns to generate vector space $\mathcal{V}$ with its dimension equal to the number of columns of X. If there exist linear dependencies among the columns of X, the dimensionality of $\mathcal{V}$ is equal to the number of linearly independent columns of X, and its dimension is smaller than the number of columns of X. See Schott (1997) for an overview of the concepts of linear algebra that are relevant for statistics.

To show that a model can be defined in terms of (restrictions on) its parameters but also (as is true in our marginal-modeling approach) in terms of restrictions on the cell probabilities, the concept of a null space is needed. The null space $\mathcal{N}$ of X is a different vector space that can be associated with matrix X. It is defined as

$$\mathcal{N} = \{y : y'X = 0\} \ .$$

The vector space $\mathcal{N}$ is the set of all vectors that are orthogonal to the columns of X. It is also often called the orthocomplement of X. Being itself a vector space, it can be generated by a set of vectors in one of its bases. One can prove that the dimensionality of the null space and the column space of a matrix sum to the number of rows of X. Let X_N be the matrix containing a basis of the null space of X, and X_B the matrix containing a basis of the column space of X. Then,

$$X_N'X_B = 0 \ .$$

An important point to realize is that a vector space can be characterized either by specifying a basis X_B, or by specifying a basis X_N of its null space: $y \in \mathcal{V}$ if and only if $y = X_B w$ for some w, if and only if $y'X_N = 0$ (or $X_N'y = 0$). We will discuss in a later paragraph how to construct bases for both vector spaces.

   With this knowledge in mind we can go back to the parametric representation of the loglinear model for independence:

$$\log(\pi) = \log(A'\pi) = X\lambda \ .$$

This relation shows that $\log(\pi)$ is an element in the vector space generated by the columns of X. Letting X_N be the matrix containing in its columns a basis of the null space of X, it follows that the loglinear model can equivalently be specified in terms of a set of constraints on the cell probabilities:

$$X_N'\log(\pi) = 0 \ .$$

   In the present example, $\mathcal{V}$ has dimension 5 whereas its null space has dimension 4. The columns of the following matrix give a basis for this null space:

$$X_N = \begin{pmatrix} 1 & 1 & 1 & 1 \\ 0 & -1 & 0 & -1 \\ -1 & 0 & -1 & 0 \\ 0 & 0 & -1 & -1 \\ 0 & 0 & 0 & 1 \\ 0 & 0 & 1 & 0 \\ -1 & -1 & 0 & 0 \\ 0 & 1 & 0 & 0 \\ 1 & 0 & 0 & 0 \end{pmatrix} \ .$$

With this choice for X_N, the constraints on the cell probabilities are

$$X_N' \log(\pi) = \log \begin{pmatrix} (\pi_{11}\pi_{33})/(\pi_{13}\pi_{31}) \\ (\pi_{11}\pi_{23})/(\pi_{13}\pi_{21}) \\ (\pi_{11}\pi_{32})/(\pi_{12}\pi_{31}) \\ (\pi_{11}\pi_{22})/(\pi_{12}\pi_{21}) \end{pmatrix} = 0 \,.$$

Independence of the row and column variable in our 3×3 table is equivalent to constraining four independent log odds ratios to be equal to zero.

Example 3: Equality of Local Odds in a 3×3 Table

In a 3×3 table with cell probabilities π_{ij}, four nonredundant local log odds can be defined as

$$\omega_{ij} = \log \left(\frac{\pi_{i,j}\pi_{i+1,j+1}}{\pi_{i,j+1}\pi_{i+1,j}} \right)$$

for $i, j = 1, 2$. Let A be the 9×9 identity matrix and define matrices C and X in the following way:

$$C' = \begin{pmatrix} 1 & -1 & 0 & -1 & 1 & 0 & 0 & 0 & 0 \\ 0 & 0 & 0 & 1 & -1 & 0 & -1 & 1 & 0 \\ 0 & 1 & -1 & 0 & -1 & 1 & 0 & 0 & 0 \\ 0 & 0 & 0 & 0 & 1 & -1 & 0 & -1 & 1 \end{pmatrix} \quad \text{and} \quad X = \begin{pmatrix} 1 \\ 1 \\ 1 \\ 1 \end{pmatrix} .$$

The hypothesis that the four local log odds are equal can be represented in parametric form as

$$C' \log(A'\pi) = X\beta \,,$$

with β the common value of the four local log odds. The columns of the following matrix U provide a basis for the null space of X:

$$U = \begin{pmatrix} 1 & 1 & 1 \\ -1 & 0 & 0 \\ 0 & -1 & 0 \\ 0 & 0 & -1 \end{pmatrix} .$$

The same hypothesis can now be represented in parameter-free form as

$$U'C' \log(A'\pi) = 0 \,.$$

Exactly which constraints are imposed on the cell probabilities can be seen from the product

$$B' = U'C' = \begin{pmatrix} 1 & -1 & 0 & -2 & 2 & 0 & 1 & -1 & 0 \\ 1 & -2 & 1 & -1 & 2 & -1 & 0 & 0 & 0 \\ 1 & -1 & 0 & -1 & 0 & 1 & 0 & 1 & -1 \end{pmatrix} ,$$

and we can represent the hypothesis concisely as

$$B' \log(A'\pi) = 0 \,.$$

Example 4: Invariance of Log Odds Ratios Over Time

Suppose that a particular dichotomous variable has been measured at three different time points, and let A, B, and C represent the three measurements. The full table ABC contains the cell frequencies n_{ijk}^{ABC} corresponding to the theoretical cell probabilities π_{ijk}^{ABC}. Remember that when this three-dimensional array is vectorized, the last subscript changes the fastest and the first subscript the slowest. Suppose now that we want to test whether the (log) odds ratio between consecutive measurements remains constant over time:

$$\frac{\pi_{11+}^{ABC}\, \pi_{22+}^{ABC}}{\pi_{12+}^{ABC}\, \pi_{21+}^{ABC}} = \frac{\pi_{+11}^{ABC}\, \pi_{+22}^{ABC}}{\pi_{+12}^{ABC}\, \pi_{+21}^{ABC}}.$$

In order to test this hypothesis, loglinear marginal modeling is needed with

$$A' = \begin{pmatrix} 1\;1\;0\;0\;0\;0\;0\;0 \\ 0\;0\;1\;1\;0\;0\;0\;0 \\ 0\;0\;0\;0\;1\;1\;0\;0 \\ 0\;0\;0\;0\;0\;0\;1\;1 \\ 1\;0\;0\;0\;1\;0\;0\;0 \\ 0\;1\;0\;0\;0\;1\;0\;0 \\ 0\;0\;1\;0\;0\;0\;1\;0 \\ 0\;0\;0\;1\;0\;0\;0\;1 \end{pmatrix}$$

$$C' = \begin{pmatrix} 1 & -1 & -1 & 1 & 0 & 0 & 0 & 0 \\ 0 & 0 & 0 & 0 & 1 & -1 & -1 & 1 \end{pmatrix}$$

and

$$X = \begin{pmatrix} 1 \\ 1 \end{pmatrix}$$

so that

$$U = \begin{pmatrix} 1 \\ -1 \end{pmatrix}$$

and

$$U'C' = \begin{pmatrix} 1 & -1 & -1 & 1 & -1 & 1 & 1 & -1 \end{pmatrix}.$$

It is easy to see that in this example the matrix product $A'\pi$ yields the cell probabilities in the marginal tables AB and BC, and that $C' \log(A'\pi)$ defines the appropriate contrasts among the logarithms of these cell probabilities. Finally, the constraint $U'C' \log(A'\pi) = 0$ corresponds to the hypothesis that the two log odds ratios are equal.

The General Parameter-free Representation of Loglinear Marginal Models

The examples above illustrate how both loglinear and loglinear marginal models can be represented by imposing constraints on the cell probabilities. Loglinear marginal models have the parameterized form

$$C' \log A' \pi = X\beta . \tag{2.4}$$

For any matrix X, a matrix U can be found whose columns contain a basis of the null space of X. Every column of U is orthogonal to all columns of X (i.e., $U'X = 0$) and the columns of U and X together span the vector space with dimensionality equal to number of cells in the frequency table. For any such matrix U (which is generally not unique), Eq. 2.4 is equivalent to

$$U'C' \log A' \pi = 0 .$$

Ordinary loglinear models are special cases of loglinear marginal models for which A and C are identity matrices of the appropriate order, and matrix U contains a basis of the null space of the design matrix X.

Here we briefly sketch (without giving a formal proof) how a basis of the null space of matrix X can be obtained. Let the $m \times k$ design matrix X ($m > k$) be of rank $r \le k$. By means of elementary column operations (Schott, 1997), a $k \times k$ matrix Q can be defined such that

$$XQ = (X_1 | 0) ,$$

with X_1 an $m \times r$ matrix of full column rank r, and 0 an $m \times (k - r)$ zero matrix. Note that this matrix Q is not uniquely defined, since it depends on the order in which the elementary column operations are carried out. The columns of matrix X_1 define a basis of the column space of X. Moreover, it is always possible to select r linearly independent rows from matrix X_1, since if this were not the case the rank of X_1 would be smaller than r. Let the $r \times r$ matrix X_{11} contain such a selection of r linearly independent rows, and let the $(m - r) \times r$ matrix X_{21} contain the remaining $m - r$ rows of X_1. Furthermore, let I_{m-r} be the $(m - r) \times (m - r)$ identity matrix. Then, the columns of matrix

$$U = \begin{pmatrix} -(X_{21}X_{11}^{-1})' \\ I_{m-r} \end{pmatrix} ,$$

after rearranging its rows in the original order, constitute a basis for the null space of X.

2.3.3 Simultaneous Modeling of Joint and Marginal Distributions: Redundancy, Incompatibility and Other Issues

In many applications, it may be necessary or interesting to simultaneously test several loglinear and loglinear marginal models for the same full table. For example, for the

3×3 table Π in the first and second example in the previous section, one might be interested in a simultaneous test of marginal homogeneity and independence. Both hypotheses are represented by different matrix constraint equations

$$U_1' C_1' \log(A_1' \pi) = 0$$

and

$$U_2' C_2' \log(A_2' \pi) = 0 \, ,$$

which can be combined in a single overall equation after defining the appropriate supermatrices:

$$\begin{pmatrix} U_1' & 0 \\ 0 & U_2' \end{pmatrix} \begin{pmatrix} B_1' & 0 \\ 0 & B_2' \end{pmatrix} \log \left[\begin{pmatrix} A_1' \\ A_2' \end{pmatrix} \pi \right] = \begin{pmatrix} 0 \\ 0 \end{pmatrix} \, .$$

This allows a simultaneous test of both hypotheses, however some caution is needed in combining the constraints of different models in this straightforward way. Sometimes when imposing simultaneous constraints on several marginal tables, or on marginal and joint distributions, several difficulties may be encountered.

A first one is that particular constraints are redundant, i.e., implied by the other ones in the set of constraints. Sometimes these redundancies are easily detected, e.g., by means of design matrices not being of full rank, but this is certainly not always so. In any case, the algorithm will not work and converge to the ML estimates if redundant constraints are specified.

Another class of problems is the specification of incompatible restrictions, i.e., restrictions which contradict each other and cannot be satisfied simultaneously. During the estimation process, such incompatabilities may be resolved by means of 'degenerate' solutions in the sense of not-intended estimated zero effects or uniform distributions, and then may lead to redundancies in the restrictions. For example, imagine a model specification for the cell entries of a successive series of turnover tables that has the (unintended) implication that the marginals of these turnover tables remain stable over time. At the same time, a model is specified for the marginals of these tables that imply a linear net change in location over time. These two models can be resolved by assuming that the linear increase or decrease in location of the marginal is zero. But then, of course, the model for the bivariate joints entries and the model for the marginals contain redundant restrictions. Finally, even if there are seemingly no problems regarding redundancy or incompatibility, difficulties may still arise in terms of applicability of standard asymptotic theory, and even with the substantive interpretation of the resulting model. Fortunately, in many cases frequently occurring in practice, and in most examples discussed in this book, no problems of these kinds occur. Below, some further details of these kinds of problems and their solutions will be presented. Although the solutions and results are only partial, they do cover important situations that occur in practice. Due to the complexity of the problems involved, it may be unrealistic to expect that, for example, a definitely conclusive test for compatibility may be attained. Further extensively discussed examples of incompatible sets of restrictions will be presented at the end of Chapter 4.

An insightful example of a combination of constraints that gives rise to surprising results is the following (Dawid, 1980; Bergsma & Rudas, 2002a, Example 7). For a $2 \times 2 \times 2$ table ABC, it is simultaneously assumed that A and B are marginally independent of each other ($A \perp\!\!\!\perp B$), but also conditionally independent given C ($A \perp\!\!\!\perp B \mid C$). Denote this model as model M_0. Furthermore, let M_1 be the model in which A and C are jointly independent of B ($AC \perp\!\!\!\perp B$), and make M_2 the model where B and C are jointly independent of A ($BC \perp\!\!\!\perp A$). Then, model M_0 is equivalent to either M_1 or M_2 or both. In other words, if model M_0 applies, exactly one of three following situations may occur:

- M_1 applies but M_2 does not;
- M_2 applies but M_1 does not, and
- M_1 and M_2 both apply.

In the last case, the three variables A, B, and C are mutually independent. On the other hand, if either M_1 or M_2 or both apply, M_0 also applies. Because it is not clear what the exact implications are from the original two restrictions in terms of the choice between M_1 and M_2, the interpretation of this combined model is not straightforward. Moreover, the dimension of the model (i.e., the number of free parameters) is not constant: if all three variables are independent, there are three free parameters; in other cases there are four. This leads to nonstandard asymptotics if the true number of free parameters is three. Fortunately, in practice it is usually not difficult to verify the absence of such problems, as discussed below.

First, consider the case of combining (compatible) restrictions on certain marginals with a loglinear model for the joint table. The above example is such a case, as restrictions were placed on the marginal table AB and a loglinear model was assumed for joint table ABC. A simple test for the absence of problems is that the restricted marginals should be a subset of the set of sufficient configurations (Bishop et al., 1975, p. 66) of the loglinear model. For the above example, in model M_0 ($A \perp\!\!\!\perp B \mid C$) the loglinear model for the full table is $\{AC, BC\}$, which means that the marginal tables AC and BC can freely be restricted (provided of course that the restrictions on the marginals themselves are compatible). Instead, above AB was restricted ($A \perp\!\!\!\perp B$), which led to some unexpected results. More generally, precise conditions using matrix formulations can be given. Suppose we restrict the linear combinations of probabilities $A'\pi$, where A is a matrix with nonnegative elements, and the loglinear model has design matrix X, i.e., we assume $\log \pi = X\beta$ for a vector of loglinear parameters β. Then, a sufficient condition for the absence of problems is that the columns of A are a linear combination of the columns of X (Lang & Agresti, 1994; Bergsma & Rudas, 2002a).

Before discussing the more general case of loglinear restrictions on nested sets of marginals (see below), the case of compatibility of fixed marginals must be discussed, as this gives some insight into the former problem. There are some obvious cases where fixed marginals are incompatible: if the AC distribution is assumed to be uniform, i.e., $\begin{pmatrix} .25 & .25 \\ .25 & .25 \end{pmatrix}$, and BC is assumed to be $\begin{pmatrix} .4 & .2 \\ .2 & .2 \end{pmatrix}$, then the C marginals in the two tables are obviously not compatible: in AC it is $(.5, .5)$ and in BC it is $(.6, .4)$. In other words, there is no joint distribution with these two bivariate marginals. The set of marginals $\{AC, BC\}$ is an example of a *decomposable* set of marginals for which compatibility can be checked easily in this way. Generally, a set of marginals is called decomposable if there is an ordering of the marginals so that, for any k, the intersection of the kth marginal with the union of the first $k-1$ marginals equals the intersection of the kth and ℓth marginals for some $\ell < k$. It is less easy to check for the compatibility of *nondecomposable* sets of marginals. An example is the nondecomposable set consisting of AB, BC, and AC. If each of the tables is restricted to be $\begin{pmatrix} 0 & .5 \\ .5 & 0 \end{pmatrix}$, then even though the one-dimensional marginals are compatible, there is no joint distribution with these bivariate marginals since the restrictions imply a perfect negative correlation for all pairs of variables, which is impossible.

The problem of compatibility of several marginal tables is very closely related to the existence of maximum likelihood estimates for loglinear models with zero observed cells, which is the reason decomposability comes in here. In particular, ML estimates of loglinear parameters for a certain loglinear model exist if, and only if, there exists a strictly positive distribution compatible with the sufficient statistics for the model. See Haberman (1973, 1974) for a rigorous treatment of the problem of the existence of ML estimates for loglinear models.

Next, more general (loglinear) constraints on possibly nested marginals will be dealt with. A first marginal (like AB) is nested in a second marginal (like ABC) if it consists of a selection of variables from the second marginal. In this way, every marginal is nested in the complete set of variables from the joint distribution. A general result was obtained by Bergsma and Rudas (2002a). The main sufficient condition for compatibility they formulated is that linear restrictions on loglinear marginal parameters are compatible if no two restricted parameters with different superscripts have indices belonging to the same variables in the subscript. In the above example of marginal and conditional independence, the constraints were $\lambda_{ij}^{AB} = 0$, $\lambda_{ij*}^{ABC} = 0$, and $\lambda_{ijk}^{ABC} = 0$. The problem arises from the first two restrictions: two corresponding loglinear parameters with subscript set $\{i, j\}$ belonging to variables A and B are restricted in the two different marginal tables AB and ABC, leading to the problems. If the compatibility condition is satisfied, then additionally the model interpretation is straightforward and standard asymptotics apply. In Section 4.5, analyses of real data are discussed where these results are relevant.

For affine restrictions on the loglinear marginal parameters, i.e., restrictions in which linear combinations of the loglinear parameters are set equal to a nonzero value, the situation is more complex. Bergsma and Rudas (2002a) showed that if the set of marginals that is restricted is *ordered decomposable*, then the above condition is sufficient to guarantee the compatibility of constraints. A set of marginals is ordered decomposable if there is an ordering such that, for any k, the maximal elements of the first k marginals in the ordering are decomposable. See Bergsma and

Rudas (2002b) and Bergsma and Rudas (2002c) for a more extended discussion and several illustrations of compatibility issues. Extensions of these results to marginal models based on cumulative and other types of logits and higher order parameters are given by Bartolucci, Colombi, and Forcina (2007). An algorithm for checking compatibility of marginal distributions with specific values is given by Qaqish and Ivanova (2006).

Finally, there is the question of the uniqueness of ML estimates. Bergsma (1997) showed that for many marginal homogeneity models, the likelihood has a unique local maximum (which then must be the global maximum). The simplest example is when the marginals are disjoint (like marginals AB and CD in contrast to marginals AC and BC): then for any practically relevant marginal homogeneity model, the likelihood has a unique stationary point that is the ML estimate.

2.3.4 ***Maximum Likelihood Estimates of Constrained Cell Probabilities

Suppose that the observed frequencies n_i, $i = 1, \cdots, I$ are multinomially or Poisson distributed with theoretical cell probabilities π_i. As shown above, when these cells satisfy a loglinear marginal model, there exist matrices A and B such that

$$h(A'\pi) = B' \log(A'\pi) = 0 .$$

The notation $h(A'\pi) = 0$ allows extension to nonloglinear marginal models as well, which are discussed in the next chapter. In general, all loglinear marginal models can be specified in such a way that all rows of matrix B' sum to zero, which means that each row represents a contrast among the logarithms of the cell probabilities. A sufficient but not necessary condition for the rows of $B' = U'C'$ to sum to zero is that the matrix X in Eq. 2.4 contains a constant column. If the rows sum to zero, it is immaterial whether we formulate the model in terms of expected cell frequencies or in terms of cell probabilities. More technically, the function h is such that for any $c > 0$, $h(cx) = h(x)$. This is an important condition simplifying maximum likelihood estimation. We say that the function h is *homogeneous*, and this issue will be discussed in more extensively in Chapter 3. For any homogeneous scalar function f, Euler's theorem says that

$$\sum_i x_i \frac{\partial f(x)}{\partial x_i} = 0 . \tag{2.5}$$

In order to test whether a particular marginal model fits the data well, first the maximum likelihood estimates of the cell probabilities must be obtained under the constraints imposed by the model. Utilizing the Lagrange multiplier method for constrained optimization as described by Aitchison and Silvey (1958) and Aitchison and Silvey (1960), these estimates are obtained by determining the saddle point of the kernel of the Lagrangian log likelihood function

$$L(\pi, \lambda, \mu) = p' \log(\pi) - \mu \left(\sum \pi_i - 1 \right) - \lambda' h(A'\pi) , \tag{2.6}$$

in which p is the vector of observed proportions, λ is a vector of unknown Lagrange multipliers, μ a Lagrange multiplier and

$$h(A'\pi) = B' \log(A'\pi) \,.$$

The term $\mu \left(\sum \pi_i - 1 \right)$ is added to incorporate the constraint that the cell probabilities sum to one. The maximum likelihood estimates are denoted as $\hat{\pi}$, $\hat{\lambda}$, and $\hat{\mu}$.

We will now give a set of equations that has the ML estimate $\hat{\pi}$ as its solution. In these equations, following the method of Bergsma (1997, Appendix A), the Lagrange multipliers will be expressed as a function of π, i.e., we effectively eliminate them from the equations, thus simplifying the problem of finding ML estimates $\hat{\pi}$. We need the Jacobian of the constraint function h, given as

$$H(x) = \frac{dh(x)'}{dx} = D_x^{-1} B \,,$$

where D_x is the diagonal matrix with vector x on the main diagonal. Note that the (i,k)th element of $H(x)$ is given by

$$\frac{\partial h_k(x)}{\partial x_i} \,.$$

Now using the shorthand

$$H = H(A'\pi) \,,$$

the chain rule for matrix differentiation leads to

$$\frac{dh(A'\pi)'}{d\pi} = AH \,.$$

The derivative of the Lagrangian function (Eq. 2.6) with respect to π then is

$$l(\pi,\lambda,\mu) = \frac{dL(\pi,\lambda,\mu)}{d\pi} = \frac{p}{\pi} - \mu - AH\lambda \,.$$

Thus, the ML estimates $(\hat{\pi},\hat{\lambda},\hat{\mu})$ are solutions to the simultaneous equations

$$l(\pi,\lambda,\mu) = 0$$
$$h(A'\pi) = 0 \,.$$

By homogeneity of h and Euler's theorem (see Eq. 2.5), $\pi'AH = 0'$. Hence, $\pi'l(\pi,\lambda,\mu) = 1'p - \mu 1'\pi = 1 - \mu = 0$, and so $\hat{\mu} = 1$. Let

$$l(\pi,\lambda) = l(\pi,\lambda,\hat{\mu}) = \frac{p}{\pi} - 1 - AH\lambda \,,$$

and we now only need to solve the simplified equations

$$l(\pi,\lambda) = 0 \tag{2.7}$$
$$h(A'\pi) = 0 \,. \tag{2.8}$$

Bergsma (1997) proposed to write the Lagrange multiplier vector λ in terms of the cell probabilities π as follows:

$$\lambda(\pi) = \left[H'A'D_\pi AH\right]^{-1} \left[H'A'(p - \pi) + h(A'\pi)\right] .$$

It can then be verified that if $\hat{\pi}$ is a solution of the equation

$$l(\pi, \lambda(\pi)) = 0 , \qquad (2.9)$$

then $(\hat{\pi}, \hat{\lambda})$, with $\hat{\lambda} = \lambda(\hat{\pi})$, is a solution of (2.7) and (2.8), and if $(\hat{\pi}, \hat{\lambda})$ is a solution of Eqs. 2.7 and 2.8, then $\hat{\lambda} = \lambda(\hat{\pi})$. Hence, by writing λ in terms of π and substituting into $l(\pi, \lambda)$, we have reduced the dimension of the problem by effectively eliminating the Lagrange multiplier λ as an independent parameter. That is, we only need to solve Eq. 2.9 in terms of π.

In general, this optimization problem cannot be solved analytically but requires appropriate numerical optimization procedures. In the next section, we describe such an algorithm, based on the likelihood equation Eq. 2.9.

If a stratified sampling procedure has been used, the definition of the Lagrangian can easily be extended to take the existence of different strata in the population into account. Here it is required that h is homogeneous relative to the stratification used (see Lang, 1996b for further details). He also showed that for inference about certain higher order loglinear marginal parameters, usually those of most interest, ignoring the fact that sampling is stratified leads to identical asymptotic inferences.

2.3.5 ***A Numerical Algorithm for ML Estimation

Bergsma (1997), building on previous work by Aitchison and Silvey (1958) and Lang and Agresti (1994), derived the following algorithm for fitting marginal models. The first step of the algorithm is to choose an appropriate starting point $\pi^{(0)}$, after which subsequent estimates $\pi^{(k+1)}$ ($k = 0, 1, 2, \ldots$) are calculated iteratively using the formula

$$\log \pi^{(k+1)} = \log \pi^{(k)} + step^{(k)} l[\pi^{(k)}, \lambda(\pi^{(k)})]$$

for an appropriate step size $step^{(k)}$. The algorithm terminates at an iteration k if $l[\pi^{(k)}, \lambda(\pi^{(k)})]$ is sufficiently close to zero. Although the algorithm looks like a linear search, it can be viewed as a form of Fisher scoring since it is based on a weighting by the inverse of the expected value of the second derivative matrix of the Lagrangian likelihood $L(\pi, \lambda)$ (Bergsma, 1997).

The algorithm of Bergsma used here and the one described by Aitchison and Silvey and Lang and Agresti are both based on the Lagrange multiplier technique, but a salient difference is that the latter searches in the product space of the probability simplex and the space of the Lagrange multipliers, whereas our algorithm searches in the lower dimensional probability simplex. In this sense, Bergsma's algorithm is simpler, and practical experience also indicates that it also performs better numerically. For example, Lang, McDonald, and Smith (1999) needed to impose additional restrictions on certain loglinear marginal models in order to achieve convergence with the Lang-Agresti algorithm, whereas convergence was easily achieved for the same models without the additional restrictions with Bergsma's simplified algorithm.

In our experience, a good choice of starting point $\pi^{(0)}$ is simply the observed cell proportions if all of them are strictly positive. If there are zero observed cells, we found that a slight smoothing towards uniformity works well, a good choice, in particular, seems to be to add .01 divided by the number of cells to each cell, and rescale so that the proportions add up to one. Starting values should always be strictly positive. For certain starting points (wildly different from the observed proportions) in certain problems, we could not reach convergence. Although there is no guarantee that any starting values lead to convergence of the algorithm, all of the manifest variable models in this book could be fitted with the default starting values of our programme. For certain latent variable models, this did not always work; further details are given in Chapter 6.

Since $l(\pi, \lambda(\pi))$ is not the derivative of an unrestricted likelihood function to be maximized, the choice of step size becomes more difficult because we do not know if a new estimate is 'better' than the previous one. However, the step size may be chosen such that an appropriate function that measures the 'distance' of the iterative estimate from the ML estimate decreases. A reasonable function is the quadratic form

$$d(\pi) = l[\pi, \lambda(\pi)]' D_\pi l[\pi, \lambda(\pi)]$$

which is zero if and only if π is a stationary point of $L(\pi, \lambda(\pi))$, and positive otherwise. In the search for an optimal value of the step size, $step^{(k)}$ is initially set equal to 1. If this results in an increase of the criterion $d(\pi)$, the step size is halved. This process of halving the step size is continued until $d(\pi)$ decreases. Unfortunately, it is not always possible to obtain a decrease of $d(\pi)$, because $l(\pi, \lambda(\pi))$ is not a gradient of $d(\pi)$. If that is the case, a 'jump' may have to be made to a different region, for example, by going back to $step^{(k)} = 1$. We could not always get convergence with this step-size halving method: for some problems, we needed to set a maximum to the step size (e.g., 0.3). The maximum permissable step size had to be found by trying out different values, but we always managed to find it fairly quickly. Note that generally speaking the smaller the maximum step size, the higher the likelihood of convergence, but the slower the algorithm potentially becomes. Concluding, the overall procedure for choosing a step size is somewhat ad hoc, but has worked well in practice for us.

A potentially serious problem with the algorithm is the possible singularity or ill-conditioning of the matrix $H'A'D_\pi AH$, which has to be inverted. If the matrix is singular at every value of π, this means that at least one constraint is redundant and needs to be removed, see also the discussion in Section 2.3.3. Another possibility is that the matrix is singular at the ML estimate $\hat{\pi}$, but not at values close to it. Our Mathematica programme then gives warnings about the ill-conditioning during the iterative process. We found that in such cases it still appears to be possible to obtain fairly good convergence of the algorithm, say, to four or five decimal places of the likelihood ratio statistic, but the algorithm will not converge any further. Fortunately, this seems to be rare for manifest data loglinear marginal models discussed in this chapter, although on the latent level (Chapter 6) we did encounter the problem.

Aside from convergence, another important issue is efficiency. Although the algorithm does involve a matrix inversion, for all of the problems in this book the matrix to be inverted is relatively small and does not form a computational bottleneck. The real computational challenge arises from computations involving the matrix A in the marginal model specification. For large tables, this matrix is large and may not be storable in computer memory. However, for marginal models the matrix consists of zeroes and ones and can be stored much more efficiently using sparse array techniques, which are implemented in computer packages such as Mathematica or MATLAB. The advantage is that the zeroes do not need to be stored, but the disadvantage is that we still need to store (the locations of) the ones. To overcome even the latter disadvantage, a programme can be written that avoids the storing of matrix A altogether, and uses its special structure to do computations directly. This approach can also lead to potentially significant speed improvements, but this all depends on the details of the programme implementation.

For either of the approaches outlined above to work, the algorithm needs to be written out in more detail, specifying the order of computations to be done. The computational bottleneck of the algorithm is the computation of $l[\pi, \lambda(\pi)]$, which can be written out fully as

$$l[\pi, \lambda(\pi)] = \frac{p}{\pi} - 1 - A \left(H \left[H'(A'D_\pi A)H \right]^{-1} \left[H'(A'(p - \pi)) + h(A'\pi) \right] \right).$$

Here, extra parentheses have been inserted to indicate the order of evaluation. There are three potentially inefficient computations involving matrix A: 1) multiplication of A by a vector, 2) computation of $A'D_\pi A$, and 3) multiplication of A' by a vector. These operations can be made (much) more efficient both by the use of sparse array techniques or doing the computations using the special structure of matrix A without creating the matrix itself. Note that the matrix $A'D_\pi A$ is typically small compared to the size of the table. For example, say we have 10 trichotomous variables, and $A'\pi$ consists of the univariate marginals, then the full table consists of $3^{10} = 59049$ cells and $A'D_\pi A$ is a 30×30 matrix with only 900 elements.

For comparison purposes, a different order of evaluation is given as follows:

$$l[\pi, \lambda(\pi)] = \frac{p}{\pi} - 1 - (AH) \left[(H'A')D_\pi (AH) \right]^{-1} \left[(H'A')(p - \pi) + h(A'\pi) \right].$$

In this case, matrix AH and its transpose have to be computed and stored. In the example with 10 trichotomous variables, its size is 59049×30. Especially for even larger tables, storing and manipulating this matrix could lead to problems in terms of computing time and space.

A final issue with the algorithm is that for loglinear marginal models, estimated marginal probabilities should not be zero, because we need to take their logarithm. However, if there are zero observed marginal probabilities, estimated probabilities may be zero as well. Since we cannot take the logarithm of zero, we advise incorporating a minimum value for the (joint) estimated probabilities in the estimation procedure. Thus, if at any point during the iterative process an estimated probability obtains a value of, say, less than 10^{-100}, we can replace its value by 10^{-100}.

Concluding this subsection, the algorithm proposed here is not without its problems, however, we have successfully applied it to fit models for tables with more than a million cells.

2.3.6 ***Efficient Computation of ML Estimates for Simultaneous Joint and Marginal Models

If we wish to simultaneously test a loglinear model and one or more loglinear marginal models, the procedure described in the previous subsection can be applied but may be (far) too inefficient, especially for large tables, and we describe a more efficient modified procedure here. This case corresponds to the simultaneous models discussed in Section 2.3.3 with either A_1 or A_2 equal to the identity matrix. The procedure is briefly outlined in Bergsma (1997, page 95) and in detail in Lang et al. (1999). It is especially important for use with the EM algorithm for loglinear latent variable models with marginal constraints (see Chapter 6). The modified algorithm gives the same iterative estimates as the algorithm described in the previous section, but computes these more efficiently.

We assume a loglinear model for π specified as

$$\log \pi = W\gamma$$

and a loglinear marginal model of the form

$$B' \log A' \pi = 0 \,, \tag{2.10}$$

where matrices A and B satisfy the regularity condition described in Section 2.3.3 that the columns of A are a linear combination of the columns of W. Using this regularity condition, we can show that $l(\pi, \lambda(\pi))$ based on the simultaneous marginal and loglinear model reduces to

$$l_W(\pi, \lambda(\pi)) = l(\pi, \lambda(\pi)) + W(W'D_\pi W)^{-1}W'(p - \pi)$$
$$= \frac{p}{\pi} - 1 - AH\lambda(\pi) + W(W'D_\pi W)^{-1}W'(p - \pi)$$

where $l(\pi, \lambda(\pi))$ in the formula is based on the marginal model in Eq. 2.10. The advantage of this formulation is that it is not necessary to compute the orthocomplement of W, which tends to be large. To illustrate, if we have a loglinear model with only first order interactions for 10 trichotomous variables, then W has size $3^{10} \times 201 = 59049 \times 201$, whereas its orthocomplement has size $3^{10} \times (3^{10} - 201) = 59049 \times 58848$, which is almost 300 times larger.

With starting values $\pi^{(0)}$ *satisfying the loglinear model*, the algorithm is analogous to what we did previously, namely, for $k = 0, 1, 2, \ldots$,

$$\log \pi^{(k+1)} = \log \pi^{(k)} - step^{(k)} l_W[\pi^{(k)}, \lambda(\pi^{(k)})] \,.$$

For $\pi^{(0)}$, the uniform distribution can be chosen. Otherwise, the same recommendations for implementation of the algorithm apply as in the previous section.

In this algorithm, in contrast to the one of the previous subsection, not one but two matrices need to be inverted at every iterative step, namely

$$Q(\pi) = W'D_\pi W$$

and

$$R(\pi) = H'A'D_\pi AH .$$

From the assumption that the columns of A are linear combinations of the columns of W, there exists a matrix U such that $A = WU$, and we can write

$$R(\pi) = H'U'W'D_\pi WUH = H'U'Q(\pi)UH .$$

Normally U has full column rank, so if H also has full column rank and Q is nonsingular, R is nonsingular. In practice, either of the matrices Q and R may be singular when evaluated at $\hat{\pi}$. For $Q(\pi)$, we found that a generalized (Moore-Penrose) inverse can be used instead of the true inverse if it doesn't exist. For $R(\pi)$, a reduction is needed in the number of constraints that are imposed, thereby reducing the number of columns of H to make it full column rank. However, in the problems of this book, we found that although the second matrix could be near singular, giving warnings by our Mathematica programme, it was still sufficiently far from singularity to allow the algorithm to converge fairly well. The main potential computational bottleneck of the present algorithm is the actual computation of $Q(\pi)$ and of $R(\pi)$ rather than their inversion. The size of these complexities are increasing with 1) the number of constraints imposed by $h(A'\pi) = 0$, which determines the number of columns of H, and 2) the number of loglinear parameters γ in the model, which determines the number of columns of W.

2.3.7 ***Large Sample Distribution of ML estimates

In the general model formulation of Eq. 2.4, several parameters may be of interest, in particular, 1) the vector of marginal probabilities $A'\pi$, 2) the loglinear marginal parameters $\phi(\pi) = C' \log A'\pi$, and 3) the vector of model parameters β. Under conditions usually met in practice, the ML estimates of these parameters are consistent estimators of the population values and have an asymptotic normal distribution (see Section 2.3.3 for exceptions). In particular, the elements of $\phi(\pi)$ must have continuous second derivatives (Lang, 1996a). Below, using results by Aitchison and Silvey (1958) (see also Lang, 1996a and Bergsma, 1997), we provide the asymptotic covariance matrices of these parameters, assuming the appropriate regularity conditions are met. We first give the asymptotic covariance matrix of the estimated cell probabilities if model defined by Eq. 2.4 is true. Although this matrix can be large and is often of little interest in itself, we can use it to calculate the asymptotic covariance matrix of the aforementioned parameters of interest using the delta method. For Poisson and multinomial sampling, the asymptotic covariance matrix of the estimated probabilities is

$$V(\hat{\pi}) = \frac{1}{N}\left(D_{\pi} - D_{\pi}AH(H'A'D_{\pi}AH)^{-1}H'A'D_{\pi} - \pi\pi'\right). \qquad (2.11)$$

By the delta method, the asymptotic covariance for the marginal probabilities is

$$V(A'\hat{\pi}) = A'V(\hat{\pi})A,$$

for the parameter vector $\hat{\phi} = \phi(\hat{\pi})$ it is

$$V(\hat{\phi}) = C'D_{A'\pi}^{-1}A'V(\hat{\pi})AD_{A'\pi}^{-1}C,$$

and since $\beta = (X'X)^{-1}X'\phi(\pi)$, for $\hat{\beta}$ it is

$$V(\hat{\beta}) = (X'X)^{-1}X'V(\hat{\phi})X(X'X)^{-1}.$$

For the residuals $p_i - \hat{\pi}_i$, to be discussed in the next subsection, the asymptotic covariance matrix is

$$V(p_i - \hat{\pi}) = \frac{1}{N}D_{\pi}AH(H'A'D_{\pi}AH)^{-1}H'A'D_{\pi}. \qquad (2.12)$$

For the perhaps more interesting residuals $\phi_{\text{obs}} - \hat{\phi}$, where $\phi_{\text{obs}} = \phi(A'p)$, the delta method yields

$$V(\phi_{\text{obs}} - \hat{\phi}) = C'D_{A'\pi}^{-1}A'V(p_i - \hat{\pi})AD_{A'\pi}^{-1}C. \qquad (2.13)$$

For stratified sampling, let $\pi^{(k)}$ be the probability vector for stratum k. Then, the asymptotic covariance matrix of the estimated probabilities is

$$V(\hat{\pi}) = \frac{1}{N}\left(D_{\pi} - D_{\pi}AH(H'A'D_{\pi}AH)^{-1}H'A'D_{\pi} - \oplus_k\pi^{(k)}(\pi^{(k)})'\right),$$

where $\oplus$ is the direct sum, defined as the block-diagonal matrix with the summed matrices as blocks. In the same way as above, the corresponding covariance matrices for other parameters are obtained using the delta method. However, we can show that the covariance matrices of loglinear parameters (i.e., the elements of $\hat{\phi}$ and $\hat{\beta}$) need not be affected by the stratification. In particular, in a two-way table formed by a stratifying variable and a response variable, the (log) odds ratios and the main loglinear effect pertaining to the response variable are unaffected, while the main loglinear effect pertaining to the stratifying variable is affected. The covariance matrix of $\hat{\pi}$ under more general types of (stratified) sampling schemes is given in Lang (1996a).

If we simultaneously impose a loglinear model $\log\pi = W\gamma$ and a loglinear marginal model $h(A'\pi) = 0$, we obtain the partitioned covariance matrix

$$V(\hat{\pi}) = V(\hat{\pi}_1) + V(\hat{\pi}_2) - V(\hat{\pi}_0) \qquad (2.14)$$

where $\hat{\pi}_1$ is the ML estimate of π under only the marginal model (see Eq. 2.11 for $V(\hat{\pi}_1)$), $\hat{\pi}_2$ is the ML estimate of π under only the loglinear model, for which

$$V(\hat{\pi}_2) = \frac{1}{N} \left(D_\pi W (W' D_\pi W)^{-1} W' D_\pi - \pi\pi' \right) ,$$

and $\hat{\pi}_0 = p$ is the unrestricted ML estimate of π for which $V(p) = D_\pi - \pi\pi'$. The partitioning holds because of the orthogonality, in the sense of asymptotic independence of ML estimates of the loglinear and marginal parameters (Bergsma, 1997, Section 5.4.2, Appendix A.3 and references therein). Again, the delta method is used to obtain expressions for the relevant asymptotic covariance matrices of other parameters.

Classical confidence intervals are easily calculated using standard errors, which we frequently provide in this book. Currently, other confidence intervals, such as those obtained by inverting the score statistic, are being developed and are now starting to become feasible for marginal models as well. Typically, such intervals are more cumbersome to compute however. The interested reader can consult (Agresti, 2002, Sections 1.4.2 and 3.1.8), Agresti and Coull (1998), Brown, Cai, and Dasgupta (1999) and Lang (2008).

2.3.8 Model Evaluation

In Section 2.1.1, the goodness-of-fit statistics G^2 and X^2 have been described that can be used to evaluate whether a given model fits the data. If the model does not fit well, insight can be gained into the reasons for this lack of fit by analyzing cell (or other) residuals, which are measures for the deviation of observed from fitted cell values. Even if the model fits well, these residuals can be used to detect certain deviations from the model that are not apparent from the overall goodness of fit.

Various types of residuals are in use. For cell i, the raw residual $p_i - \hat{\pi}_i$, where p_i is the observed proportion in cell i, depends strongly on the size of $\hat{\pi}_i$ and is therefore of limited use. A measure that adjusts for the size of $\hat{\pi}_i$ is the *standardized residual*, which is defined as

$$e_i = \sqrt{N} \frac{p_i - \hat{\pi}_i}{\sqrt{\hat{\pi}_i}}.$$

The e_i are related to the Pearson statistic by $\sum e_i^2 = X^2$. Thus, they show, for every cell, exactly how much it contributes to a large value of X^2. Pearson residuals may be useful for marginal probabilities as well, although in that case their squares do not add up to the X^2 statistic, so a bit more care has to be taken in their interpretation.

One drawback of standardized residuals is that their variance is smaller than 1, so a comparison with the standard normal distribution is not appropriate. The *adjusted residual* proposed by Haberman (1974) is defined as the raw residual $p_i - \hat{\pi}_i$ divided by its standard error. Because its mean is 0 and variance is 1, it is better suited for comparison with the standard normal than the standardized residual. Denoting the adjusted residuals by r_i, the definition is

$$r_i = \frac{p_i - \hat{\pi}_i}{se(p_i - \hat{\pi}_i)}.$$

The values of the standard errors $se(p_i - \hat{\pi}_i)$ are given by the square roots of the diagonal entries of the right hand side of Eq. 2.12.

Other adjusted residuals may be considered as well. Perhaps most interesting are the adjusted residuals of marginal loglinear parameters, defined as

$$\frac{\phi_{obs,j} - \hat{\phi}_j}{se(\phi_{obs,j} - \hat{\phi}_j)}.$$

The denominator is obtained as the square root of the corresponding diagonal element of the right-hand side of Eq. 2.13.

3

Nonloglinear Marginal Models

Despite the great flexibility of loglinear models, loglinear marginal models are not always the most appropriate way of analyzing marginal categorical data. Investigators might prefer or need to compare averages, dispersion measures, association coefficients, etc. rather than (specifically) odds and odds ratios. Many of these statistics cannot be formulated within the loglinear framework. Therefore, in this chapter, nonloglinear marginal modeling will be discussed. It will be shown how to extend the maximum likelihood inference procedures and algorithms of the previous chapter to include nonloglinear models. A few concrete examples of the appropriate matrix manipulations for defining these nonloglinear indices will be presented to give the reader a clear picture of how to implement these nonloglinear procedures. Sometimes these matrix manipulations may become rather complicated. However, we have developed R and Mathematica procedures (found on the book's website) that do many of these manipulations automatically. In this way, the examples on the book's website can be adapted in an easy and flexible way for the researcher's specific purposes. But first, as in the previous chapter, several motivating empirical examples will be introduced. The same data from the previous chapter will be used, answering similar research questions but by means of different coefficients. The interested reader may want to compare more closely the results obtained in this chapter with the results from the previous loglinear analyses.

3.1 Comparing Item Characteristics for Different Measurement Levels

In the body satisfaction example of the previous chapter, how the marginal satisfaction distributions varied among the body part items and between men and women was investigated. Now the focus will be on two of the most important characteristics of one-way distributions, viz. central tendency and dispersion. It will first be assumed that body satisfaction is measured as a categorical interval-level variable; next, satisfaction will be treated as a categorical ordinal variable; and finally, as just

W. Bergsma et al., *Marginal Models: For Dependent, Clustered, and Longitudinal Categorical Data*, Statistics for Social and Behavioral Sciences, DOI: 10.1007/978-0-387-09610-0_3, © Springer Science+Business Media, LLC 2009

a nominal scale variable. In this way, it will be shown how in the presence of dependent observations, averages and dispersions of variables can be compared at all levels of measurement.

3.1.1 Interval Level of Measurement

The means of the body satisfaction items were reported in Table 2.5. Going from the least to the most satisfying body aspect, the order was: thighs, buttocks, hips, legs, weight, figure, and body build. Before attaching too much weight to this order, a researcher must make sure that the mean differences are not due to sampling error. Marginal modeling methods have to be used to perform these tests, given the dependencies between the marginal observations. The hypothesis that all item means are equal in the population has to be rejected: $G^2 = 30.49, df = 6 \ (p = .000, X^2 = 22.21)$. It is interesting to note, in line with results from the previous chapter, that the ML estimate of the common mean under this equal means the hypothesis is 3.320, while the overall mean, directly calculated on the data in Table 2.5, inappropriately assuming independent observations equals 3.317. The difference between 3.320 and 3.317 is certainly not large, but it is not due to rounding errors: there is a real difference in parameter estimates due to ignoring the dependencies among the observations. In other applications, these kinds of differences may be much larger than in the present case.

The observed marginal means can be inspected to investigate the mean item differences. By far the largest difference between two successive items in the satisfaction order given above is between legs and weight: $3.3023 - 3.4186 = -.1145$. So there are thighs, buttocks, hips, and legs on the one hand, and weight, figure, and body build on the other. Further detailed significance tests can be carried out to test the mean differences within and between these two clusters.

Regarding the mean differences between the sexes, men are on average somewhat more satisfied with their body than women. The overall mean satisfaction score for all items, calculated in the naive way directly based on the means in Table 2.7, is 3.735 for men, while for women the overall mean equals 3.118. This general tendency is also found for all separate body parts, but the observed differences are larger for one item than for another. One way to test whether these differences in mean are significant is to test the hypothesis that all item means are the same among men and the same among women. The test outcomes for the hypothesis that all item means are equal among men are $G^2 = 8.64, df = 6 \ (p = .195, X^2 = 5.63)$, with an estimated common item mean of 3.738; the analogous outcomes for women are $G^2 = 42.51$, $df = 6 \ (p = .000, X^2 = 27.41)$, with an estimated common item mean of 3.125 (note again that these estimated common subgroup means are a bit different from the corresponding observed subgroup means based directly on Table 2.7). The simultaneous hypothesis that all means are equal among the men and that at the same time all means are equal among the women yields $G^2 = 8.64 + 42.52 = 51.16, df = 6 + 6 = 12$ with $p = .000$ and therefore has to be rejected. Looking at the separate test results for men and women, the cause of the rejection of the simultaneous hypothesis seems to be that the item means are different among women, but not among men. However, as noted before, the test for the men has less power than the test for the women. If the larger sample size of the women is applied to the proportional satisfaction distribution for the men, the test outcomes for the equality of the means among the men is $G^2 = (204/97) \times 8.64, = 18.17, df = 6$ ($p = .006$, $X^2 = 11.84$), a significant result. On the other hand, from the observed means in Table 2.7 it is seen that the item means vary less among men than among women. New data may allow more definite conclusions about the equality of the item means among the men.

Allowing for variation among the item means, also for the men, an interesting model is the no-three-factor interaction model: is the difference between the means of a particular item j and another particular item j' the same for men and women? Or, formulated the other way around: are the mean differences between men and women the same for all items? Application of this no-three-factor interaction model gives an estimated common item mean difference between men and women of .543 with an estimated standard error of .099. However, this model does not fit the data: $G^2 = 27.87, df = 6$ ($p = .000$, $X^2 = 18.55$). To evaluate how specific or exceptional a particular item is with respect to sex differences, the residuals of the equal mean item difference model are useful. The raw residuals are equal to the observed item mean differences minus the estimated constant difference of .543. Larger than the estimated constant difference are the observed differences for thighs (raw residual of .413), followed by legs (.334), buttocks (.201) and hips (.098). Negative raw residuals are found for weight ($-.035$), figure ($-.208$), and body build ($-.291$). According to the adjusted residuals (not reported here), all residual differences are significant except for hips and weight. The general conclusions are that men are on average more satisfied with all their body parts than women; however, for particular parts, (e.g., their thighs), they are especially much more satisfied than women, while for other parts, (e.g., their body build), there is much less difference.

Besides these mean differences, there may be differences in the variation of satisfaction. Perhaps with some body parts, everybody is more or less equally (dis)satisfied, but regarding other parts, the respondents may differ a lot in their answers. Assuming that *Satisfaction* is measured at the interval level, the variance is the most commonly used dispersion statistic. The maximum likelihood estimates of the relevant standard deviations $\hat{\sigma}$ were reported in Tables 2.5 and 2.7

There are small differences among the item variances in the whole sample. Satisfaction with weight has the largest variance ($\hat{\sigma}^2 = 1.326$), followed by thighs (1.208), buttocks (1.195), legs (1.194), hips (1.156), figure (1.048), and finally, body build (1.020). The hypothesis that all item variances are the same in the whole sample has to be rejected at the .05, but not at the .01 level: $G^2 = 15.44, df = 6$ ($p = .017$, $X^2 = 13.06$). The common estimated variance $\hat{\sigma}^2$ equals 1.212 and $\hat{\sigma} = \sqrt{1.212} = 1.101$. The naive overall variance computed directly from Table 2.5 has a somewhat different value of 1.177.

Comparing the variances of men and women (Table 2.7), it is seen that within both groups the observed item variances vary only a little bit, and that the observed variances of the women are all somewhat larger than for the men. The test outcomes for the hypotheses that all item variances are equal within each of the two subgroups are as follows: for men: $G^2 = 8.35$, $df = 6$ ($p = .214$, $X^2 = 6.18$); for women: $G^2 = 8.59$, $df = 6$ ($p = .198$, $X^2 = 6.73$); for both men and women: $G^2 = 8.35 + 8.59 = 16.95$, $df = 6 + 6 = 12$, $p = .152$. The hypotheses that the item variances are equal among the men and among the women need not be rejected. The identical item variance among the men is estimated as 1.102, and among women as 1.150. An even more extreme hypothesis, in which all item variances are the same and moreover identical for men and women, is also acceptable: $G^2 = 18.24$, $df = 13$ ($p = .149$, $X^2 = 15.55$). The common item variance estimate among men and women equals 1.140 (and $\hat{\sigma} = 1.068$).

Despite its many attractive properties, a well-known disadvantage of the variance is its sensitivity to outliers. To overcome this problem, other measures of variation for interval data have been proposed, among them Gini's mean absolute difference g, discussed in Chapter 1. If the g's are estimated for the whole sample, the order of the items in terms of decreasing item variation is essentially the same for $\hat{g}$ as for $\hat{\sigma}^2$: weight ($\hat{g} = .635$), thighs (.607), buttocks (.603), legs (.601), hips (.587), body build (.546), and figure (.545). The differences in $\hat{g}$ are not very large. The mean of the $\hat{g}$'s, simply and directly computed from these observed values, is .589: the expected difference between two randomly drawn observations from an item distribution is roughly .6 for all items, just over half a category. Nevertheless, the hypothesis that all g's are the same in the population has to be rejected: $G^2 = 18.39$, $df = 6$ ($p = .005$, $X^2 = 15.39$). The estimated common constant $\hat{g}$ under this hypothesis is .605, and only body build and figure have significant (negative) adjusted residuals under this model (not reported here).

Also the comparisons of the $\hat{g}$'s for men and women lead to essentially the same conclusions as the comparisons of the variances. The observed g's (not reported here) for the women are consistently a bit higher than for the men, and within the subgroups there is not much variation. The test outcomes for the hypotheses that all item g's are equal within each of the two subgroups are as follows: for men, $G^2 = 9.84$, $df = 6$ ($p = .132$, $X^2 = 7.48$); for women, $G^2 = 9.66$, $df = 6$ ($p = .140$, $X^2 = 7.76$); for both men and women, $G^2 = 9.84 + .9.68 = 19.52$, $df = 6 + 6 = 12$, $p = .077$. The hypotheses that the g's are equal among the men and among the women need not be rejected. The common item variance g for the men is estimated as .553, and for women as .592. The most restrictive assumption that all g's are equal for all items and for men and women is also acceptable: $G^2 = 21.23$, $df = 13$ ($p = .069$, $X^2 = 17.73$), with an estimate for g of .583. Looking at the (significant) adjusted residuals for this model (not reported here), it is suggested that among the men, satisfaction with thighs has less (and with weight, more) variation than $\hat{g} = .583$; among the women, the adjusted residuals point to less variation for body build.

When the categorical data are truly interval data, classical approaches such as t- and F-tests and (M)ANOVA models may form well-known alternatives to the categorical marginal-modeling approach (but assuming continuous population variables). Under which conditions the marginal-modeling approach has more power than the classical approach, or vice versa, is not immediately clear. However, the approach advocated here offers in the first place a much more general and flexible framework for comparing dependent observations and categorical distributions that can also be used in nonstandard, non-textbook situations; and secondly, it does not rely on underlying normal distributions: an underlying (product)multinomial distribution is all that is required (see also the discussion on marginal models for continuous data in Chapter 7).

3.1.2 Ordinal Level of Measurement

The use of means and variances above implied that *Satisfaction* was treated as a variable measured at the interval level. The respondents, however, may not have reacted to the *Satisfaction* scale in terms of an equal distance interval scale, but treated the scale numbers merely in terms of more or less satisfaction. In other words, *Satisfaction* might be an ordinal rather than an interval-level variable. As explained in Chapter 1, the median will not be used as the location statistic for ordinal variables. Instead, the comparative coefficient L_{ij} will be used.

When the matrix of all observed $\hat{L}_{ij}$'s (not reported here) is inspected for the whole sample, an interesting pattern emerges. If the items are put in the order thighs–buttocks–hips–legs–weight–figure–body build, then all $\hat{L}_{ij}$'s are negative when item i appears earlier in the sequence than item j (note that this is the same order as found for the means). Moreover, the larger the distance between the two items in the sequence, the larger the (absolute) value of the L_{ij}'s is. This points to a quasi-symmetrical pattern, and such a pattern suggests that the BT-model as described in Chapter 1 might be valid for these data. More precisely, it suggests the validity of the hypothesis that all $l_{ij} = 0$ (Agresti, 1990, p. 438-39; Fienberg & Larntz, 1976). The test results for the hypothesis that indeed $l_{ij} = 0$ are $G^2 = 16.52$, $df = 15$ ($p = .348$, $X^2 = 16.12$), and there appears to be no reason to reject the null hypothesis that all $l_{ij} = 0$.

The degrees of freedom have been calculated, following Agresti, by noting that there are $7 \times 6/2 = 21$ observed $\hat{L}_{ij}$'s minus six remaining parameters l_i to estimate (Agresti, 2002, p. 436). In our example, however, the matrix of observed $\hat{L}_{ij}$'s is not of full rank. There is a high degree of (multi)collinearity in this matrix and there are only 18 independently observed $\hat{L}_{ij}$'s. This would suggest $18 - 6 = 12$ rather than the theoretical 15 degrees of freedom. Furthermore, when the BT model was fitted, even more collinearity was found: it was only necessary to set nine appropriately chosen l_{ij}'s to zero rather than the 12 predicted from the rank of the observed L_{ij}'s. This suggests that the 'true' degrees of freedom equals 9, which would yield the p-value .057. However, calculating the number of restrictions that are needed to fit the model in this empirical manner makes the number of degrees of freedom a random variable, based on the nature of the sample data rather than fixed and based on the model itself and a given design. Not much more can be said regarding these different amounts of 'empirical' degrees of freedom other than basing the p-value on the theoretical, fixed 15 degrees of freedom yields a conservative test. On the other hand, the estimates of interaction effects l_{ij} in the saturated model are all very small: their absolute values are all smaller than .006 and almost none of them are significant. In the end, it was concluded that the BT-model with all $l_{ij} = 0$ is acceptable, given the data.

This degrees-of-freedom problem occurs in all following analyses using L_{ij} (and W_{ij} in the analyses below), both in the whole sample and in the of subgroups men and women. From now on, only the theoretically correct number of degrees of freedom will be reported, but in drawing conclusions, the outcomes of conditional tests will be mainly decisive, comparing hierarchically nested, more - and less - restricted models, and in this way doing away with the multicollinearity problem in the L_{ij}'s.

The estimates of l_i in the BT-model with $l_{ij} = 0$ order the items according to the same sequence as above on the basis of the observed $\hat{L}_{ij}$'s (between parentheses $\hat{l}_i$ and its standard error): thighs ($-.079; .017$), buttocks ($-.068; .019$), hips ($-.033; .017$), legs ($-.010; .018$), weight ($.033; .021$), figure ($.068; .017$), and body build ($.089; .018$). The later an item appears in the sequence, the higher its location toward the end of the satisfaction scale and the more people that are satisfied with it. More precise and exact statements can be made for all item comparisons on the basis of $\hat{l}_i$. For example, comparing satisfaction with thighs to satisfaction with body build: $\hat{L}_{T,Bb} = \hat{l}_T - \hat{l}_{Bb} = -.079 - .089 = -.168$, meaning that if one first randomly draws an observation from the satisfaction distribution of thighs and then secondly from the satisfaction distribution of body build, it is estimated that the probability that the thigh draw has a higher satisfaction score than the body build draw is .166 lower than the probability that the thigh draw has a lower satisfaction score than the body build draw.

The hypothesis that in the BT-model all $l_i = 0$ amounts to the hypothesis that all ordinal item locations are the same and that, given the validity of the BT model, $L_{ij} = 0$. The test statistics for this hypothesis are $G^2 = 34.21, df = 21$ ($p = .034$, $X^2 = 25.89$). The outcomes of the conditional test of $L_{ij} = 0$ against the (not further restricted) BT-model are $G^2 = 34.21 - 16.52 = 17.69, df = 6, p = .007$. It seems best to put our faith in the unrestricted BT-model and use the $\hat{l}_i$'s to compare the ordinal items in terms of their location.

Regarding the subgroup comparisons, one should first investigate whether or not the BT-model fits within the subgroups of men and women, i.e., test the hypotheses that all $l_{ij} = 0$ for men and for women. In both subgroups (separately and simultaneously), the BT-model can be accepted. For men, the test outcomes are $G^2 = 9.52, df = 15$ ($p = .849, X^2 = 7.39$), and for women they are $G^2 = 5.29, df = 15$ ($p = .989, X^2 = 4.97$). Inspection of the $\hat{l}_i$'s (not reported here) leads to conclusions very much in line with the results obtained above when comparing the item means. For example, conditionally testing the hypothesis $l_i = 0$, given that the BT-model is true, leads to the similar conclusion that the hypothesis can be accepted among the men ($G^2 = 12.95 - 9.52 = 3.43, df = 6, p = .753$), but not among the women: ($G^2 = 44.79 - 5.29 = 39.50, df = 6, p = .000$). Given the smaller sample size for men, the best decision is probably to accept for these data that there may be location differences for both men and women. It then becomes interesting to investigate the model of no-three-factor interaction, that is, the model of constant men-women location differences for all ordinal items.

To be able to compare the locations of the items of men and women with each other, the model in Eq. 1.1 has to be extended:

$$L_{ikjl} = l_{ik} - l_{jl} + l_{ikjl}$$

where subscripts i and j refer to *Items* and k and l to *Sex*.

Altogether there are $14(14-1)/2 = 91$ pairwise comparisons L_{ikjl} (or 21 for men, 21 for women, and 49 (= 7×7) for men-women comparisons). The first interesting thing about the observed L_{ikjl}'s (not reported here) pertaining to the pairwise men-women comparisons (L_{i1j2}) is that they are all positive: no matter which items are being compared, men are always more satisfied than women. Note that this is in agreement with the means reported in Table 2.7, where the lowest mean for the men (3.65 for body build) is still higher than the highest mean among the women (3.13 for figure).

For being able to make the men-women comparisons between all items, the BT-model should be valid, i.e., hypothesis $l_{ikjl} = 0$ should be valid. The degrees of freedom for the test of this hypothesis are 91 (paired comparisons) minus $14 - 1 = 13$ parameters to be estimated. The test outcomes are $G^2 = 34.23$, $df = 78$ ($p = 1.000$, $X^2 = 26.93$). Given this test outcome and the sizes of $\hat{l}_{ikgl}$, the BT-model will be accepted. Looking at the item differences ($l_{i1} - l_{i2}$) for men and women in this model, the biggest difference is found for thighs ($.182 - (-.275) = .457$) and the smallest for body build ($.048 - (-.001) = .050$)). The hypothesis $l_{i1} - l_{i2} = l_{j1} - l_{j2}$, that all men-women item differences are the same, has to be rejected on the basis of the conditional test, given the validity of the BT-model (with $l_{ikjl} = 0$): $G^2 = 53.30 - .34.23 = 19.09$, $df = 6$, $p = .004$. The estimated constant item difference between men and women in this rejected model equals .265 (s.e. $= .019$). According to the adjusted residuals for this constant difference model (not reported here), the difference between men and women is significantly larger for thighs and legs (men tend to be much more satisfied about these body parts than women) and significantly smaller for body build and figure (men tend to be just a little more satisfied with these bodily aspects than women), conclusions that are much in agreement with the overall conclusions of the item mean comparisons between men and women.

For the investigation of the variation in an ordinal variable, the coefficient W_{ij} was introduced in Chapter 1. Inspection of the observed $\hat{W}_{ij}$'s in the whole sample (not reported here) shows that the pairwise differences in ordinal variation are very small. All absolute values are less than .009, except for the pairwise comparisons with weight, which are around .025 (note that weight had the largest standard deviation in Table 2.5). For being able to meaningfully compare the W_{ij}'s with each other, the BT-model should be valid with the crucial hypothesis $w_{ij} = 0$. The BT-model gives the test outcomes: $G^2 = 18.07$, $df = 15$ ($p = .259$, $X^2 = 16.78$). The $\hat{w}_i$'s estimated under this BT-model again point to very small differences in variation. A conditional test that all variances are the same, given the BT-model, that is, $\hat{w}_i = 0$ need not be rejected: $G^2 = 27.60 - 18.07 = 9.53$, $df = 6$, $p = .146$.

For the men-women comparisons, an analogous equation may be set up as for the ordinal location:

$$W_{ikjl} = w_{ik} - w_{jl} + w_{ikjl} \; .$$

The BT-model with $w_{ikjl} = 0$ need not be rejected: $G^2 = 34.41$, $df = 78$ ($p = 1.000$, $X^2 = 25.19$). Looking at the parameter estimates $\hat{w}_{ik}$ in the (modified) BT-model, the ordinal variation seems to be the same for all items and for men and women. Carrying out a conditional test of the equality of all ordinal variations among items and simultaneously between men and women, given the BT-model, yields $G^2 = 43.03 - 34.41 = 8.62$, $df = 13$, $p = .801$. There is no reason to reject the hypothesis of equal ordinal variation among items and between men and women: none of the adjusted residuals even approximates statistical significance, except weight among men. This one significant adjusted residual equals 3.64. The satisfaction distribution of weight among men is the only item that shows a variation that deviates from all the others and is bigger than all other ordinal variations.

3.1.3 Nominal Level of Measurement

Completely ignoring the ordered nature of the categories of *Satisfaction* means treating *Satisfaction* as a purely nominal-level variable. For nominal-level variables, the terms location and variation have only a somewhat analogous meaning compared to ordinal and interval-level variables. The relevant concept for nominal-level variables is diversity or dispersion in the sense of how evenly the research elements are dispersed over the nominal categories. Even the mode, which is sometimes denoted as a location statistic, is closely related to the concept of diversity since it indicates in which category most elements are concentrated. Because the mode suffers from the same discontinuity problem as the median and as hypotheses about the comparison of modes are rare in practice, the mode will be ignored here. Substantive research questions that come perhaps closest to location hypotheses are questions about comparing the density or the relative frequency of a particular category, such as: are more women 'completely satisfied' with their figure than men? Actually, such questions have been discussed in the previous chapter and will be taken up again in a somewhat different form in the next subsection.

In this subsection, research questions will be dealt with that pertain directly to dispersion in nominal characteristics, such as 'How do cities compare regarding religious or racial dispersion' or more directly, in terms of our body satisfaction example 'How do men and women compare regarding their satisfaction dispersion?' Two often-used measures for variation in nominal-level variables are the index of diversity D and the information index H, which are both discussed in Chapter 1. If we order the body satisfaction items in the whole sample (Table 2.5) according to these two measures of nominal variation, the first thing to note is that these two rankings are exactly the same. The ranking is also completely identical to the order of the items on the basis of $\hat{\sigma}^2$, $\hat{g}$, and $\hat{W}_{ij}$; the differences that do occur pertain to very small differences in size. Just to give an idea of the sizes of the coefficients D and H, the item order regarding nominal variation is (between parentheses the coefficients D and H, respectively): weight (.757; 1.496), buttocks (.750; 1.476), thighs (.746; 1.465), legs (.743; 1.460), hips (.735; 1.445), body build (.713; 1.383), and figure (.694; 1.363).

Although these coefficients do not vary much among the items, the hypothesis that all D's are the same has to be rejected ($G^2 = 24.75$, $df = 6$, $p = .000$, $X^2 = 23.40$), as is the hypothesis of equal H's ($G^2 = 23.39$, $df = 6$, $p = .001$, $X^2 = 18.97$).

It is also found (again) that the variation among women is a bit larger than among men, both regarding D and H. The test outcomes for the hypotheses that all item variances are the same among men and are the same among women are as follows:

- for D:
  - among men $G^2 = 9.81$, $df = 6$ ($p = .133$, $X^2 = 7.80$) with $\hat{D} = .699$
  - among women $G^2 = 15.07$, $df = 6$ ($p = .020$, $X^2 = 13.23$) with $\hat{D} = .744$
  - simultaneous $G^2 = 9.81 + 15.07 = 24.88$, $df = 12$, $p = .015$
- for H:
  - among men $G^2 = 8.37$, $df = 6$ ($p = .212$, $X^2 = 6.62$) with $\hat{H} = 1.338$
  - among women $G^2 = 14.10$, $df = 6$ ($p = .029$, $X^2 = 11.00$) with $\hat{H} = 1.455$
  - simultaneous $G^2 = 8.37 + 14.10 = 22.47$, $df = 12$, $p = .033$

The test results obtained for the women and for the simultaneous analysis of both groups show that the equal H and equal D hypotheses are acceptable at the .01 level and have to be rejected at the .05 level; the equal variation hypotheses are acceptable for men, but then it has to be remembered that the sample size for men is much lower than for women. Allowing for the possibility that the item H's and the item D's are not constant among the men and not constant among the women, an interesting question is whether the men-women differences are the same for all items. In other words, is the no-three-factor interaction model valid in the population? The test outcomes for D are $G^2 = 9.93$, $df = 6$ ($p = .127$, $X^2 = 8.90$) with an estimated difference between men and women for all items of $\hat{D}_m - \hat{D}_w = -.049$ (s.e. $= .019$); for H, $G^2 = 9.40$, $df = 6$ ($p = .153$, $X^2 = 7.56$) with an estimated difference between men and women for all items of $\hat{H}_m - \hat{H}_w = -.117$ (s.e. $= .051$). The hypotheses of common item differences in terms of the dispersion coefficients D and H can be accepted. The common item difference between men and women is such that men show somewhat less variation than women; the difference is small but significant (unlike the results for and $\hat{\sigma}^2$, $\hat{g}$ and $\hat{W}_{ij}$).

3.2 Comparing Associations

In some respects, the distinction that is made here (and before) between comparing distributions and comparing associations is an artificial one. In general, association measures indicate how much two or more distributions differ from each other, usually how different they are from the situation of statistical independence (or how far from marginal homogeneity when dependent observations are involved). In that sense, the difference between the means of the satisfaction with body items i and j, for example, is a measure of the association between the two items and when investigating whether this mean difference is the same for men and women, associations are being compared. Further, the whole previous chapter on loglinear marginal modeling is essentially about estimating odds ratios and comparing odds ratios. Therefore,

Table 3.1. Association between *Sex Role Attitude* and *Sex* for Parents and Children. Source NKPS, see text

Respondent's Status		Parent		Child	
Sex		Men	Women	Men	Women
Sex Role	1. Nontraditional	195 (.304)	427 (.344)	329 (.434)	724 (.643)
	2, 3. Traditional	446 (.696)	816 (.656)	429 (.566)	402 (.357)

as will become obvious, the hypotheses and models in this subsection are closely related to those of the previous sections and chapters. However, they will be formulated in terms of comparisons of (dependent) bivariate distributions and by means of coefficients that are commonly called 'association measures'.

To illustrate the marginal analyses for comparing associations coefficients, the data of the NKPS from Chapter 2 will be used again. First, the relationship between *Sex* and *Sex Role Attitude* will be dealt with. The pertinent data are given in the last two columns of Table 2.9. The relative distribution of *Sex Role Attitude* in the group of men is given by the probability vector $(.375, .474, .152)$ for the three consecutive response categories. In the group of women, this distribution is $(.486, .399, .115)$. Using the proportions' difference ε as a measure of association, the main difference between men and women is for the nontraditional category 1, $\hat{\varepsilon}_1 = .375 - .486 = -.111$ (s.e. $= .016$); for the moderately traditional category 2, $\hat{\varepsilon}_2 = .075$ (s.e. $= .016$); and for the traditional category 3, $\hat{\varepsilon}_3 = .036$ (s.e. $= .011$). One of the three $\hat{\varepsilon}$'s is of course redundant, since their sum is necessarily equal to zero. Men appear to be somewhat more traditional than women. When computing the standard errors of the $\hat{\varepsilon}$'s, one should take into account that the comparisons of men and women involve (partially) dependent observations, as explained in the previous chapter. The same is true for the significance tests. Using the appropriate marginal-modeling approach, it was found that all three $\hat{\varepsilon}$'s are statistically significant, given the estimated standard errors and the chi-square tests (for the smallest one $\hat{\varepsilon}_3$ and testing the hypothesis $\varepsilon_3 = 0$: $G^2 = 10.59$, $df = 1$ ($p = .001$, $X^2 = 10.78$). Not surprisingly, the hypothesis that all ε's are 0 must be rejected. This last hypothesis is identical to the marginal homogeneity model for men and women where men and women have the same proportional distribution of *Sex Role Attitude;* the test results are accordingly exactly the same as found before in Chapter 2: $G^2 = 46.91$, $df = 2$ ($p = .001$, $X^2 = 46.33$).

In the previous chapter, it was found that (significant) men-women differences regarding *Sex Role Attitude* existed only for children, not for parents. In these loglinear analyses, the men-women differences are expressed according to a multiplicative model, using odds and odds ratios. These differences can, however, also be expressed in terms of an additive model using ε's. Table 3.1 provides the relevant data (obtained from Table 2.10). For reasons of simplicity, *Sex Role Attitude* has been collapsed into two categories, viz. 1 *vs* 2 and 3 (note that the biggest $\hat{\varepsilon}$ is for this dichotomy). Among the parents, there is hardly any difference between men and women regarding *Sex Role Attitude* (Table 3.1): $\hat{\varepsilon}_P = .304 - .344 = -.039$, while among children the difference is much bigger: $\hat{\varepsilon}_C = .434 - .643 = -.209$. For testing the significance of $\hat{\varepsilon}_P$ and $\hat{\varepsilon}_C$ separately, the usual standard tests can be applied to the data in Table 3.1 because the respondents within the subtable for parents and the subtable for children have been independently observed given the sampling design of the NKPS module used here. The test results for $\hat{\varepsilon}_P$ are: $G^2 = 2.98$, $df = 1$ ($p = .085$, $X^2 = 2.955$), and for $\hat{\varepsilon}_C$ are: $G^2 = 80.39$, $df = 1$ ($p = .000$, $X^2 = 80.23$). The hypothesis that *Sex* has no influence on the *Sex Role Attitude* can be accepted for the parents, but not for the children. However, from the fact that for parents the tests results are not significant and for children they are, it cannot be concluded that the two ε's are different from each other. Therefore, a direct test for the hypothesis $\varepsilon_P = \varepsilon_C$ is needed, and this test does involve (partially) dependent (matched) observations, and a marginal-modeling approach is now necessary. Hypothesis $\varepsilon_P = \varepsilon_C$ must be rejected ($G^2 = 26.99$, $df = 1$ $p = .000$, $X^2 = 27.04$), indicating that the parents' sex differences regarding traditionalism are different from the children's.

Another set of questions that was investigated in the previous chapter concerned the relationship between the two related characteristics of *Sex Role Attitude* and *Marriage Attitude* (Table 2.12). The loglinear analyses showed that the association between the two characteristics is the same for parents and children. Now it is asked here whether this result also holds for other association measures. Treating *Sex Role Attitude* and *Marriage Attitude* as interval-level variables, product moment correlations ρ can be compared for parents ($r_P = .453$, s.e. $= .019$), and for children ($r_C = .440$, s.e. $= .022$). The marginal-modeling test for the hypothesis $\rho_P = \rho_C$ yields $G^2 = 0.19$, $df = 1$ ($p = .663$, $X^2 = 0.19$). Also in terms of ρ, there is no reason to assume that the association between the two traditional attitude variables is different for parents and children. The estimated common value of ρ is $r = .447$ with a standard error of $.015$.

Treating the two variables more realistically as ordinal-level variables, γ might be a good choice for measuring the association. The estimated values are $\hat{\gamma}_P = .647$ and $\hat{\gamma}_C = .725$. The test statistics for the hypothesis $\gamma_P = \gamma_C$ are $G^2 = 5.57$, $df = 1$ ($p = .018$, $X^2 = 5.54$). Whether to accept the equal γ's null hypothesis or not depends on the significance level chosen ($.01$ or $.05$). The common γ is estimated as $\hat{\gamma} = .681$ (s.e. $= .017$).

Completely ignoring the order of the categories leads to the use of association coefficients for nominal-level variables, two of which will be given here. One is Goodman and Kruskal's τ, which can be interpreted as the proportion of explained variation when variation is measured in terms of the dispersion (or concentration) index D. Predicting (arbitrarily) the scores on *Marriage* from *Sex Role Attitude*, the results are: $\hat{\tau}_P = .174$ (s.e. $= .015$) and $\hat{\tau}_C = .201$ (s.e. $= .022$). The hypothesis that the relationship is of equal strength for parents and children in terms of the τ coefficient $\tau_P = \tau_C$ can be accepted ($G^2 = 0.88$, $df = 1$ $p = .349$, $X^2 = .88$), with a common $\hat{\tau} = .186$ (s.e. $= .013$). Similar results are obtained for the association coefficient U based on explained variation in terms of the entropy (or uncertainty) coefficient H. Predicting again *Marriage* from *Sex Role Attitude*, $\hat{U}_P = .122$ (s.e. $= .010$), $\hat{U}_C = .129$ (s.e. $= .012$), and the test results for the hypothesis $U_P = U_C$ are

$G^2 = .19$, $df = 1$ ($p = .661$, $X^2 = .125$). The common estimate is $\hat{U} = .125$ (s.e. $= .008$). So, only in terms of γ, there may be a difference in association between parents and children.

The main purpose of presenting all the above examples was to show to the reader the extreme flexibility of the marginal-modeling approach advocated here. How to exactly specify the nonloglinear models and how to obtain the maximum likelihood estimates is the subject of the remainder of this chapter.

3.3 Maximum Likelihood Estimation

For fitting nonloglinear marginal models, essentially the same algorithm as described in the previous chapter can be used, but with a more complex constraint function and associated Jacobian. In Sections 3.3.1 and 3.3.3, the matrix specification of the model constraints for nonloglinear marginal models will be presented, and in Section 3.3.4, more details for the implementation of the algorithm will be given.

3.3.1 Generalized exp-log Specification of Nonloglinear Marginal Models

The formulation of marginal models and the notational system used in the previous chapter can be extended to include nonloglinear marginal models as well. Let π be the vectorized theoretical cell probabilities of the joint table. Remember that when vectorizing a matrix or a multidimensional array, here the first index changes the fastest and the last index changes the slowest. As seen in Chapter 2, but now with a slightly more general and flexible notation, for appropriately defined matrices A, C, and X, loglinear marginal models can be represented as

$$\phi(A'\pi) = X\beta \, ,$$

where $\phi(A'\pi)$ is a vector of marginal loglinear parameters defined as

$$\phi(A'\pi) = C' \log(A'\pi) \, .$$

In parameter-free form, the loglinear marginal model is

$$B'\phi(A'\pi) = 0 \, ,$$

where B is an orthocomplement of X. This representation of loglinear marginal models can be generalized to represent general nonloglinear marginal models by the formula

$$\theta(A'\pi) = X\beta \, ,$$

which, analogously to loglinear marginal models, can be represented in parameter free form as

$$B'\theta(A'\pi) = 0 , \qquad\qquad (3.1)$$

again with B an orthocomplement of X. The main issue now is the specification of θ.

The approach proposed here for the representation of θ and nonloglinear marginal models is an extension of the notational system developed by Grizzle et al. (1969); Forthofer and Koch (1973), Kritzer (1977) and Bergsma (1997). Although these authors (except for the latter one) developed their system for use with the weighted least squares (or GSK) method, it can also be applied within the context of ML estimation and testing. First, the developed (recursive) system will be described in general terms and then several examples will be provided to illustrate its application to concrete estimation and testing problems. By means of these examples, the great versatility of this recursive notational system will be shown.

A General Recursive Representation for Nonloglinear Coefficients

First, the general representation of a vector of coefficients $\theta(x)$ is considered. To formulate marginal models as above, x needs to be replaced by $A'\pi$ in order to obtain the vector of marginal coefficients $\theta(A'\pi)$ that defines all relevant marginal probabilities to which the ensuing transformations will be applied. For the nonloglinear coefficients θ considered in this book, it is always possible to find a sequence of scalar functions $t_1, t_2, \cdots, t_n$ and a sequence of associated weight matrices $W_1, W_2, \cdots, W_n$ of appropriate order so that a sequence of transformations $u_0(x), u_1(x), \cdots, u_i(x), \cdots u_n(x)$ of x is defined that satisfies

$$u_0(x) = x$$
$$u_1(x) = W_1' t_1(u_0(x))$$
$$\vdots$$
$$u_i(x) = W_i' t_i(u_{i-1}(x))$$
$$\vdots$$
$$u_n(x) = W_n' t_n(u_{n-1}(x)) .$$

Then,

$$\theta(x) = u_n(x) .$$

This system was formulated by Bergsma (1997), who only used the exponential and logarithmic functions for the t_i, and applied the term 'exp-log notation'. Although these functions are most commonly needed, for some coefficients other functions are needed as well, and hence we use the term 'generalized exp-log notation' for this very flexible notational system.

For the parameters in a loglinear model

$$\phi(x) = C' \log(x) ,$$

one scalar function $t_1(x) = \log(x)$ and one weight matrix $W'_1 = C'$ must be defined, i.e., $n = 1$. For many coefficients, the scalar functions t_i will either be the (natural) logarithm $t_i(x) = \log(x)$, the exponential function $t_i(x) = \exp(x)$, or the identity function $t_i(x) = x$. In some applications, more 'unusual' functions may be needed, but the estimation procedure discussed later requires all those functions to be continuous and differentiable. A vector of coefficients θ that can be represented by the recursively defined $u_n(x)$ of the above form is said to be *depth-n representable*. All coefficients dealt with in this chapter are depth-n representable for some n, and loglinear coefficients are depth-1 representable. Representations of this kind need not be unique, however.

To make the above general exposition more concrete, a few examples will be presented that also illustrate the flexibility of the recursive notation to represent various nonloglinear coefficients and models. There will be four examples. In the first three, it will be shown how to represent various coefficients and in the last one, a model is defined using the coefficient specification. To simplify the ensuing mathematical expressions, no use will be made of the superscript notation to represent joint or marginal cell probabilities in these examples, but we will simply use a vector π as the argument in $\theta(\pi)$.

Example 1: Representation of ε in a 2×2 Table

Let $\pi' = (\pi_{11}, \pi_{12}, \pi_{21}, \pi_{22})$ be the (row-wise) vectorized form of a 2×2 contingency table. The ε coefficient (with the column variable taken as the response and the row variable as the explanatory variable) can be represented in the following way. First define the matrix

$$W'_1 = \begin{pmatrix} 1 & 0 & 0 & 0 \\ 1 & 1 & 0 & 0 \\ 0 & 0 & 1 & 0 \\ 0 & 0 & 1 & 1 \end{pmatrix},$$

so that, with $t_1(x) = x$,

$$W'_1 t_1(\pi) = \begin{pmatrix} \pi_{11} \\ \pi_{1+} \\ \pi_{21} \\ \pi_{2+} \end{pmatrix}.$$

Then define

$$W'_2 = \begin{pmatrix} 1 & -1 & 0 & 0 \\ 0 & 0 & 1 & -1 \end{pmatrix},$$

and $t_2(x) = \log(x)$ so that

$$W'_2 t_2(W'_1 \pi) = \begin{pmatrix} \log(\pi_{11} / \pi_{1+}) \\ \log(\pi_{21} / \pi_{2+}) \end{pmatrix}.$$

Finally, with

$$W_3' = \begin{pmatrix} 1 & -1 \end{pmatrix}$$

and $t_3(x) = \exp(x)$, the desired ε coefficient follows:

$$\varepsilon = W_3' t_3(W_2' t_2(W_1' t_1(\pi))).$$

In this recursive representation of a single ε-coefficient, the three scalar functions needed are $t_1(x) = x$, $t_2(x) = \log(x)$ and $t_3(x) = \exp(x)$.

Example 2: Representation of the Association Coefficient γ

Suppose a researcher wants to compute the association coefficient γ in a 2×3 table. Let vector

$$\pi' = (\pi_{11}, \pi_{12}, \pi_{13}, \pi_{21}, \pi_{22}, \pi_{23})$$

contain the cell probabilities from this 2×3 table. Now, define matrices W_1, W_2, W_3, W_4 and W_5 as follows:

$$W_1' = \begin{pmatrix} 1 & 0 & 0 & 0 & 1 & 0 \\ 1 & 0 & 0 & 0 & 0 & 1 \\ 0 & 1 & 0 & 0 & 0 & 1 \\ 0 & 1 & 0 & 1 & 0 & 0 \\ 0 & 0 & 1 & 1 & 0 & 0 \\ 0 & 0 & 1 & 0 & 1 & 0 \end{pmatrix},$$

$$W_2' = \begin{pmatrix} 1 & 1 & 1 & 0 & 0 & 0 \\ 0 & 0 & 0 & 1 & 1 & 1 \\ 1 & 1 & 1 & 1 & 1 & 1 \end{pmatrix},$$

$$W_3' = \begin{pmatrix} 1 & 0 & -1 \\ 0 & 1 & -1 \end{pmatrix},$$

and

$$W_4' = \begin{pmatrix} 1 & -1 \end{pmatrix}.$$

With $t_1(x) = \log(x)$ and $t_2(x) = \exp(x)$, it follows that

$$u_2(\pi) = W_2' t_2(W_1' t_1(\pi))$$
$$= \begin{pmatrix} \pi_{11}\pi_{22} + \pi_{11}\pi_{23} + \pi_{12}\pi_{23} \\ \pi_{12}\pi_{21} + \pi_{13}\pi_{21} + \pi_{13}\pi_{22} \\ \pi_{11}\pi_{22} + \pi_{11}\pi_{23} + \pi_{12}\pi_{23} + \pi_{12}\pi_{21} + \pi_{13}\pi_{21} + \pi_{13}\pi_{22} \end{pmatrix}.$$

Let Π_C be the probability of concordance and Π_D the probability of discordance in the table. Then Π_C and Π_D are given by twice the first and second element in the vector above, and $\Pi_C + \Pi_D$ by twice the third element. Now with $t_3(x) = \log(x)$ and $t_4(x) = \exp(x)$,

$$u_4(\pi) = W_4' t_4(W_3' t_3(u_2(\pi))) = \frac{\Pi_C}{\Pi_C + \Pi_D} - \frac{\Pi_D}{\Pi_C + \Pi_D},$$

so that

$$\gamma = u_4(\pi).$$

Example 3: Representation of the Uncertainty Coefficient U

The main reason for introducing this example is to show that other functions than the logarithmic and exponential functions can be incorporated into the recursive representation in a straightforward (although not always easy to see) manner as a transformation function t_k. In Chapter 1, the proportional reduction in variation index U was introduced. This coefficient can be written as

$$U = -\frac{\sum_i \sum_j \pi_{ij} \log \pi_{ij}}{\sum_j \pi_{+j} \log \pi_{+j}} + \frac{\sum_i \pi_{i+} \log \pi_{i+}}{\sum_j \pi_{+j} \log \pi_{+j}} + 1 \ .$$

Suppose you have a 2×2 table with

$$\pi' = (\pi_{11}, \pi_{12}, \pi_{21}, \pi_{22}) \ .$$

Then let $t_1(x) = x$ and

$$W_1' = \begin{pmatrix} 1 & 0 & 0 & 0 \\ 0 & 1 & 0 & 0 \\ 0 & 0 & 1 & 0 \\ 0 & 0 & 0 & 1 \\ 1 & 1 & 0 & 0 \\ 0 & 0 & 1 & 1 \\ 1 & 0 & 1 & 0 \\ 0 & 1 & 0 & 1 \end{pmatrix} \ .$$

With $t_2(x) = x \log x$, it follows that

$$t_2(W_1' t_1(\pi)) = \begin{pmatrix} \pi_{11} \log \pi_{11} \\ \pi_{12} \log \pi_{12} \\ \pi_{21} \log \pi_{21} \\ \pi_{22} \log \pi_{22} \\ \pi_{1+} \log \pi_{1+} \\ \pi_{2+} \log \pi_{2+} \\ \pi_{+1} \log \pi_{+1} \\ \pi_{+2} \log \pi_{+2} \end{pmatrix}$$

and from

$$W_2' = \begin{pmatrix} -1 & -1 & -1 & -1 & 0 & 0 & 0 & 0 \\ 0 & 0 & 0 & 0 & -1 & -1 & 0 & 0 \\ 0 & 0 & 0 & 0 & 0 & 0 & -1 & -1 \end{pmatrix}$$

it follows that

$$u_2(\pi) = W_2' t_2(W_1' t_1(\pi)) = \begin{pmatrix} -\sum_i \sum_j \pi_{ij} \log \pi_{ij} \\ -\sum_i \pi_{i+} \log \pi_{i+} \\ -\sum_j \pi_{+j} \log \pi_{+j} \end{pmatrix} \ .$$

All coordinates of $u_2(\pi)$ are now positive, so applying the transformation $t_3(x) = \log(x)$ on these coordinates is allowed. Then from

$$W_3' = \begin{pmatrix} 1 & 0 & -1 \\ 0 & 1 & -1 \end{pmatrix}$$

it follows that

$$u_3(\pi) = \begin{pmatrix} \log\left(-\sum_i \sum_j \pi_{ij} \log \pi_{ij}\right) - \log\left(-\sum_j \pi_{+j} \log \pi_{+j}\right) \\ \log\left(-\sum_i \pi_{i+} \log \pi_{i+}\right) - \log\left(-\sum_j \pi_{+j} \log \pi_{+j}\right) \end{pmatrix}$$

$$= \begin{pmatrix} \log \dfrac{\sum_i \sum_j \pi_{ij} \log \pi_{ij}}{\sum_j \pi_{+j} \log \pi_{+j}} \\ \log \dfrac{\sum_i \pi_{i+} \log \pi_{i+}}{\sum_j \pi_{+j} \log \pi_{+j}} \end{pmatrix}.$$

With $t_4(x) = \exp(x)$, $t_5(x) = x + 1$,

$$W_4' = \begin{pmatrix} -1 & 1 \end{pmatrix}$$

and

$$W_5' = \begin{pmatrix} 1 \end{pmatrix},$$

we obtain

$$u_5(\pi) = U.$$

Example 4: Testing Stability of ε Over Time

Using the representation of coefficients described above, it will now be shown how to specify a complete model. Suppose a dichotomous variable is measured at three time points, and let π denote the vectorized joint table of cell probabilities π_{ijk}. A researcher might be interested in the question how well one can predict an observation at time point t from observations at time point $t - 1$, and whether this quality of prediction (measured by ε) remains constant over time. Let ε_{12} and ε_{23} be the ε coefficients computed in the marginal tables of the first and second measurement, and the second and third measurement, respectively. In both cases, the row variable is the explanatory variable for predicting the column variable. For testing the null hypothesis $\varepsilon_{12} = \varepsilon_{23}$, nonloglinear marginal modeling is necessary.

First, one needs to specify the matrix for producing the marginal tables 1×2 and 2×3. With

$$A_{12}' = \begin{pmatrix} 1 & 1 & 0 & 0 & 0 & 0 & 0 & 0 \\ 0 & 0 & 1 & 1 & 0 & 0 & 0 & 0 \\ 0 & 0 & 0 & 0 & 1 & 1 & 0 & 0 \\ 0 & 0 & 0 & 0 & 0 & 0 & 1 & 1 \end{pmatrix}.$$

$A_{12}'\pi$ produces the 1×2 marginal and with

$$A'_{23} = \begin{pmatrix} 1\,0\,0\,0\,1\,0\,0\,0 \\ 0\,1\,0\,0\,0\,1\,0\,0 \\ 0\,0\,1\,0\,0\,0\,1\,0 \\ 0\,0\,0\,1\,0\,0\,0\,1 \end{pmatrix}$$

$A'_{23}\pi$ produces the 2×3 marginal. Therefore,

$$A' = \begin{pmatrix} A'_{12} \\ A'_{23} \end{pmatrix}$$

so that $A'\pi$ produces all the marginals of interest. Now let the matrices W_1, W_2, W_3 and the functions t_1, t_2, and t_3 be the same as in Example 1 above. Then, define

$$\tilde{W}'_1 = \begin{pmatrix} W'_1 & 0 \\ 0 & W'_1 \end{pmatrix}$$

$$\tilde{W}'_2 = \begin{pmatrix} W'_2 & 0 \\ 0 & W'_2 \end{pmatrix}$$

$$\tilde{W}'_3 = \begin{pmatrix} W'_3 & 0 \\ 0 & W'_3 \end{pmatrix}.$$

Here, $\tilde{W}'_k$ is called the direct sum of W_k and W_k. With these definitions, and

$$\tilde{u}_0(x) = x,$$

and for $k = 1, 2, 3$,

$$\tilde{u}_k(x) = \tilde{W}'_k t_k(\tilde{u}_{k-1}(x)),$$

it follows that

$$\tilde{u}_3(A'\pi) = \begin{pmatrix} \varepsilon_{12} \\ \varepsilon_{23} \end{pmatrix}.$$

To complete the model specification, define

$$B' = \begin{pmatrix} 1 & -1 \end{pmatrix}$$

and the model of equality of epsilons is specified as

$$h(A'\pi) = B'\theta(A'\pi) = B'\tilde{u}_3(A'\pi) = 0.$$

The most difficult aspect of the model specification appears to be the specification of the coefficient vector θ. Once that has been done, the specification of the model constraint is relatively straightforward. The book's website provides general algorithms for specifying these coefficients.

3.3.2 Compatibility and Redundancy of Restrictions

In the previous chapter, compatibility issues concerning the marginal restrictions have been discussed extensively for loglinear marginal models. But of course, the same kinds of difficulties may be encountered in nonloglinear marginal models. However, although the same types of solutions probably apply, it is much more difficult in the nonloglinear case to prove general conditions than it is for the loglinear marginal models. Therefore, just a few specific points will be mentioned.

An important issue is that matrix B in the general model in Eq. (3.1) should be of full column rank. If it is not, then there are redundant restrictions that should be removed for the determination of the correct number of degrees of freedom, and for our algorithm to converge.

Even if the matrix B has full column rank, there may still be redundant restrictions. For example, for dichotomous marginals, one should avoid simultaneously restricting the mean and the variance since the variance is a function of the mean. For a dichotomous variable with scores 0 and 1, if the mean is specified to be equal to $\frac{1}{2}$ and the variance is specified to be equal to $\frac{1}{4}$, then both restrictions imply a uniform distribution and one of them is redundant. For dichotomies, one should generally avoid simultaneous restrictions on the mean and the variance, or on any two coefficients, since there is only one degree of freedom.

For three variables, the correlation matrix should be positive definite. Hence, we cannot fit the model where the three marginal correlations are all equal to $-\frac{2}{3}$. The requirement of a positive definite correlation matrix is a lack of variation independence of the marginal correlations, and this may occur with other parameters as well, although it is difficult to give general rules.

In Chapter 4, we discuss extensively such problems that may occur in the specification of certain markovian models.

3.3.3 Homogeneous Specification of Coefficients

For readers who want to define their own coefficients, it is important that these coefficients are specified in an homogeneous way for our estimation procedure to work. Therefore, homogeneity will be briefly discussed in this subsection. Homogeneity of coefficients in the present context was described by Bergsma (1997, Section 4.3.2 and Appendix D). Lang (2005) extended this to a general class of sampling schemes including stratified sampling.

The coefficient θ is called *homogeneous* if $\theta(cx) = \theta(x)$ for any positive constant c. The loglinear (marginal) parameters discussed in the previous chapter are, except for the intercept term, examples of homogeneous functions. For example, in a 2×2 table with cell probabilities π_{ij} and corresponding expected frequencies $m_{ij} = N\pi_{ij}$, it does not matter whether the odds ratio is defined as either

$$\frac{\pi_{11}\pi_{22}}{\pi_{12}\pi_{21}}$$

or as

$$\frac{m_{11}m_{22}}{m_{12}m_{21}}.$$

For the coefficient of agreement κ (see Section 1.3.4),

$$\kappa = \frac{\sum_i \pi_{ii} - \sum_i \pi_{i+}\pi_{+i}}{1 - \sum_i \pi_{i+}\pi_{+i}},$$

replacing π_{ij} by the corresponding expected frequency m_{ij} yields a completely different (and erroneous) value for the coefficient. This shows that this formulation of κ is not homogeneous.

Note that homogeneity is not a property of a coefficient, but rather of its functional form or representation. For example, since $\sum_{i,j} \pi_{ij} = 1$, kappa can also be represented as

$$\kappa = \frac{\sum_{i,j} \pi_{ij} \sum_i \pi_{ii} - \sum_i \pi_{i+}\pi_{+i}}{\left(\sum_{i,j} \pi_{ij}\right)^2 - \sum_i \pi_{i+}\pi_{+i}}$$

This functional form *is* homogeneous, because, if the π_{ij} are now replaced by the corresponding m_{ij}, the same correct value will be obtained.

As discussed in the next subsection, if a coefficient is homogeneous, it has a simpler formula for its asymptotic covariance matrix, and maximum likelihood estimation is simplified because the constraint $\sum \pi_i = 1$ will be automatically satisfied by ML estimates. The programs presented on the book's website require a homogeneous specification.

Homogeneous specification is generally easy to achieve by reformulating the original model in terms of the cell probabilities π_{ij}. This can be done using the system described in the previous subsection in the following way. Suppose that a non-homogeneous parameter θ is originally defined in terms of the expected frequencies m_i, then homogeneity is obtained by redefining the base function u_0 as

$$u_0(m) = \frac{1}{m_+}m = \exp(\log m - \log 1'm) = \exp\left[(I_t - 1)\log\left(\frac{I_t}{1'}\right)m\right]$$

so that $u_0(m) = u_0(\pi)$, and hence also $\theta(m) = \theta(\pi)$. Measures that are already naturally written as a homogeneous function, for example the logit function π_1/π_2 or the odds ratio, do not need this redefinition of u_0. As previously stated, loglinear marginal parameters (except for the intercept term) are naturally written in homogeneous form.

3.3.4 ***Algorithm for Maximum Likelihood Estimation

The ML fitting algorithm that is described in Chapter 2 can be used to fit nonloglinear marginal models as well. The nonloglinear constraints are defined by Eq. (3.1) so that

$$h(A'\pi) = B'\theta(A'\pi) = 0,$$

To derive the Jacobian H belonging to these constraints, the Jacobian of θ is needed. If $\theta(x)$ is a depth-n representable parameter, then its Jacobian can be computed as follows. Let

$$U_k(x)' = \frac{du_k(x)}{dx'}$$

be the Jacobian of u_k. Then

$$U_0(x)' = I \,,$$

and, for $k = 1, 2, \ldots, n$, the chain rule for differentiation yields

$$U_k(x)' = W_k' D[\dot{t}_k(u_{k-1}(x))]U_{k-1}(x)' \,,$$

where $\dot{t}_k$ is the first derivative of t_k and $D[.]$ represents a diagonal matrix with the elements of the vector in square brackets on the main diagonal. Define

$$T(x)' = U_n(x)'$$

as the Jacobian of $\theta(x)$. The Jacobian of the constraint function then is

$$H(x)' = \frac{dh(x)}{dx'} = B'T(x)'$$

The ML algorithm described in the previous chapter can now be implemented straightforwardly, using $H = H(A'\pi)$.

In spite of the more complex model formulations, in practice we did not find nonloglinear marginal models more difficult to fit than loglinear marginal models. In fact, possibly because typically nonloglinear marginal models involve fewer constraints than loglinear ones, the former seem to be slightly easier to fit in the sense of it being easier to find a step size and fewer necessary iterations for convergence. Note that all the recommendations on choosing a starting point and a step size discussed in the previous chapter apply to nonloglinear models as well.

3.3.5 ***Asymptotic Distribution of ML Estimates

Under the assumption that $h(A'\pi)$ is homogeneous, has continuous second derivatives, and has full rank Jacobian for the population distribution, $\hat{\pi}$ has an asymptotic normal distribution. Its mean is then the population value of π and covariance matrix given in Section 2.3.7, where H is now calculated as described above. These results are based on the work of Aitchison and Silvey (1958), which were applied to the nonloglinear marginal models of this chapter in Bergsma (1997). The formal derivations and regularity conditions can be found in Lang (2004) and Lang (2005). Given the regularity conditions on h, all the results of Sections 2.3.7 and 2.3.8 also apply to nonloglinear marginal models, with H appropriately calculated.

4

Marginal Analysis of Longitudinal Data

In the previous chapters, the foundations have been laid for a comprehensive full ML estimation approach for the handling of marginal categorical data. The approach was illustrated by means of research questions that necessitated the use of marginal modeling, but in a relatively simple form. In the next three chapters, it will be shown that marginal modeling can be used to answer important research questions that require more complex analyses and involve longitudinal data (Chapter 4), structural equation modeling and (quasi-)experimental designs (Chapter 5), and latent variable models (Chapter 6). Except for the extension to latent variable models, no really new principles will be involved. However, that does not mean that the application of marginal modeling in the following chapters is always easy and straightforward. The next chapters will try to explain in an accessible way how to translate particular substantive research questions into the sometimes complex language of the marginal-modeling approach. The data and the programs on our website will help the readers to carry out the analyses themselves and in this way learn by doing.

This chapter will illustrate extensively the extreme usefulness of marginal models for the analysis of longitudinal data, especially for panel data. Because panel data are longitudinal in the strict sense, namely obtained by investigating the same research units repeatedly over time, dependent observations arise. In many cases, marginal modeling turns out to be the best choice for handling these dependencies when searching for the appropriate answers to important research questions. Of course, the use of panel data does not require marginal-modeling methods for all research questions. When the researcher is interested in the exact nature of the dependencies and wants to know whether and how previous behavior, beliefs or attitudes determines later behavior, beliefs or attitudes, standard conditional or transition analyses will provide the answers. Random coefficient and autocorrelation models have also proven to be very useful to account for the dependencies among the over time observations. However, in comparison to marginal models, these latter approaches usually make strict assumptions about the nature of the dependencies and essentially, somewhat different research questions are answered than when using marginal models. More on this will be said in the last chapter.

W. Bergsma et al., *Marginal Models: For Dependent, Clustered, and Longitudinal Categorical Data*, Statistics for Social and Behavioral Sciences, DOI: 10.1007/978-0-387-09610-0_4, © Springer Science+Business Media, LLC 2009

Duncan identified a number of important and often occurring types of research questions for which marginal analysis of panel data provides the best answer (Duncan, 1979; Duncan, 1981; Hagenaars, 1990). These typical panel questions or key hypotheses in panel analysis are about:

- The comparisons over time of net (aggregate) and of gross (individual) changes in one characteristic
- The comparisons over time of net and of gross changes in two or more characteristics
- The investigation of changes in association
- The comparison of changes among subgroups

The comparisons of net changes in one characteristic over time will be dealt with in Section 4.2, including the comparison of net changes in subgroups and the comparison of associations between the changing characteristic and a characteristic that is stable over time. The comparison of gross changes over time in one characteristic is the topic of Section 4.3. In Section 4.4, changes in two or more characteristics are studied, including net changes in two or more characteristics (Section 4.4.1), changing associations among nonfixed characteristics (Section 4.4.2), gross changes in two or more characteristics (4.4.3), and simultaneous hypotheses about net and gross changes (4.4.4).

Time will always be treated in this chapter as discrete, despite the fact that continuous time models for categorical data have been in existence for a long time. A well-known example of such a continuous time model is the continuous time markov process for discrete characteristics (Coleman, 1964). The reason that such continuous time models are not covered in this book is not because it is thought that social and behavioral processes evolve in an inherently discrete way over time with jumps at specific observation moments. The main reason for treating time discretely is that there is usually not enough information to do otherwise. In most longitudinal studies, the observations are discrete, carried out at just a few particular points in time and, most importantly, only provide information about those particular moments of observation. In such studies, empirical information about the exact nature of the underlying continuous process is simply lacking. What is essentially done when applying continuous time models to such data is to fill in hypothetically the missing information about the continuous process. Usually it is assumed — very often unrealistically — that the changes are continuously, smoothly, and identically distributed over a certain period. However, misspecification of the underlying process often biases the results very strongly. Therefore, a pragmatic stance has been taken here where the consequences of the underlying, possibly continuous and smooth processes are investigated at particular discrete points in time (Hagenaars, 1990, p. 16, 22).

Before turning to marginal models for panel data, first longitudinal data in the form of trend data will be discussed. Trend data result from repeated, independent cross-sectional samples from the same population. Although, in general, trend data can be analyzed in standard nonmarginal ways because the observations over time are independent of each other, the comparison of net changes in more than one characteristic does require a marginal-modeling approach (Hagenaars, 1990, Section 6.4).

4.1 Trend Data

To indicate the kinds of research questions that make use of trend data but do require the use of marginal modeling, an example is taken from Hagenaars (1990). According to conventional wisdom, Dutch society was characterized for a long time (certainly in the sixties and seventies of the previous century) as leftist, progressive, and very permissive. But gradually this image (or reality?) started changing. Against this background, Hagenaars investigated whether '... in reaction to the permissive society of the 1960s and early 1970s and as a consequence of the worsening economic situation, a stronger right wing conservative ideology was developing in the late 1970s stressing the importance of an orderly society in which law and order prevailed and everybody had to pull their own weight. A growing concern about crime and about the possible misuses of social security provisions were thought to be indicative of this rising conservative ideology' (Hagenaars, 1990, p. 289). Three hypotheses and consequences following from the rising conservative ideology were tested. First, it was investigated whether indeed during the period of 1977-1981 the concern about crime and abuse of social security increased when considered separately. This turned out not to be true. Second, the hypothesis was tested that these two different aspects of the new conservative ideology became more closely related to each other over time, as indicated by an increasing individual level association between the two items. This hypothesis was confirmed (it will be looked into again in Section 4.1.2). Because the data used came from a trend study, the analyses for testing these two hypotheses only involved independent observations, and nonmarginal methods were used. However, it was also investigated (thirdly) whether the marginal fluctuations in concern about crime and abuse of social security over time occurred in a parallel fashion, being supposedly brought about in the same way by the same kinds of events that affected general conservative feelings. For this, marginal-modeling methods are required because the marginal distribution of concern about crime involves the same respondents as the marginal distribution of concern about abuse. Hagenaars (1990) used conditional MH tests involving (quasi-)symmetry models. Below, these analyses will be replicated using our much more flexible unconditional marginal approach.

4.1.1 Comparing Net Changes in More Than One Characteristic

The full cross-classification of the variables is given in Table 4.1. This table will be referred to as Table TCS (*Time* × *Concern about crime* × *Concern about abuse social security*). The relevant marginal distributions of *Concern about crime* (C) and *Concern about abuse of social security* (S) are presented in Table 4.2. The data come from the Amsterdam Continuous Survey, which was conducted by the University of Amsterdam. It consisted of a series of approximately semiannual investigations that started in January 1972, with a new sample from the Dutch population being drawn at each moment of observation (Van der Eijk, 1980).

As Hagenaars showed, and as may be clear from inspection of the distributions in Table 4.2, the marginal distributions of the *Concern for abuse of social security* and

Table 4.1. *Concern about abuse of social security (S)* and *Concern about crime (C)* at ten time points. Response alternatives: 1 = no (big) problem; 2 = big problem; 3 = very big problem

		Time									
		1	2	3	4	5	6	7	8	9	10
S	C	No-77	Ja-78	Jn-78	No-78	Ma-79	Oc-79	Ap-80	Oc-80	De-80	Jn-81
1	1	12	26	28	38	38	25	33	27	45	26
1	2	26	37	29	39	35	38	35	37	30	52
1	3	35	44	32	46	35	22	19	12	19	20
2	3	6	9	4	16	11	12	13	18	22	18
2	2	61	74	80	80	62	96	98	101	84	127
2	3	86	82	75	77	72	82	77	71	87	95
3	1	11	11	9	15	15	11	8	9	11	11
3	2	37	39	35	51	55	56	37	62	43	60
3	3	249	265	303	222	268	267	247	265	238	186
Total		523	587	595	584	591	609	567	602	579	595

Table 4.2. Marginal tables (%) for *Concern about abuse of social security (S)* and *Concern about crime (C)* at ten time points. Response alternatives: 1 = no (big) problem; 2 = big problem; 3 = very big problem

	Time									
	1	2	3	4	5	6	7	8	9	10
S=1	14	18	15	21	18	14	15	13	16	17
S=2	29	28	27	30	25	31	33	32	33	40
S=3	57	54	58	49	57	55	52	56	50	43
100 % =	523	587	595	584	591	609	567	602	579	595
C=1	6	8	7	12	11	8	10	9	14	19
C=2	24	26	24	29	26	31	30	33	27	40
C=3	71	67	69	59	64	61	61	58	59	51
100 % =	523	587	595	584	591	609	567	602	579	595

the *Concern about crime* vary considerably over time but not according to a nice, smooth (curvi)linear pattern. At the same time, the ups and downs of the time curves of the two items are remarkably parallel (Hagenaars, 1990, p. 292). This parallelism was not caused by general fluctuations in concern. The Amsterdam study contained a number of other concern issues (such as abortion or nuclear energy) and the marginal distributions of those items did not show a similar parallel pattern over time. The concerns about abuse of social security and about crime have something special in common.

This parallelism will be investigated more formally by means of loglinear marginal models applied to Table 4.2. As analogously done before, Table 4.2 will be referred to as if it were a normal *TIR* (*Time* × *Item* × *Concern Response*) table, where T has ten categories 1 through 10, I has two (1. abuse of social security and 2. crime), and R three (1. no (big) problem 2. big problem 3. very big problem). Models for this table have to estimated using the marginal-modeling approach because the upper half of Table 4.2 refers to the same respondents as the lower half. Of interest are several logit models in which T and I are considered the independent variables influencing the concern response R, conditioning on the distribution of TI. The most parsimonious logit model $\{TI,R\}$ for table *TIR* implies that the response distribution of concern is exactly the same for the two items and also exactly the same at each point of time. Model $\{TI,TR\}$ assumes that the distribution of R changes over time, but at each particular point in time, the concern distribution is exactly the same for both items, whereas model $\{TI,IR\}$ allows for differences in concern between the two items, but not for net changes over time. These three models $\{TI,R\}$, $\{TI,TR\}$, and $\{TI,IR\}$ represent very strict forms of the parallelism noted above and have to be rejected ($p = .000$) (using here and elsewhere in this chapter the marginal-modeling algorithms from Chapters 2 and 3).

However, parallelism is still possible in a somewhat weaker and more realistic form. It may well be that the concern distribution is different for the two items and that, moreover, the concern distribution changes over time, but then in such a way that the changes over time are the same for the two items. In other words, the differences between the two item distributions will remain the same over time. Expressing these differences in terms of loglinear parameters, this weaker form of parallelism leads to the no-three-variables interaction model $\{TI,IR,TR\}$. Although model $\{TI,IR,TR\}$ fits the observed data much better than the more parsimonious models, it still does not fare very well: $G^2 = 39.79$, $df = 18$ ($p = .002$, $X^2 = 39.25$). On the other hand, inspection of the adjusted residual frequencies for model $\{TI,IR,TR\}$ shows that almost all residuals are small; the only serious exceptions are the residuals involving $T = 9$. A similar conclusion follows from inspection of the three-variable interaction parameters $\hat{\lambda}_{t\,ir}^{TIR}$ in saturated model $\{TIR\}$: all estimates are very small, except those for $T = 9$, but then the highest and significant estimates are still just $\hat{\lambda}_{9\,11}^{TIR} = -.153$ and $\hat{\lambda}_{9\,12}^{TIR} = .115$.

From all this and from the parameter estimates in models $\{TI,IR,TR\}$ and $\{TIR\}$ (not further reported here), the substantive conclusions are first, that people were much more concerned about crime than about abuse of social security money.

Second, the concern with both items changed (irregularly) over time, but in a like, parallel fashion, except in December 1980 ($T = 9$). For some reasons, that may be further investigated, the odds of finding crime no (big) problem rather than a big problem increased at $T = 9$ more than the similar odds for the concern of abuse of social security. Just to mention possible explanations to look for, perhaps some hideous crime was reported in the newspapers about that time or there was a debate on crime issues in Parliament that was extensively covered by the newspapers.

From a methodological and statistical point of view, comparisons were made among the conditional analyses reported in Hagenaars (1990, Section 6.4), the appropriate marginal analyses carried out above, and the (not reported) inappropriate analyses where the loglinear models were directly applied to Table 4.2, ignoring the dependencies among the observations. No general patterns could be found regarding the size of test statistics and standard errors for the three approaches. The most important lesson to learn from this, as noted before, is that for multiway tables, ignoring dependencies among the observations is not a simple matter of systematically having more or systematically having less statistical power.

4.1.2 Simultaneous Tests for Restrictions on Association and Net Change: Modeling Joint and Marginal Tables

The data and variables in Table 4.1 can also be used to illustrate the usefulness of our proposed marginal-modeling approach for combining models for joint and marginal tables. Next to the (confirmed) hypothesis of the parallelism of the time curves of the two concern items, there was also the hypothesis of an increasing positive individual level association between C and S. This latter hypothesis can be tested applying ordinary, nonmarginal loglinear models to the full table TCS (Table 4.1). Assuming a linear increase over time in the association between C and S, the three-variable interaction parameters in saturated model $\{TCS\}$ are restricted in the following way

$$\lambda_{t\,i\,j}^{TCS} = \mu_{i\,j}^{CS}T_t, \tag{4.1}$$

where the parameters $\mu_{i\,j}^{CS}$ sum to 0 over any subscript, and the equidistant scores assigned to T_t range from -4.5 to $+4.5$. The test statistics for the restriction in Eq. 4.1 are $G^2 = 32.00$, $df = 32$ ($p = .466$, $X^2 = 32.09$): there is no reason to reject the assumption of a linear increase over time in the (loglinear) association of C and S. From the parameter estimates (not reported here, but see Table 4.3), it is seen that the association is positive. Hagenaars reported a value of $G^2 = 31.43$ for the model in Eq. 4.1 instead of $G^2 = 32.00$. The small difference with the presently obtained value is caused by the fact that the semiannual studies were only roughly held every sixth month and Hagenaars used the exact timing of the studies when assigning scores to T (Hagenaars, 1990, p. 298-299). To keep things as simple as possible, equidistant scores were used here, but all differences between the two ways of assigning scores are substantively negligible.

It is also possible to test both hypotheses about the expanding conservative ideology thesis at the same time, i.e., the simultaneous validity of the hypotheses of the

Table 4.3. Loglinear parameters $\hat{\lambda}_{ij\,t}^{SC|T}$ for relationship between *Concern about abuse of social security (S)* and *Concern about crime (C)* at first ($T = 1$) and last ($T = 10$) points in time; standard errors between parentheses; see also text

		$C = 1$	$C = 2$	$C = 3$
$T = 1$	$S = 1$	.779 (.083)	−.175 (.070)	−.605 (.068)
	$S = 2$	−.471 (.098)	.573 (.067)	−.102 (.065)
	$S = 3$	−.308 (.091)	−.399 (.067)	.707 (.059)
$T = 10$	$S = 1$	1.012 (.082)	−.001 (.067)	−1.011 (.073)
	$S = 2$	−.189 (.089)	.254 (.060)	−.066 (.062)
	$S = 3$	−.823 (.102)	−.253 (.069)	1.076 (.065)

parallel time curves and of the linear increase in association. This involves a simultaneous test of the hypothesis for the joint Table 4.1 and the hypothesis for marginal Table 4.2. A disadvantage of such a simultaneous test is that a misspecification in one part may also influence the outcomes for the other part, but an advantage is that if the complete model is true, estimates are obtained that are optimal under the given complete model. Using our marginal-modeling approach, it is rather straightforward to test model $\{TI,IR,TR\}$ for marginal table TIR (Table 4.2), and at the same time the restriction in Eq. 4.1 for the entries of the joint table TCS (Table 4.1). No incompatibilities exist between the two sets of restrictions that follow from the two hypotheses. The test statistics for the simultaneous tests of both sets of restrictions are $G^2 = 71.00$, $df = 50$ ($p = .027$, $X^2 = 69.52$). The degrees of freedom are simply the sum of the degrees of freedom for the separate models ($32 + 18 = 50$). This value $G^2 = 71.00$ is just a bit smaller than the sum of the two separate test statistics obtained before: ($32.00 + 39.79 = 71.79$). The low p-value for the simultaneous test is only caused by the outliers in marginal table TIR for the parallelism hypothesis.

In the simultaneous model, the linear restriction in Eq. 4.1 yields estimates for the relationship between C and S that produced the expected increasing positive association over time, in terms of odds ratios, as shown in Table 4.3 for the first and last time points. From these estimates, it can be derived that the extreme log odds ratio involving cells $RC = [11, 13, 31, 33]$ at $T = 1$ equals 2.399, and the odds ratio itself equals 11.008. This log odds ratio linearly increases from each time point to the next by an amount of .169 (or the odds ratio by a factor of 1.184), leading to an extreme log odds ratio at $T = 10$ of 3.922 and an odds ratio of 50.486.

At least two aspects of the assumed crystallization process of a conservative ideology are confirmed. There was in the period concerned an increasing individual level association between the concerns about crime and abuse of social security. Moreover, the two concerns were affected over time in a parallel fashion by ongoing events. Marginal modeling makes it possible to test both aspects simultaneously by modeling both the joint and a marginal table simultaneously.

Further, this example makes clear that in the context of a trend study where we have observations from independent samples over time, marginal modeling might also be required to answer pertinent research questions. Marginal modeling is not only very well-suited for the analysis of the truly longitudinal (panel) but may also be needed for the analysis of trend data (or for that matter, for the analysis of one-shot surveys, as shown in the two previous chapters).

4.2 Panel Data: Investigating Net Changes in One Characteristic

It is often said that the strongest point of a panel design is the fact that the researcher is able to investigate individual (gross) changes over time. And, of course, this is an outstanding feature of panel data. But, nevertheless, in many research reports using panel data, a lot of analyses and key hypotheses are actually about the net and not about gross changes over time. The investigation of net changes using panel data requires marginal-modeling methods because of the strict longitudinal character of the data. This section will discuss how to analyze the net changes in one particular characteristic — in both the total population and among subgroups. It will also be shown how these subgroup comparisons of net changes can be interpreted as changes in association between the one changing characteristic and a stable, fixed variable, such as *Gender*. By means of two examples, one on net changes in political orientation and one on growth curves of the use of marijuana, several different aspects of net change comparisons will be explained.

4.2.1 Overall Net Changes; Cumulative Proportions; Growth Curves

Net Changes in Political Orientation: Cumulative Proportions

The variable *Political orientation* used in Table 2.1 was also measured in a panel study as part of the American National Election Studies. A group of 408 individuals was interviewed at three points in time about their political orientation in the years 1992, 1994, and 1996. The data concerning the marginal distributions of *Political orientation* at the three time points are reported in Table 4.4.

An important research question is whether and how the overall mood of the country changed over the three years along the liberal-conservative dimension. In other words, do the marginal distributions in Table 4.4 show any (net) changes in political orientation among the three time points, and if so, in what direction? Since these marginals come from a panel study, they involve the same persons and therefore dependent observations and marginal modeling must be applied to answer the research question. A good starting point is the hypothesis that the three marginals are exactly the same. Treating Table 4.4 as if it were a conventional *T(ime)* × *P(olitical orientation)* table, the marginal homogeneity (MH) hypothesis (i.e., the no-net-change hypothesis) can be represented as loglinear independence model $\{T,P\}$ for table TP. Application of the appropriate marginal-modeling approach to take the dependencies among the observations of the panel data into account yields the following test

Table 4.4. *Political orientation* in the United States: 1992, 1994, 1996. Source: U.S. national election studies, 1992–1997

	1992	1994	1996	MH
1. Extremely liberal	11	6	6	9.17
2. Liberal	48	43	36	42.52
3. Slightly liberal	51	58	69	58.78
4. Moderate	113	95	98	103.60
5. Slightly conservative	92	83	86	86.90
6. Conservative	77	107	98	91.83
7. Extremely conservative	16	16	15	15.21
Total	408	408	408	408

statistics for the MH hypothesis: $G^2 = 27.66$, $df = 12$ ($p = .006$, $X^2 = 26.11$). The expected frequencies under MH are reported in the last column of Table 4.4. Note that the expected frequencies under MH, as seen before, are not simply the average of the three marginals. The MH model fits the data poorly. Inspection of the adjusted residuals of the MH model (see Table 4.5), and of the estimated two-variable parameters of saturated model $\{TP\}$ (not reported here), made it clear that the observed marginal frequency of the conservative Category 6 in 1992 was significantly larger than its expected frequency under MH, while the observed marginal frequencies of the slightly liberal Category 3 in 1996 and the conservative Category 6 in 1994 were significantly smaller than their expected values. But no simple interpretable pattern explaining the lack of fit of the MH model was immediately seen.

Therefore, another approach was tried, also illustrating the great flexibility of the marginal-modeling approach. The variable *Political orientation* actually consists of two dimensions: ideology (liberal/conservative) and intensity (extreme, normal, slightly), plus for both dimensions the same neutral category 'moderate/middle of the road' (as it was originally indicated). Now one might wonder whether the net changes that obviously occurred took place along the ideology or the intensity dimension, or whether it was the specific interaction of the two dimensions that caused the net changes. Because the middle Category 4 belongs to both dimensions, it is first investigated whether a model will fit in which the marginal probabilities for Category 4 do not change over time. If the short-hand notation is used in which the restrictions are indicated directly in terms of the marginal probabilities in table TP for Category 4 at each point in time, this partial marginal homogeneity model implies:

$$\pi_{14}^{TP} = \pi_{24}^{TP} = \pi_{34}^{TP} \tag{4.2}$$

There is no reason to reject the hypothesis of no net change in neutral Category 4: $G^2 = 3.83$, $df = 2$ ($p = .147$, $X^2 = 3.83$).

Marginal homogeneity in the ideology dimension implies, in addition to no net change in the neutral category, that there are no net changes over time in the total proportion of liberals and conservatives:

$$\pi_{11}^{TP} + \pi_{12}^{TP} + \pi_{13}^{TP} = \pi_{21}^{TP} + \pi_{22}^{TP} + \pi_{23}^{TP} = \pi_{31}^{TP} + \pi_{32}^{TP} + \pi_{33}^{TP}$$

$$\pi_{15}^{TP} + \pi_{16}^{TP} + \pi_{17}^{TP} = \pi_{25}^{TP} + \pi_{26}^{TP} + \pi_{27}^{TP} = \pi_{35}^{TP} + \pi_{36}^{TP} + \pi_{37}^{TP}$$

(4.3)

A simultaneous test of the validity of all restrictions in Eqs. 4.2 and 4.3 yields the following test outcomes: $G^2 = 6.75, df = 4$ ($p = .150, X^2 = 6.69$). The hypothesis can be accepted that there is no net change along the ideology dimension. A more specific test of Eq. 4.3 by means of the conditional test of the model with the restrictions in Eqs. 4.2 and 4.3 against the model containing only the restrictions in Eq. 4.2 confirms that the marginal proportions of liberals and conservatives did not change over time: $G^2 = 6.75 - 3.83 = 2.92, df = 4 - 2 = 2, p = .232$.

But then perhaps, given that the MH model as such did not hold, the intensity of the political orientation has changed. The hypothesis of no net changes in intensity involves the following restrictions:

$$\pi_{11}^{TP} + \pi_{17}^{TP} = \pi_{21}^{TP} + \pi_{27}^{TP} = \pi_{31}^{TP} + \pi_{37}^{TP}$$

$$\pi_{12}^{TP} + \pi_{16}^{TP} = \pi_{22}^{TP} + \pi_{26}^{TP} = \pi_{32}^{TP} + \pi_{36}^{TP}$$

(4.4)

$$\pi_{13}^{TP} + \pi_{15}^{TP} = \pi_{23}^{TP} + \pi_{25}^{TP} = \pi_{33}^{TP} + \pi_{35}^{TP}$$

The restrictions in Eqs. 4.4 can also be accepted: the simultaneous test outcomes for the restrictions in Eqs. 4.2 and 4.4 are $G^2 = 9.74, df = 6$ ($p = .136, X^2 = 9.68$), and the conditional test outcomes for the comparison of the model with the simultaneous restrictions in Eqs. 4.2 and 4.4 against the model with only the restrictions in Eq. 4.2 yields $G^2 = 9.74 - 3.83 = 5.91, df = 6 - 2 = 4$ ($p = .206$). Finally, the model in which all marginal homogeneity restrictions in Eqs. 4.2, 4.3 and 4.4 are assumed to be simultaneously valid need not be rejected: $G^2 = 12.62, df = 8$ ($p = .126$, $X^2 = 12.58$).

On the one hand, from a substantive point of view, this is a disappointing result: we had hoped to come up with a parsimonious and clear interpretation of the failure of the original MH model for Table 4.4 in terms of net changes in ideology or intensity or both. But there is no need to assume overall net changes in either ideology or intensity. So, it must be the interaction between ideology and intensity and the net changes in specific categories of *Political orientation* that are responsible for the failure of the original MH model. Which categories these are was indicated in the beginning of this section, namely those exhibiting the largest adjusted residual frequencies.

On the other hand, from a methodological and statistical point of view, this is an interesting example that illustrates the flexibility of the marginal-modeling approach. For example, when testing all restrictions in Eqs. 4.2, 4.3 and 4.4 simultaneously, several kinds of dependencies among the observations have to be taken into account: dependencies arising from the panel character of the data, but also others originating from the fact that the simultaneous set of restrictions in Eqs. 4.3 and 4.4 involve partially overlapping categories of *Political orientation*. This use of partially overlapping categories can easily be extended to the general analyses of cumulative

Table 4.5. Adjusted residuals for MH model *Political orientation* Table 4.1

	1992	1994	1996
Extremely liberal	.93	−1.70	−1.70
Liberal	1.56	.14	−1.85
Slightly liberal	−1.71	−.16	2.26
Moderate	1.64	−1.49	−.97
Slightly conservative	.92	−.70	−.16
Conservative	−2.98	3.04	1.23
Extremely conservative	.30	.30	−.08

Table 4.6. Adolescents' *Marijuana use*. Source: U.S. National Youth Survey (Elliot, Huizinga, & Menard, 1989)

	Age (*A*)									
Marijuana use (*M*)	13	%	14	%	15	%	16	%	17	%
1. Never	220	91.7	195	81.3	169	70.4	158	65.8	140	58.3
2. Once a month	15	6.3	27	11.3	41	17.1	41	17.1	52	21.7
3. More than once a month	5	2.1	18	7.5	30	12.5	41	17.1	48	20.0
Total	240	100	240	100	240	100	240	100	240	100

and global odds that are often advocated for the analyses of ordinal data (Agresti, 2002; Molenberghs & Verbeke, 2005). Because monotonicity of the local odds is a stricter form of ordinality than monotonicity of the cumulative or global odds, and because we believe that the theoretical, substantive expectations about ordinal relationships almost always refer to the strict ordinal form rather than to weaker forms, only local odds are dealt with in this book. However, as may be clear from the above, cumulative and global odds can be handled just as easily.

Marijuana Use: Loglinear Growth Curves

Another kind of application of a net change analysis, and one that is more typical of traditional growth curves studies and actually exemplifies a categorical growth rate model, is provided by the data on the use of marijuana in Table 4.6. The data are taken from the U.S. National Youth Survey (Elliot, Huizinga, & Menard, 1989). The panel study started in 1976 when the 240 children were 13 years old and ended in 1980 when they were 17 years old. The data in Table 4.6, containing the marginal distributions at the different ages, have previously been analyzed by Lang et al. (1999). The growth in marijuana use will be investigated below, not only in terms of net changes in the whole marginal distributions, analogously to the above analyses of the political orientation in the United States, but also in terms of net changes in summary characteristics of these distributions — more specifically, their location and dispersion. All analyses will be carried out at the population level (i.e., population averaged). Often growth curve analyses are carried out at the subject-specific level by means of random coefficient models. As remarked before, individual level analyses will generally

Table 4.7. Estimated parameters $\hat{\lambda}_{ij}^{AM}$ for saturated model {MA}, Table 4.6; * significance level $p < .01$

Marijuana use (M)	Age (A)				
	13	14	15	16	17
1. Never	.908*	.204*	−.201*	−.350*	−.562*
2. Once a month	−.094	−.089	.067	−.015	.131
3. More than once a month	−.814*	−.115	.133	.364*	.431*

lead to different results and the average of the individual growth curves will not necessarily coincide with the overall population growth curve, certainly not for logistic growth curves. Which approach to use depends mainly on the research question. For example, if in a clinical setting one wants to know what is expected to happen to the drug use of a particular adolescent, the individual growth curves may be the most relevant. If a government wants to know what will happen to drug abuse at the societal level, the marginal analyses proposed here may be the most appropriate (see Rule 6 in Firebaugh, 2008). Or emphasizing a different aspect, if one thinks that random coefficient models capture the true underlying individual process, but one is interested in the overall results of these individual processes, one should compute the relevant marginals from these random coefficient models. However, if at the individual level processes than those implied by the standard random-effect model are valid (an autocorrelation process, for example), the thus obtained marginal results may be biased. The marginal approach advocated here captures the consequences at the population level without making any assumptions about the underlying individual processes.

The data in Table 4.6 (here further referred to as if it were a standard table MA) show a definite increase in marijuana use. Not surprisingly, the MH hypothesis, loglinear model {M,A} applied to Table 4.6, has to be rejected: $G^2 = 90.12$, $df = 8$ ($p = .000$, $X^2 = 71.04$). The estimated parameters for the relationship between M and A in saturated model {MA} are presented in Table 4.7. These estimates indicate an ordinal, monotonically increasing relationship between Age and $Marijuana$ use — more precisely, a decrease of category 'never use' over the years and a corresponding increase of category 'more than once a month'. The largest increase appears to occur from 13 to 14 years.

Perhaps more parsimonious models than the saturated model might be found for this overall trend. An obvious candidate to consider is the loglinear linear-by-linear (or uniform association) model in which M is treated as an interval-level variable and the trend over time is assumed to be linear: the odds of belonging to category i of M rather than to $i + 1$ increase for all i in the same way by a constant factor over time. For this model we have

$$\lambda_{ij}^{MA} = \rho M_i A_j \tag{4.5}$$

in which A_j and M_i are fixed scores for the levels of A and M, respectively. However, this linear-by-linear model does not fit the data: $G^2 = 21.07$, $df = 7$ ($p = .004$, $X^2 = 16.75$). There may be at least two reasons for this negative result. First, $Marijuana$

Table 4.8. Estimated parameters $\hat{\lambda}_{i\ j}^{AM}$ for the parabolic model; data in Table 4.6

Marijuana use (M)	Age (A)				
	13	14	15	16	17
1. Never	.827	.251	−.162	−.414	−.503
2. Once a month	0	0	0	0	0
3. More than once a month	−.827	−.251	.162	.414	.503

use may not be an equidistant interval-level variable, as assumed in the linear-by-linear model, or second, the time trend may not be linear. First, treating M as just a nominal-level variable, but still assuming linear trends over time for each category of M, leads to a nominal-by-interval (or row association) model. In this model, the odds of belonging to category i of M rather than to $i+1$ increase or decrease by a constant factor over time, but not necessarily for all i in the same way or even in the same direction. More formally, this model states

$$\lambda_{i\ j}^{MA} = \rho_i^M A_j . \tag{4.6}$$

The test statistics of this row association model are $G^2 = 16.11$, $df = 6$ ($p = .013$, $X^2 = 13.22$), and for the conditional test, comparing the linear-by-linear model against the row association model: $G^2 = 21.07 - 16.11 = 4.96$, $df = 7 - 6 = 1$ ($p = .026$). It is not very clear what to decide about the nominal or interval level of M given these borderline test outcomes.

Investigation of the second possible reason for the failure of the linear by linear model, viz., that the trend over time is not linear provides a much less ambiguous picture. Inspection of the estimates in Table 4.7 already shows that the increase over time is not the same between all successive years. Also, from a substantive point of view, one might expect the largest increase in marijuana use in the first years of adolescence. Perhaps a model in which the equal interval character of M is maintained, but the trend over time is assumed to be a parabolic curve that describes the data for the ages 13 to 17 well. In such a parabolic model, the differences over time between the log odds are not constant but decrease with a constant factor. Under this model, we have

$$\lambda_{i\ j}^{MA} = \rho_1 M_i A_j + \rho_2 M_i A_j^2 \tag{4.7}$$

for given scores M_i and A_j. This turns out to be the case. The parabolic model fits the data well ($G^2 = 5.60$, $df = 6$, $p = .469$, $X^2 = 5.52$) and it fits significantly better than the linear-by-linear model. The estimates of the relevant loglinear parameters of this parabolic model for the relationship between A and M are reported in Table 4.8. Accepting the results in Table 4.8 as our final estimates, it appears that the use of marijuana systematically increases over the years, but in such a way that the difference (in $\hat{\lambda}$) between two consecutive years becomes gradually less and less (by .162 for each next consecutive pairs of years, to be exact).

The above loglinear analyses described the growth curve of marijuana use in terms of the whole marginal distributions of M at the several age groups. However, often the interest lies in the changes in a particular characteristic of the distributions, such as changes in average or dispersion. Analyses of these latter kinds will be explained below, assuming different levels of measurement for marijuana use M.

Growth Curves: Changes in Location

By considering M as an interval-level variable, the mean can be used as a central tendency measure. To compute the means of M at the several ages, equidistant scores 1, 2, and 3 have been assigned to its categories. For the successive ages t, the mean scores $\hat{\mu}_t$ of M in Table 4.6 are as follows:

t	$\hat{\mu}_t$
13	1.104
14	1.263
15	1.421
16	1.513
17	1.617

Note that there are no restrictions placed on the means (yet) and therefore, as in saturated loglinear models, the means can be obtained directly and simply from Table 4.6. The mean use of marijuana increases consistently over the years. With μ_t denoting the mean response at time t, the difference δ between the means at times $t+1$ and t is given by

$$\delta_{t+1} = \mu_{t+1} - \mu_t .$$

The estimates of δ_{t+1} with their corresponding z-scores are presented in the first two rows of Table 4.9. Again, these differences can simply be computed from the data and means in Tables 4.6 and 4.9 because no restrictions are placed on the differences over time. However, when computing the standard errors of these differences, the dependencies among the observations have to be taken into account, and the procedures outlined in Section 2.3.6 have to be applied. All four mean differences are positive and clearly significant, and the overall hypothesis that all means are equal, that is, that $\delta_{t+1} = 0$ for all t has to be rejected: $G^2 = 79.61$, $df = 4$ ($p = .000$, $X^2 = 57.98$).

The $\hat{\delta}_{t+1}$-scores in the upper part of Table 4.9 suggest that the mean increase in marijuana use is somewhat bigger in the categories 13-14-15 than in the later years of 15-16-17, which is a pattern somewhat different from the outcomes of the loglinear analysis. However, this difference is not significant. The test for a linear increase in the means, or (what amounts to the same) the test that all successive δ's have the same value

$$\mu_t = \mu_* + \beta A_t \quad \text{or}$$
$$\delta_{t+1} = \delta_* (= \beta)$$

can be accepted: $G^2 = 3.11$, $df = 3$ ($p = .374$, $X^2 = 3.14$). Moreover, this linear model need not be rejected in favor of a model with a parabolic/quadratic increase in the means, i.e., with a linearly increasing or decreasing difference δ_{t+1}:

Table 4.9. Parameter estimates of changes in mean ($\hat{\delta}$) and changes in variances ($\hat{\delta}^{\sigma}$) and their corresponding z-scores under the saturated model; see text, data in Table 4.6

	13–14	14–15	15–16	16–17
$\hat{\delta}_{t+1}$	0.16	0.16	0.09	0.10
$z = \hat{\delta}_{t+1}/\mathrm{se}(\hat{\delta}_{t+1})$	4.42	4.15	2.40	2.54
$\hat{\delta}^{\sigma}_{t+1}$	.21	.15	.10	.04
$z = \hat{\delta}^{\sigma}_{t+1}/\mathrm{se}(\hat{\delta}^{\sigma}_{t+1})$	3.82	2.93	2.12	1.26

$$\mu_t = \mu_* + \beta A_t + \gamma A_t^2 \quad \text{or}$$
$$\delta_{t+1} = \delta_* + \alpha A_t .$$

The conditional test of the parabolic model against the linear model yields $G^2 = 3.11 - 0.57 = 2.55$, $df = 3 - 2 = 1$ ($p = .110$), and the more parsimonious linear model can be accepted. The estimated constant difference between the estimated means $\hat{\delta}_{t+1} = \hat{\delta}_*$ in the linear model equals .131 with a standard error of .015.

So, application of the additive model in which differences between means are investigated leads to the conclusion that the relationship between A and M is linear. This conclusion about the linear shape of the growth curve of means is different from that of the earlier application of the multiplicative model, which required a parabolic relationship between A and M to obtain a well-fitting model. In terms of differences in mean use, there is a similar increase in marijuana use from one age category to the next. In terms of odds ratios, the odds of increased use grow systematically faster in the younger age groups than in the older ones. A result like this is to be expected because of the initially small percentage of users. An increase in proportion of regular users from .02 to .05 is, in terms of odds ratios, about the same as an increase from .20 to .40. In terms of differences, the increase from .02 to .05 is of course the same as an increase from .20 to .23. In the end, it is all about how to measure effects and about whether a causal effect increases differences or odds ratios. In this example, perhaps the loglinear representation is more adequate; in other situations mean differences may make more sense.

The differences between the means might also have been tested by means of standard statistical procedures such as the t-test for dependent samples. However, these standard procedures are based on the assumption of normality, which never holds for categorical data and certainly not for the skewed data in Table 4.6. The methods applied here do not assume normality of observed score distributions, but only require that the observations in the full joint table are multinomially distributed. As such, they have broader applicability than the methods based on normal theory. Whether or not there might be any loss or gain of power in our categorical marginal approach compared to the standard procedures must be further investigated.

Using the mean as a measure of central tendency implies that variable M is treated as an interval-level variable. However, variable M is perhaps better seen as just an

Table 4.10. Parameter estimates and z-scores under the BT model; see text, data in Table 4.6

	$BT-A$				
t	1	2	3	4	5
$\hat{l}_t$	$-.186$	$-.078$	.028	.083	.152
$z = \hat{l}_t/\mathrm{se}(\hat{l}_t)$	-10.01	-4.62	1.85	4.84	7.42

	$BT-M$				
t	1	2	3	4	5
$\hat{l}'_t$	-1.246	$-.286$	.278	.514	.740
$z = \hat{l}'_t/\mathrm{se}(\hat{l}'_t)$	-7.16	-2.81	3.63	6.08	7.83

ordinal-level variable. In Chapter 1, the ordinal location coefficients L_{ij} and L'_{ij} were introduced and applied. L_{ij} refers to the additive variant subtracting the probability of scoring higher from the probability of scoring lower on X_i compared to X_j, while L'_{ij} indicates the log-multiplicative variant in which the logarithms of these probabilities are subtracted. For the substantive interpretation of these ordinal location coefficients, it is important that the BT model with parameters $l_{ij} = 0$ or $l'_{ij} = 0$ can be accepted, as argued before in Chapter 3. When applied to the data in Table 4.6, the BT variant fits the data almost perfectly. The test results for L_{ij} (BT-A variant) are $G^2 = 1.16$, $df = 6$ ($p = .979$, $X^2 = 1.15$) and for L'_{ij} (BT-M variant) $G^2 = 1.80$, $df = 6$ ($p = .937, X^2 = 1.75$). Due to multicollinearity, the empirical degrees of freedom found for both models was 3; see Chapter 3 for a discussion of the empirical degrees of freedom. The relevant location parameters $\hat{l}_t$ and $\hat{l}'_t$ for each period of time t under the BT model are reported in Table 4.10.

There is a clear increase in the ordinal central tendency of M. The hypothesis that there are no location differences in the population, or more precisely that in the BT model the location parameters l (the parameters l', respectively) are all equal to each other, leads to exactly the same expectations about the data in the BT-A and in the BT-M model: $p(X_i > X_j) = p(X_i < X_j)$, and consequently to the same test outcomes in BT-A and BT-M. This hypothesis of equal l's or equal l''s has to be rejected ($p = .000$).

As above for the means and the log odds, (curvi)linear models can be fitted for the ordinal location measures to estimate the growth over time. Because the required specifications are completely analogous to the specifications for the loglinear parameters and for the mean differences, they will not be repeated here in equation form, but just referred to as linear and parabolic models. It is interesting to see that a linear BT-A model (for L_{ij} and l) fits the data well and, in that sense, behaves in the same way as the growth curve for the differences between the means. For the BT-M model (for L'_{ij} and l'), however, a parabolic curve is needed, just as for the log odds. More specifically, the test results for the linear BT-A model are as follows. The conditional test outcomes for the linear BT-A model, comparing it to the BT-A model without further restrictions, are $G^2 = 5.12 - 1.16 = 3.96$, $df = 3$ ($p = .266$).

Compared to the parabolic model, the conditional test statistics for the linear BT-A model are $G^2 = 5.12 - 2.00 = 3.12$, $df = 1$ ($p = .077$). The BT-A model with a linear growth curve for the ordinal locations can be accepted, and it is not necessary to assume a parabolic growth curve. The estimated location parameters $\hat{l}_t$ in the linear BT-A model increase by a constant amount of .085 from $-.170$ at age 13 to .170 at age 17: if a randomly drawn 17-year old is compared with a randomly drawn 13-year old, the probability that the older one has a higher score on marijuana use than the younger one is .34 ($= .170 + .170$) higher than the probability that the older one has a lower score.

For the BT-M model, the following results are obtained. The conditional test outcomes of the linear BT-M model against the unrestricted BT-M model are $G^2 = 17.18 - 1.80 = 15.38$, $df = 3$ ($p = .002$), and the linear model has to be rejected. The parabolic BT-M model does not fit significantly worse than the unrestricted BT-M: $G^2 = 4.02 - 1.80 = 2.22$, $df = 2$ ($p = .330$). The conditional test statistics for testing the linear model against the parabolic model yields $G^2 = 17.18 - 4.02 = 13.16$, $df = 1$ ($p = .000$), and the parabolic model has to be preferred. The $\hat{l}_t$ estimates in the parabolic BT-M model are as follows:

t	$\hat{l}_t$
13	-1.148
14	-0.355
15	0.219
16	0.574
17	0.710

The ordinal (log)locations increase systematically, but with diminishing differences between the successive age periods of .219 (or the odds ratios diminish by a factor $e^{.219} = 1.244$). Again, when comparing a 17-year old with a 13-year old and interpreting the odds rather than the log odds, the probability that the older one has a higher score on marijuana use than the younger one is 6.41 ($= e^{(1.148 + .710)}$) times higher than the probability that the older one has a lower score.

Growth Curves: Changes in Dispersion

Besides in the growth in central tendency, a researcher might be interested in growth in dispersion. If M is regarded as an interval-level variable, the variance σ^2 is an appropriate coefficient. When analyzing changes in both variances and means, one should keep in mind that, for asymmetrical distributions, the values of the mean and the variance are not independent of each other. The covariance between the sample mean and variance is equal to μ_3/N, with μ_3 being the third central moment of the population distribution. Certainly for the skewed data in Table 4.6, a growth in means will be necessarily connected to a growth in variance. Variances are in this comparative sense perhaps the most useful if the distributions are either symmetric or if the means of the distributions to be compared are the same. This is certainly not the case here. Nevertheless, the variances highlight aspects of the marijuana use distributions that are different from the means. A second point to keep in mind is that in this and the next paragraphs, the growth curve will be described in terms of differences between variances. There is no natural, compelling way of working with the variance differences for comparative purposes. Differences between the means appear naturally within the context of regression analysis, and differences between log odds appear naturally within the loglinear model. But there seems to be no reason not to prefer the ratios of the variances above their differences, or to choose working with the differences or ratios of the standard deviations rather than the variances. Although for this data, the essential conclusions about the (curvilinear) shape of the growth curves for the dispersion remain the same if one of the alternative approaches will be applied, this need not be true in general.

The variance estimates $\hat{\sigma}_t^2$ of *Marijuana use* for the successive ages in Table 4.6 are as follows:

t	$\hat{\sigma}_t^2$
13	.135
14	.344
15	.494
16	.592
17	.636

The successive differences of the variances denoted as $\hat{\delta}_{t+1}^\sigma$ are reported in the lower part of Table 4.9 with

$$\delta_{t+1}^\sigma = \sigma_{t+1}^2 - \sigma_t^2 .$$

Using the marginal-modeling approach, the hypothesis that there is no change in variance

$$\delta_{t+1}^\sigma = 0$$

for all t can be tested. This hypothesis has to be rejected ($p = .000$). A borderline test outcome is obtained for the model with a linear increase in variance

$$\delta_{t+1}^\sigma = \delta_*^\sigma$$

with $G^2 = 9.05$, $df = 3$ ($p = .029$, $X^2 = 9.14$). However, a parabolic growth model for the variances

$$\delta_{t+1}^\sigma = \delta_*^\sigma + \alpha A_t$$

fits perfectly and significantly better than the linear model. The conditional test results for the linear against the parabolic model are $G^2 = 9.05 - .003 = 9.05$, $df = 3 - 2 = 1$ ($p = .002$). According to this parabolic model, the estimated variances increase as follows

t	$\hat{\sigma}_t^2$
13	.135
14	.342
15	.493
16	.592
17	.636

with a diminishing difference between two consecutive variances of $\hat{\alpha} = -.054$. As is clear from these results and the data in Table 4.6, this increase of variance over the years in use of marijuana is mainly due to the distributions' shifts towards the higher scores and the marginal distributions becoming less skewed in this way.

Nominal level variation coefficients, such as the diversity index D or the uncertainty (entropy) coefficient U, have a different definition of variation compared to σ^2, as explained in Chapter 1. Nevertheless, for the marijuana use distributions, the substantive results are the same. Using similar testing procedures as for σ_t^2, it was concluded that the growth curves of both D and U behave in a similar, parabolic way over time as σ_t^2.

It is perhaps good to remind the reader (again) of the fact that the necessary matrix operations for defining these coefficients and their restrictions are rather complicated: on the book's webpage it will be shown how to do this exactly.

The ordinal variation coefficients W and W' behave very differently from the other variation coefficients dealt with so far. The combined hypothesis that the BT model is true, and that all W's and all W''s respectively are the same, can be accepted: $G^2 = 3.07$, $df = 4$ ($p = .547$, $X^2 = 3.02$). Given the big changes in the distributions of M over the years, this is a strange result. Obviously, with three categories and very skewed distributions, and because of the deletion of all kinds of tied observations and pairs in the computation of W and W', the ordinal variance test does not have much power. The variation coefficients W and W' appear not to be very suited if the ordinal variable(s) have just a few categories.

4.2.2 Subgroup Comparisons of Net Changes

All analyses of the growth curves for marijuana so far have been carried out for the whole sample. However, researchers are very often interested in knowing if and how growth curves differ for particular subgroups in the population. In this subsection, how to carry out such marginal analyses will be illustrated using the marijuana data that have now been split according to the sex of the respondents in Table 4.11.

The percentages in Table 4.11 and the mean responses in Fig. 4.1 suggest that boys and girls show more or less the same (curvilinear) increase in marijuana use over the years as found in the total sample. It is also clear that boys are heavier users than girls at all ages. In principle, all analyses of the previous subsection might be repeated for both subgroups. However, for practical reasons, the analyses of the subgroup growth curves will be carried out using only loglinear models and differences among the means.

Table 4.11. *Marijuana use* and *Age* for boys ($N = 119$) and girls ($N = 121$). Source: U.S. National Youth Survey

	Age (A)									
	13	%	14	%	15	%	16	%	17	%
Boys' *Marijuana use* (B)										
1. Never	106	89.1	89	74.8	78	65.5	73	61.3	65	54.6
2. Once a month	9	7.6	17	14.3	20	16.8	20	16.8	22	18.5
3. More than once a month	4	3.4	13	10.9	21	17.6	26	21.8	32	26.9
Girls' *Marijuana use* (G)										
1. Never	114	94.2	106	87.6	91	75.2	85	70.2	75	62.0
2. Once a month	6	5.0	10	8.3	21	17.4	21	17.4	30	24.8
3. More than once a month	1	.8	5	4.1	9	7.4	15	12.4	16	13.2

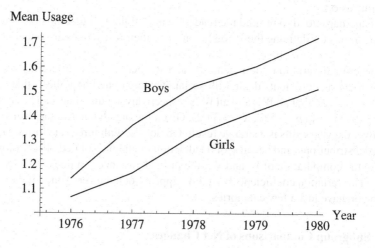

Fig. 4.1. *Marijuana use*: means of boys and girls for the ages 13 to 17, data in Table 4.11

A first obvious question to ask about these data is whether or not the marijuana use of boys and girls evolves in the same way when they grow older: are the growth curves parallel over time? Starting with the loglinear analysis, one may look at the three-variable interaction parameters in saturated model {SAM} for Table 4.11, referring to this table as usual as if it were a standard SAM ($Sex \times Age \times Marijuana$ use) table, where S has the categories $B(oys)$ and $G(irls)$. After computation of the standard errors in the appropriate way, and taking the dependencies into account, it turned out that none of the three-variable interaction parameter estimates is significant and all have rather small values: none of their absolute values is larger than .14. A test of the parallelism of the growth curves for boys and girls is given by logit model $\{SA,SM,AM\}$ for table SAM, conditioning on the distribution of SA. In this model, the level of the growth curves of boys and girls may differ from each other (because of the presence of λ_{s*m}^{SAM}), but the curves are parallel, indicated by the presence of λ_{*am}^{SAM} in combination with the absence of λ_{sam}^{SAM}. Model $\{SA,SM,AM\}$ can certainly be accepted with $G^2 = 5.17, df = 8$ $(p = .739, X^2 = 5.12)$. It may be concluded that, in the population, the effects of age on marijuana use are the same for both boys and girls. To see whether or not the parallel growth curves both follow a parabolic pattern, the λ_{*am}^{SAM} parameters in model $\{SA,SM,AM\}$ for table SAM are restricted as in Eq. 4.7. The thus restricted model $\{SA,SM,AM\}$, with identical growth curves for boys and girls, fits the data excellently with $G^2 = 11.94, df = 14$ $(p = .612, X^2 = 10.78)$. It does not fit worse than model $\{SA,SM,AM\}$ without further constraints $(G^2 = 11.94 - 5.17 = 6.77, df = 14 - 8 = 6$ $p = .343)$, and better than the linear growth curve model $(G^2 = 24.88 - 11.94 = 12.94, df = 15 - 14 = 1, p = .000)$. The relevant parameter estimates for describing the growth curve are, for all practical purposes, the same as the ones reported in Table 4.8.

Contrary to the parabolic growth curve for the log odds, the growth curve of the means followed a linear pattern in the whole sample. This is also true in both subgroups. First, it was investigated whether the mean differences in M over time are the same for boys and girls. Indicating the difference between two successive means for boys as $\delta_{t+1}^B = \mu_{t+1}^B - \mu_t^B$, and for girls as $\delta_{t+1}^G = \mu_{t+1}^G - \mu_t^G$, the hypothesis to be tested for all t is

$$\delta_{t+1}^B = \delta_{t+1}^G.$$

This same difference hypothesis can be accepted: $G^2 = 3.50, df = 4$ $(p = .477, X^2 = 3.38)$. At all ages, the difference between boys and girls in mean marijuana use is the same and therefore no significant difference between the shapes of the mean growth curves for boys and girls exist. Are the growth curves for the means in both subgroups also linear? The model with the same differences over time for boys and girls in combination with a linear increase over time

$$\delta_{t+1}^B = \delta_{t+1}^G = \delta_*^{BG}$$

can also be accepted: $G^2 = 5.99, df = 7$ $(p = .541, X^2 = 5.92)$. This linear model does not perform worse than the same difference model per se $(G^2 = 5.99 - 3.50 = 2.49, df = 7 - 4 = 3$ $(p = .477)$, and also not worse than the same difference model with a parabolic increase over time $(G^2 = 5.99 - 4.55 = 1.45, df = 7 - 6 = 1$ $(p = .229)$. The statistically significant increase of the means among both boys and girls for all ages t to $t+1$ is estimated as $\hat{\delta}_*^{BG} = .123$ $(s.e. = .012)$. An almost identical increase was obtained for the whole sample, namely $\hat{\delta}_* = .131$.

To what systematic differences in marijuana use this leads to between boys and girls will be discussed in the next subsection.

4.2.3 Changes in Associations

One of Duncan's typical panel questions mentioned in the beginning of this chapter concerned the comparison of associations over time. The data and discussions in Subsection 4.2.2. can also be interpreted in this light. The point of view taken in this last subsection has been comparison of the net changes in marijuana use between boys and girls over time. However, exactly the same data and models can be viewed from the angle of comparing the boy-girl differences in marijuana use at several points in time, in other words, as the comparison over time of the association between a changing characteristic (viz. marijuana use) and a stable characteristic (viz. gender). For example, the question of whether or not the net differences over time in marijuana use are the same for boys and girls is identical to the question whether or not the boy-girl differences at each age category remain the same for all ages, which in its turn is just another way of asking whether or not the association between marijuana use and gender is the same at all ages.

The estimation of separate logit models for the effects of S on M in each age category separately does not require marginal-modeling methods, but as soon as comparisons are made among the age groups, marginal modeling is required because these marginals involve the same respondents. Above, it was found that model $\{SA,SM,AM\}$ fits the data in Table 4.11 very well, and it was concluded that the growth curves of boys and girls were similar (parallel). But from these results, it can also be concluded that the boy-girl differences are the same for all five age categories. Turning directly to the model that was chosen as the final model, viz. model $\{SA,SM,AM\}$ with the parabolic growth curve, the (significant) effects $\hat{\lambda}_{s*m}^{SAM}$ of S on M in that model are as follows:

M	Boys	Girls
1	$-.149$	$.149$
2	$-.054$	$.054$
3	$.203$	$-.203$

Hence, the local odds ratio for the 2×2 subtable involving boys and girls and Categories 1 and 2 of M equals $\exp(-.149+.054-.149-(-.054)) = \exp(.190) = .827$. From analogous computations, it follows that the local odds ratio for boys and girls and Categories 2 and 3 of M equals $.598$, and the odds ratio for boys and girls and Categories 1 and 3 equals $.495$. These associations between S and M in terms of odds ratios are the same for all age categories, and imply that at each age, the odds that one is a regular user ($M = 3$) rather than a nonuser ($M = 1$) are $1/.495 = 2.022$ times larger for boys than for girls.

Also the differences between means discussed in the previous subsection can be seen as association measures, similar to unstandardized regression coefficients. The model with restriction $\delta_{t+1}^{B} = \delta_{t+1}^{G}$ was accepted, and implied similar mean growth curves for boys and girls. But at the same time, from exactly this same model, it can be concluded that the differences between boys and girls in mean marijuana use are the same for all age categories. In the preferred model from the previous subsection with $\delta_{t+1}^B = \delta_{t+1}^G$ and the additional constraint of a linear growth curve for the means, the constant boy-girl differences in marijuana use are estimated as $\hat{\mu}_B - \hat{\mu}_G = .113$ ($s.e. = .045$). For each age category, the mean M score for boys is .113 higher than for girls.

Measures of association other than odds ratios or mean differences can, of course, be used to investigate the stable association hypothesis for the relationship S-M at each point in time, for example, gamma (γ). For each age category, the following values are observed (standard errors between parentheses):

t	$\hat{\gamma}_t$ (se)
13	.335 (.214)
14	.400 (.140)
15	.247 (.124)
16	.207 (.119)
17	.200 (.111)

All these γ's are positive, indicating higher marijuana use for boys than for girls. However, only in age categories 14 and 15 are the $\hat{\gamma}$'s significant at the .05 level. The hypothesis that $\gamma = 0$ for all age categories need not be rejected: $G^2 = 7.65$, $df = 5$ ($p = .176$, $X^2 = 7.47$). However, the model in which all five γ's are supposed to be equal also fits the data well ($G^2 = 2.17$, $df = 4$ $p = .705$, $X^2 = 2.14$) and (at the .05 level) better than the model with $\gamma = 0$: $G^2 = 7.65 - 2.17 = 5.49$, $df = 5 - 4 = 1$ ($p = .019$). An almost perfectly fitting model is the model in which a linear increase in the γ's is assumed. However, the equal γ model does not need to be rejected in favor of this linear γ model ($G^2 = 2.17 - .95 = 1.22$, $df = 4 - 3 = 1$ $p = .270$). Therefore, the equal γ model may well be the best choice. The (constant) association for all age groups between S and M is then estimated as $\hat{\gamma} = .238$ ($s.e. = .097$). Had the model with γ linearly changing over time been accepted, the interesting result would have been obtained that the differences between boys and girls in terms of γ would have become smaller over time, from $\hat{\gamma} = .376$ at age 13 to $\hat{\gamma} = .193$ at age 17. Conclusions about the possible net changes in the strength of the relationship between S and M do indeed depend here on the association measure chosen.

The main purpose of this section has been to illustrate that questions about net changes over time are part and parcel of panel studies, and that one needs marginal-modeling methods to answer these questions. Our marginal-modeling approach turned out to be very flexible and capable of handling a large variety of research hypotheses about net changes in many different forms. This is not to deny that the formulation of the appropriate matrices by means of the generalized exp-log notation is often complex and difficult, especially when the models are not loglinear. The matrices for these examples put on the book's webpage should enable the researchers to construct them easily and correctly in agreement with their own research purposes.

4.3 Gross Changes in One Characteristic

4.3.1 Comparing Turnover Tables for Different Periods

The analyses of the previous section on net changes in panel studies involved the comparisons of one-way marginals. However, the marginal-modeling approach of the previous section can be easily extended to cover what is often seen as the main strength of panel studies — the analysis of individual, gross changes in a partic- ular characteristic. The comparison of gross changes is another one of Duncan's key hypotheses in panel analysis that requires marginal modeling when one needs to compare individual gross changes over time. Rather than comparing one-way marginals, now higher-way marginals in the form of turnover tables have to be com- pared. Although the analysis of one particular transition table is a form of standard conditional analysis — in what way are the values on time $t + 1$ dependent on the scores on t — the comparison of several such transition tables requires marginal- modeling methods.

Comparisons of gross changes are often of great theoretical interest. For example, a political scientist might expect that the voters' intended party choices become more and more crystallized and stable in the course of an election campaign and that per- haps this crystallization process will be stronger among young rather than old people. Another related hypothesis might be that, in times of political conflicts, voters have stronger and more stable partisan opinions than in times of political peacefulness. In general, these kinds of hypotheses must be investigated by means of panel data comparing the individual changes in successive turnover tables, e.g., comparing the transition table from time t to $t + 1$ with the transition table from $t + 1$ to $t + 2$. Funda- mental work in this area, usually in the context of markov chain modeling was done in the 1950's by statisticians like Anderson, Goodman, Madansky. A summary and extensions of this work is provided by Bishop et al., 1975, Chapter 7. This section further extends their approach.

Below, a number of illustrations will be provided regarding different types of research questions about gross changes. In a way, these examples are rather straight- forward extensions of the approach explained in the previous section. However, some extra difficulties will arise that have to do with possible incompatibilities or redun- dancies among the imposed constraints. These difficulties are special cases of the redundancies or incompatibilities discussed in Section 2.3.3, and will be dealt with further when discussing the particular examples and, in a more general way, in the last section of this chapter.

By way of example, the data underlying Table 4.4 will be used. Table 4.4 con- tained the one-way marginals for *Political orientation* in 1992 (P_1), 1994 (P_2), and 1996 (P_3). The full $7 \times 7 \times 7$ cross-classification table $P_1 P_2 P_3$ is presented on the book's website. From this full cross-classification, three marginal two-way turnover tables can be formed, viz. $P_1 P_2$ (with, in shorthand notation, cell probabilities π_{ij}^{12}), $P_2 P_3$ (with entries π_{jk}^{23}), and $P_1 P_3$ (with entries π_{ik}^{13}) (tables not presented here, but see Table 4.12 below).

Now assume that a researcher wants to investigate whether the individual changes from 1992 to 1994 are similar to the changes from 1994 to 1996. As usual, the marginal tables of interest can be put together into one new table. In this case, it means that a new table is defined, consisting of the two marginal tables 1992–1994 and 1994–1996. If the rows of these two marginal tables refer to the earlier time point, and the columns to the later time point, this new table can be denoted as a *TRC* table, where *T* refers to the period of observation with $T = 1$ to the cells in marginal turnover table 1992–1994 and $T = 2$ to marginal turnover table 1994–1996. *R* refers to *Political orientation* as the row variable with seven categories in both marginal tables, and *C* to *Political orientation* as the seven-category column variable.

The hypothesis that the two marginal turnover tables 1992-1994 and 1994-1996 are the same is represented by loglinear model {*T,RC*} for table *TRC*. However, unlike the discussions so far in this book, this nice and elegant way of defining the marginal restrictions in terms of loglinear models for 'quasi-tables', such as *TRC*, causes some difficulties now. Depending on the research design (e.g., number of waves or kinds of turnover tables to be compared) and the nature of the particular marginal model fitted, straightforwardly defining the marginal restrictions in terms of the seemingly appropriate loglinear model for the 'quasi-table' may not work well because the loglinear model may imply redundant or incompatible restrictions and suggest the wrong degrees of freedom. These difficulties occur when the successive turnover tables to be compared have overlapping marginals. In our example, when turnover table 1992–1994 is compared with turnover table 1994–1996, the column variable in the first turnover table is the same variable as the row variable in the latter one. This difficulty will be further explained below.

Loglinear model {*T,RC*} for table *TRC* is meant to imply that the two marginal turnover tables P_1P_2 and P_2P_3 are completely identical. That is,

$$\pi_{ij}^{12} = \pi_{ij}^{23} . \tag{4.8}$$

Further, and important for the discussions below, note that marginal table P_1P_2 is equal to marginal turnover table *RC* for $T = 1$, and that in the same vein marginal turnover table P_2P_3 is equal to *RC* for $T = 2$. Therefore, $\pi_{ij}^{12} = \pi_{ij1}^{RC|T}$ and $\pi_{ij}^{23} = \pi_{ij2}^{RC|T}$; moreover, λ_{ij}^{12} from the saturated model for table P_1P_2 is identical to the conditional loglinear parameter for the relationship $R - C$ when $T = 1$: $\lambda_{ij}^{12} = \lambda_{ij1}^{RC|T}$ and similarly, $\lambda_{ij}^{23} = \lambda_{ij2}^{RC|T}$.

Determination of the number of degrees of freedom for testing the equality of the two turnover tables looks simple. Calculated in the standard way, loglinear model {*T,RC*} applied to table *TRC* has 48 degrees of freedom. The same result is obtained starting from the (identical) restrictions in Eq. 4.8. Because *Political orientation* has seven categories, and because in each turnover table the proportions sum to 1, Eq. 4.8 implies $(7 \times 7) - 1 = 48$ independent restrictions and, therefore 48 degrees of freedom left for testing Eq. 4.8. However, because of the overlapping variables in the successive turnover tables, one has to be very careful about computing the number of degrees of freedom in this direct, simple way because, in general, the overlapping variables and marginals in the successive turnover tables constrain the cell frequencies and there may be fewer degrees of freedom than expected.

Nevertheless, and perhaps somewhat surprisingly, for this particular table and this particular model, the determination of the independent restrictions and the calculation of the number of degrees of freedom happens to be correct. But this is certainly not true in general. If *Political orientation* had been measured at four points in time rather than three, and if whether the turnover probabilities were the same in the three successive turnover tables P_1P_2, P_2P_3 and P_3P_4 had been tested, the number of degrees of freedom would seemingly be $2 \times 48 = 96$ when calculated similarly as before for the three-wave panel. However, for the four-wave panel study, the correct number of degrees of freedom is only 90.

This is perhaps more easily seen if the restrictions in model $\{T,RC\}$ (or equivalently in Eq. 4.8) are explicitly formulated in terms of restrictions on the loglinear parameters for the relevant marginal tables — first, in terms of table *TRC* for three waves only. There are two sets of restrictions needed, one on the two-way effects in the marginal tables P_1P_2 and P_2P_3, and another on the one-way effects in the one-way marginal tables P_1, P_2, and P_3:

$$\lambda_{ij}^{12} = \lambda_{ij}^{23}$$

$$\lambda_i^1 = \lambda_i^2 = \lambda_i^3 .$$

(4.9)

Note that the second set of loglinear restrictions on the one-way marginals is equivalent to the restriction that all marginal distributions are equal:

$$\pi_i^1 = \pi_i^2 = \pi_i^3.$$

In general, for an arbitrary set of $I \times I$ tables, equal corresponding odds ratios in combination with equal marginal row distributions and with equal marginal column distributions make the two tables completely identical. In this case, because of the overlapping marginal in the two successive turnover tables, all one-way marginal distributions become equal to each other. The set of restrictions on the two-variable parameters in Eq. 4.9 implies $(7-1)(7-1) = 36$ independent restrictions; the number of restrictions on the one-variable parameters is $2 \times 6 = 12$. Altogether, there are $36 + 12 = 48$ degrees of freedom, as found above. However, if analogous restrictions (as in Eq. 4.9) would have been imposed but now for the four wave case on the three successive marginal turnover tables, the two-variable restrictions would have implied $2 \times 36 = 72$ independent restrictions and the number of independent restrictions on the one-way parameters would be $3 \times 6 = 18$, a total 90 independent restrictions, rather than the naively calculated 96 degrees of freedom.

Because the difficulties in determining the correct number of degrees of freedom and finding an appropriate set of independent restrictions stem here from the overlapping one-way marginals and the restrictions on one-variable parameters, yet another look at these one-variable parameters may be illuminating. Model $\{T,RC\}$ applied to table *TRC* for three waves implies that the loglinear parameters for (conditional) marginal turnover table *RC* for $T = 1$ (marginal table P_1P_2) are identical to the corresponding loglinear parameters for (conditional) marginal turnover table *RC* for $T = 2$

(marginal table P_2P_3). Besides the restrictions on the (conditional) two-variable parameters ($\lambda_{ij}^{12} = \lambda_{ij}^{23}$), the following restrictions on the one-variable parameters are implied in our marginal notation:

$$\lambda_{i*}^{12} = \lambda_{i*}^{23}$$
$$\lambda_{*i}^{12} = \lambda_{*i}^{23}.$$

These two sets of restrictions on the one-variable parameters given the number of categories imply $2 \times 6 = 12$ independent restrictions (in addition to the restrictions on the two-variable parameters). Because of the overlapping marginals in the model, in which the two successive marginal turnover tables are completely identical, λ_{i*}^{23} will be equal to λ_{*i}^{12}. But this still leaves $2 \times 6 = 12$ independent restrictions. However, model $\{T,RC\}$ implies the following restrictions on the one-variable parameters in the successive turnover tables for four waves:

$$\lambda_{i*}^{12} = \lambda_{i*}^{23} = \lambda_{i*}^{34}$$
$$\lambda_{*i}^{12} = \lambda_{*i}^{23} = \lambda_{*i}^{34}.$$

There seem to be $(2 \times 6) \times 2 = 24$ independent restrictions on the one-variable parameters. However, because of the overlapping marginals in combination with model $\{T,RC\}$ for table TRC, $\lambda_{i*}^{23} = \lambda_{*i}^{12}$ and $\lambda_{i*}^{34} = \lambda_{*i}^{23}$. Therefore, only $3 \times 6 = 18$ independent restrictions on the one-variable effects remain rather than 24.

The model of completely identical turnover tables for 1992–1994 and 1994–1996 (Eq. 4.8) just fits the data at the .05 level: $G^2 = 64.60$, $df = 48$ ($p = .055$, $X^2 = 56.6$). Although it is a borderline result, it is a bit surprising because, as seen in the previous section, the MH model for no net changes in the one-way marginals in Table 4.4 had to be rejected ($p = .006$). The overall test for the present model of identical successive turnover tables, i.e., for the equality of association *and* marginal distributions has obviously less power regarding the specific hypothesis of no net change than the MH test used before. On the other hand, the estimated turnover table $t - (t + 1)$ under Eq. 4.8 contains many small values for the cell frequencies and two zero cells. This sheds doubt on the approximation of the theoretical chi-square distribution by G^2(and uncertainty about the number of degrees of freedom to use).

A weaker, but theoretically more interesting, assumption than completely identical cell entries in the two successive turnover tables is the hypothesis of equal transition probabilities in both marginal turnover tables. Transition probabilities occupy a central place in panel analysis, much more so than the joint probabilities. Equal transition probabilities in marginal turnover tables P_1P_2 and P_2P_3 mean:

$$\pi_{j|i}^{2|1} = \pi_{j|i}^{3|2}. \tag{4.10}$$

In Bishop et al. (1975, Chapter 7, Section 7.2.4) it is shown that the equal transition probability model corresponds to loglinear model $\{TR,RC\}$ applied to table TRC. Bishop et al., following Goodman and others, investigate the equal transition probabilities model for successive turnover tables in the context of the extra assumption of first-order stationary markov chains. In our marginal-modeling approach, no such extra assumptions are needed about the underlying dependency process over time (although it is of course possible to add such an extra restriction). Model $\{TR,RC\}$ essentially implies that the end state at $t + 1$ (C) is independent of the period concerned (T), given the starting state at t (R). Model $\{TR,RC\}$ applied to table TRC for our three-wave example has 42 degrees of freedom, when computed in the naive, simple way. However, this number is correct. Even better, it is also correct if the equal transition model for successive turnover tables is imposed on a panel study with more than three waves. This can be easily seen when the restrictions are formulated in terms of restrictions on the loglinear parameters of the successive marginal turnover tables P_1P_2 and P_2P_3 (and then analogously to four and more waves). Model $\{TR,RC\}$ implies

$$\lambda_{ij}^{12} = \lambda_{ij}^{23}$$
$$\lambda_{*j}^{12} = \lambda_{*j}^{23}$$

Now the restrictions must be made on the one-variable parameters in the two-way tables and not on the one-variable parameters in the one-way marginals: in the equal transition model, the one-way marginals p_i^2 and p_i^3 (or λ_i^2 and λ_i^3) need not be equal to each other. However, if the row percentages are the same in tables P_1P_2 and P_2P_3, then the average conditional log distributions of the column variable will be the same in the two tables and therefore $\lambda_{*j}^{12} = \lambda_{*j}^{23}$.

The hypothesis of equal successive transition tables (Eq. 4.8) need not be rejected: $G^2 = 43.28$, $df = 42$ ($p = .417$, $X^2 = 35.24$). Moreover, and important in the light of the sparse data in the estimated transition table (see Table 4.12), the conditional test for the restrictions in Eq. 4.8 against the restrictions in Eq. 4.10 turns out in favor of the model of equal transition probabilities: $G^2 = 64.60 - 43.28 = 21.32$, $df = 48 - 42 = 6$ ($p = .002$). Note further that accepting Eq. 4.10 is not in contradiction with the rejection of the MH hypothesis for the one-way marginals.

A still weaker hypothesis about the similarity of the successive turnover tables P_1P_2 and P_2P_3 is that only the corresponding local odds ratios in the two tables are the same, i.e., loglinear model $\{TR,TC,RC\}$ for table TRC is true. In terms of restrictions for the loglinear parameters in the two marginal turnover tables, only the restrictions on the two-way effects in Eq. 4.9 are applied: $\lambda_{ij}^{12} = \lambda_{ij}^{23}$. Because no one-way (overlapping) marginals or one-variable effects are involved in the restrictions, there are no problems in computing the correct number of degrees of freedom in the usual way (and the same is true for four or more waves). Not surprisingly in the light of the above results, this equal odd ratio model also fits the data well with $G^2 = 34.77$, $df = 36$ ($p = .527$, $X^2 = 28.22$). However, on the basis of the conditional test, the model of equal odds ratios need not be preferred above the more stringent equal transitions model ($G^2 = 43.28 - 34.77 = 8.51$, $df = 42 - 36 = 6$, $p = .203$).

A practical problem when estimating this equal odds ratio model was that some of the estimated frequencies in the pertinent marginal tables were zero. This also happened in the other two models $\{T,RC\}$ and $\{TR,RC\}$, but because there the restrictions involved direct equality restrictions on the (transition) probabilities, no

Table 4.12. Estimated transition probabilities $\pi_{t+1|t}$ for *Political Orientation* under the model in Eq. 4.10. Source: U.S. National Election Studies, 1992–1997

	1. Extr.lib.	2. Lib.	3. Sl.lib.	4. Mod.	5. Sl.cons.	6. Cons.	7. Extr.cons.	Total
1. Extr.lib.	.418	.318	.077	.119	0	0	.066	1.000
2. Lib.	.030	.542	.239	.139	0	.049	0	1.000
3. Sl.lib.	.015	.128	.549	.193	.103	.011	0	1.000
4. Mod.	0	.044	.154	.514	.197	.077	.014	1.000
5. Sl.cons.	0	.005	.049	.201	.470	.258	.016	1.000
6. Cons.	0	.006	.019	.054	.181	.666	.073	1.000
7. Extr.cons.	0	0	0	.064	.074	.542	.319	1.000

convergence problems were encountered during the estimation process. However, odds ratios involving zero cells are not defined and the algorithm did not converge. It was decided to replace all empty observed cells in the full observed table $P_1P_2P_3$ by a small constant $c = 10^{-80}$. The results above have been obtained by this replacement of the observed empty cells.

Among the three models, the equal transition model in Eq. 4.10 appeared to be the best choice. The estimated transition probabilities are presented in Table 4.12.

Close inspection of the outcomes in Table 4.12 reveals precisely how people have changed from t to $t + 1$. It can be seen how stable the respondents are or to which categories people move, once they have chosen for a particular category. For example, the conservatives (Category 6) are the most stable: 66% of those who classified themselves as conservative the first time did so the second time. The least stable are the extreme conservatives (Category 7) with a relative stability of 32%. A test of whether these relative stabilities are in fact identical in the population is possible by introducing the extra restriction that all conditional probabilities on the main diagonal $\pi_{i\,i}^{2|1}$ $(= \pi_{i\,i}^{3|2})$ are the same for all categories i into the equal transition model. Testing the equal transition model with the extra assumption of equal relative stabilities for all categories yields a borderline result ($p = .044$), but the conditional test outcomes of the extra restricted equal transition model against the equal transition model *per se* are unequivocally against the validity of the extra restriction: $G^2 = 65.90 - 43.28 = 22.62$, $df = 48 - 42 = 6$ ($p = .001$). In the extra restricted model, the transition probabilities on the main diagonal were estimated as $\hat{\pi}_{i\,i}^{2|1} = .532$ for all i. Looking at the adjusted residual frequencies for this model (not reported here, but see also the main diagonal entries in Table 4.12), this extra assumption is not problematic for the first four categories (liberal/moderate), but the residuals are significant (in absolute values: > 1.96) for the last three categories (conservatives). However, the interpretation is not simply in terms of conservatives having smaller or larger relative stability than the liberals: the significant residuals are negative for Categories 5 and 7, but positive for Category 6. How to explain this is not immediately clear.

These kinds of analyses and inspection of the transition probabilities can be carried further, e.g., by imposing (quasi-)symmetry in the turnover tables. But rather than showing this (these are rather natural and easy extensions) a note of warning is appropriate. It is well-documented, although in practice often forgotten, that even large, systematically looking differences in transition probabilities may occur purely as artifacts of different category sizes in combination with very small amounts of random (measurement) error, without any true changes or differences in transitions taking place. This might also be a partial explanation of the erratic results obtained regarding the equal relative stability model. Latent variable models are required to investigate this phenomenon (Bassi, Hagenaars, Croon, & Vermunt, 2000; Hagenaars, 1990, 2005). How to introduce latent variables in marginal models will be discussed in Chapter 6 of this book.

4.3.2 Comparing Summary Measures of Gross Change

The investigation of the net changes in the previous section focused not only on the comparison of the whole marginal distributions but also on some characteristics of these distributions. In the same vein, the analyses of gross changes can be carried out using summary statistics for the gross changes. From the very first developments of panel methods forward, such summary statistics have been proposed, especially by Paul Lazarsfeld and his associates (Lazarsfeld & Fiske, 1938, Lazarsfeld, Berelson, & Gaudet, 1948, see also many unpublished references by Lazarsfeld c.s., Hagenaars, 1990, pp. 147, 148, 200). In the past, these summary statistics have been mostly used in a descriptive way. Marginal modeling methods provide a flexible and general framework for statistical inferences regarding these statistics.

For a characteristic with ordered categories such as *Political orientation*, it makes sense to ask whether the gross changes in a two-way turnover table are more in the direction of becoming more conservative or, on the contrary, of becoming more liberal. To answer this question, one must compare for a randomly chosen respondent the probability that the respondent belongs to the upper diagonal cells with the probability of belonging to the lower diagonal cells in the turnover table. For entries π_{ij} in the two-way turnover table, the sum $\sum_{i<j} \pi_{ij}$ has to be compared with $\sum_{i>j} \pi_{ij}$ by taking their difference, resulting in coefficient Ω or their log ratio yielding Ω'. The observed $\hat{\Omega}$'s and $\hat{\Omega}'$'s are reported in Table 4.13 for the three two-way turnover tables P_1P_2, P_2P_3, and P_1P_3 that have been formed from the data on *Political orientation* discussed before (tables were not presented). In this way, it can be seen whether there is an overall tendency in the gross changes towards more liberalism or more conservatism between successive time points, e.g. t and $t+1$.

As can be seen in Table 4.13, there is a slight but statistically significant tendency towards a more conservative attitude in terms of the gross changes in the first period from 1992 to 1994, while in the second period from 1994 to 1996, there is no clear average gross movement towards a more liberal or a more conservative attitude. The tendency from the beginning (1992) to the end (1996) is about the same as for the first period, and if anything, a little bit stronger. Further hypotheses about the Ω's and Ω''s (which would truly require a marginal-modeling approach) have not been tested, as the above results were clear enough.

Table 4.13. *Political Orientation*; observed Ω, standard errors and test results; see text source: U.S. National Election Studies, 1992–1997

	P_1P_2	P_2P_3	P_1P_3
$\sum_{i<j}\pi_{ij}$	.301	.203	.294
$\sum_{i>j}\pi_{ij}$	.211	.223	.186
$\hat{\Omega}'_{ij}$	.358	-.092	.457
$\hat{\Omega}_{ij}$	.091	-.020	.108
$s.e.(\hat{\Omega}_{ij})$	.035	.032	.034
$z\,(\hat{\Omega}_{ij})$	2.56	-0.61	3.14

Note that the coefficients Ω and Ω' look a lot like the coefficients L and L' used to determine the net changes in ordinal location in the marginal distributions. Nevertheless, they are not the same. The main difference between the two sets of coefficients resides in the way the units, whose scores are to be compared, are sampled: L and L' compare the scores of two units sampled independently from the two marginal distributions, whereas Ω and Ω' compare the scores of a single unit sampled from the bivariate distribution. As a consequence, the L coefficients may lead to different conclusions than those based on Ω and Ω'.

Another summary statistic of the turnover table that can be used at all levels of measurement is the raw stability coefficient $\sum_i \pi_{ii}$, that is, the sum of the cell entries at the main diagonal of the two-way turnover table. It has a nice and simple interpretation as the percentage of people that have the same score at two points in time. However, it has the disadvantage that it is strongly dependent on chance stability: even when the respondents' responses at the two time points are statistically independent of each other, $\sum_i \pi_{ii}$ can attain very high values. The coefficient used most often to correct for this spurious amount of stability is the agreement coefficient κ, discussed in Chapter 1, where 'agreement' denotes 'stability'.

Consider the panel data in Table 4.14 which gives the labor force participation of 1,583 women in five consecutive years. Table 4.15 contains the values of the observed uncorrected and the observed corrected proportion of stable respondents for the four transitions between consecutive years.

All observed $\hat{\kappa}$'s presented in Table 4.15 are significantly different from 0. It is more interesting to see whether the response stability remains constant over time, i.e., is the same for all successive turnover tables

$$\kappa_{12} = \kappa_{23} = \kappa_{34} = \kappa_{45}.$$

This hypothesis has to be rejected at the .05 but not at the .01 level: $G^2 = 10.02$, $df = 3$, $(p = .018, X^2 = 9.82)$. The estimate of the common κ equals $\hat{\kappa} = .716$. Comparison of this value .716 with the observed $\hat{\kappa}$'s in Table 4.15 suggests an increase in κ over time, and it may well be a linear increase. The hypothesis of a linear increase of κ

$$\kappa_{12} - \kappa_{23} = \kappa_{23} - \kappa_{34} = \kappa_{34} - \kappa_{45}$$

Table 4.14. Women's labor participation in five consecutive years. Source: Heckman and Willis, 1977

				1971	
1967	1968	1969	1970	Yes	No
Yes	Yes	Yes	Yes	426	38
Yes	Yes	Yes	No	16	47
Yes	Yes	No	Yes	11	2
Yes	Yes	No	No	12	28
Yes	No	Yes	Yes	21	7
Yes	No	Yes	No	0	9
Yes	No	No	Yes	8	3
Yes	No	No	No	5	43
No	Yes	Yes	Yes	73	11
No	Yes	Yes	No	7	17
No	Yes	No	Yes	9	3
No	Yes	No	No	5	24
No	No	Yes	Yes	54	16
No	No	Yes	No	6	28
No	No	No	Yes	36	24
No	No	No	No	35	559

Table 4.15. Proportion of stable respondents and observed and linear $\hat{\kappa}$ (see text) for women's labor participation in five consecutive years; Table 4.14

Transition	$\sum_i \hat{\pi}_{ii}$	Observed $\hat{\kappa}$	SE Observed $\hat{\kappa}$	Linear $\hat{\kappa}$
1967-68	.845	.687	.018	.681
1968-69	.852	.703	.018	.704
1969-70	.857	.714	.018	.728
1970-71	.880	.759	.016	.751

need not be rejected ($G^2 = 1.15$, $df = 2$ $p = .564$, $X^2 = 1.15$), and is also to be preferred when compared to the previous no difference model: $G^2 = 10.02 - 1.15 = 8.88$, $df = 3 - 2 = 1$ ($p = .003$). The values of the four $\hat{\kappa}$ coefficients under this model of linear increase are reported in Table 4.15 showing a constant, statistically significant increase over time of .023 (s.e. $= .008$). It is concluded that there is linear increase in the proportion of women, corrected for chance stability, that remained in or remained out of the labor force for two consecutive years when considering all successive turnover tables $t - (t+1)$ from 1967 to 1971.

4.3.3 Extensions; Net *Plus* Gross Changes; Multiway Turnover Tables; Subgroup Comparisons

The analyses carried out so far in this section on gross changes in one characteristic can be extended in several important ways. A first possibility is to *combine hypotheses on gross change with hypotheses on net changes.* There are many research situations where such combined hypotheses are meaningful. For example, in social mobility research, the kernel of the investigations is formed by turnover tables in which the occupations or classes people belong to at time one (beginning of the career or (grand)parents) are cross-classified with occupation or class at time two (later in the career or (grand)children). In these investigations, hypotheses are often formulated about structural mobility forced by the differences between the marginals and simultaneous hypotheses about free circulation mobility or fluidity that is not enforced by the marginal differences. There have been numerous discussions about the meanings of structural and free mobility and how to measure and distinguish them (see among many others, Luijkx, 1994). The marginal-modeling approach advocated here, which makes it possible to combine hypotheses about net and gross changes, may well provide a better solution to these problems than the solutions found so far in the social mobility literature. An empirical application of simultaneously testing hypotheses about net and gross changes will be provided at the end of the next section, albeit in a different context than social mobility.

A second extension concerns the *dimensionality of the turnover tables.* In all analyses above, only two-way turnover tables for successive periods have been compared, but of course also three- and higher-way tables can be simultaneously investigated. For example, in a five-wave panel study, marginal loglinear models can be used for the comparison of turnover tables, $T_1 T_2 T_3$, $T_2 T_3 T_4$, and $T_3 T_4 T_5$. The principles involved do not differ from the above discussions about the comparisons of two-way turnover tables, and their applications to higher-way tables is straightforward.

Thirdly, just as for the net changes, *subgroup comparisons* may be interesting for gross changes. For example, the successive turnover tables may be set up separately for boys and girls. All analyses carried out above for the whole sample can also be done for these two subgroups, under the conditions that these models are completely identical or different in one or more respects for boys and girls. Once more, the principles, models, and examples presented so far in this section can be applied in a simple, direct manner to problems involving the comparisons of gross changes for subgroups.

Finally, as with the comparison of net changes, these same subgroup comparisons can be viewed from the standpoint of *association* between the gross changes in changing characteristics on the one hand, and on the other hand a stable characteristic such as *Gender.* This comes close to setting up a causal model for the various variables, which is the topic of the next chapter.

Table 4.16. *Marijuana* and *Alcohol use* data ($N = 208$). Source: U.S. National Youth Survey

	13	%	14	%	15	%	16	%	17	%
					$T(ime)$ $(= age\ t)$					
I(tem) = 1 (Marijuana)										
R(esponse)										
1. Never	193	92.8	172	82.7	149	71.6	139	66.8	120	57.7
2. Once a month	12	5.8	23	11.1	37	17.8	40	19.2	48	23.1
3. More than once a month	3	1.4	13	6.3	22	10.6	29	13.9	40	19.2
Total	208	100%	208	100%	208	100%	208	100%	208	100%
I(tem) = 2 (Alcohol)										
R(esponse)										
1. Never	145	69.7	93	44.7	78	37.5	54	26.0	38	18.3
2. Once a month	60	28.8	98	47.1	101	48.6	98	47.1	87	41.8
3. More than once a month	3	1.4	17	8.2	29	13.9	56	26.9	83	39.9
Total	208	100%	208	100%	208	100%	208	100%	208	100%

4.4 Net and Gross Changes in Two Related Characteristics

The applications and examples in the above sections showed that there are many important kinds of research questions that require the study of changes in just one particular characteristic over time. At the same time, researchers also often consider theories that must be tested by similtaneously comparing changes in two or more characteristics. For example, a researcher of drug use may have theoretical reasons to assume that adolescents show the same patterns over time when getting addicted to whatever drug they start using, or on the contrary, adopt different patterns for different drugs. To find empirical answers to many such questions, marginal models have to be applied. Only instances involving just two characteristics will be discussed here, but extensions to three or more follow easily. The example will be on drug use, more specifically on the use of marijuana and alcohol where the alcohol data were collected from the same adolescents as the marijuana data (Table 4.6). Because of additional missing data for the alcohol data, the sample size decreased from $N = 240$ to $N = 208$.

First, it will be shown how to investigate the net changes in two related characteristics (Section 4.4.1). The pertinent marginal models are very similar to the ones presented in Section 4.1 on trend data, but now the panel character of the data causes additional dependencies that have to be taken into account. Second, it will be shown how to compare the associations between two characteristics over time; researchers may want to know whether, for example, associations increase over time (Section 4.4.2). Thirdly, it might be important to know whether and how the gross changes over time differ for the two characteristics (Section 4.4.3). Fourth, and finally, how to simultaneously test hypotheses about net and gross changes for two characteristics will be shown. From this overview and these examples, the readers must be able to formulate and test their own marginal models regarding changes in two or more

Table 4.17. Conditional loglinear parameters $\hat{\lambda}_{t\,k\,i}^{TR|I}$ for *Marijuana* and *Alcohol use* data (N=208)

	$T(ime)$ ($= age\ t$)				
	13	14	15	16	17
Marijuana					
R(esponse)					
1. Never	1.003	.220	-.209	-.374	-.640
2. Once a month	-.106	-.123	.067	.050	.113
3. More than once a month	-.896	-.097	.143	.324	.527
Alcohol					
R(esponse)					
1. Never	1.260	.223	-.083	-.537	-.863
2. Once a month	.208	.104	.005	-.112	-.205
3. More than once a month	-1.468	-.327	.078	.649	1.068

characteristics (of course helped by the data and matrix formulations presented on the book's website).

4.4.1 Net Changes in Two Characteristics

The one-way marginals for marijuana and alcohol use are presented in Table 4.16. This table will be referred to as if it were a standard *ITR* table, where *I* refers to *Item* (1 = marijuana; 2 = alcohol), *T* represents *Time* (with the five age categories 1 through 5), and *R* is the *Response* variable (with the three frequency-of-use categories 1 through 3). Symbol *T* will now be used for the *Age* categories to avoid confusion with the use of alcohol.

The data in marginal Table 4.16 will be analyzed by means of loglinear models and by applying different models for the response means μ. A first impression of the growth curves for marijuana and alcohol use can be gained from the relevant parameters of saturated loglinear model {*ITR*} (Table 4.17) and the observed means (Fig. 4.2). The loglinear parameters in Table 4.17 are the conditional parameters $\hat{\lambda}_{t\,k\,i}^{TR|I}$ that are computed from the $\hat{\lambda}_{*t\,k}^{ITR}$ and $\hat{\lambda}_{it\,k}^{ITR}$ estimates of model {*ITR*} as

$$\lambda_{t\,k\,i}^{TR|I} = \lambda_{*t\,k}^{ITR} + \lambda_{it\,k}^{ITR}.$$

They represent the loglinear relationship between age and use, i.e., between *T* and *R* for *Marijuana use* ($\hat{\lambda}_{t\,k\,1}^{TR|I}$) and *Alcohol use* ($\hat{\lambda}_{t\,k\,2}^{TR|I}$), respectively.

From Tables 4.16 and 4.17 and Fig. 4.1, it is clear that there are more alcohol than marijuana users in all age groups and that both alcohol and marijuana use increase over time, the use of alcohol somewhat more than marijuana. It is also reassuring to see that the additional nonresponse following from the inclusion of the use of alcohol does not seem to seriously biases the results, at least not for the use of marijuana. The growth curve for marijuana in Fig. 4.2 is similar to the growth curve in Fig. 4.1.

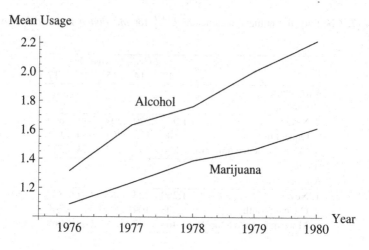

Fig. 4.2. *Alcohol* and *Marijuana Use*

Formal loglinear analyses confirm these findings. Several models that contradict these descriptive conclusions in one or more respects have to be rejected ($p = .000$): model $\{IT,R\}$ for table ITR, in which the use of alcohol and marijuana is supposed to be at the same level and constant over time; model $\{IT,IR\}$, in which changes over time are allowed but no differences between alcohol and marijuana use; and model $\{IT,IR\}$, with stable use over time but at a different level for alcohol and marijuana. A parsimonious model that allows for different levels of the use of marijuana and alcohol and for changing use over time (ie., model $\{TI,IR,TR\}$) for table ITR fits the data somewhat better. Note that in this model, the marijuana-alcohol differences are supposed to be the same at each age category and the age differences in use the same for alcohol and marijuana. In other words, the growth curves for the log odds of R for alcohol and marijuana are parallel, but do not coincide and one may be above or below the other. The test results for model $\{TI,IR,TR\}$ are $G^2 = 17.32$, $df = 8$ ($p = .027$, $X^2 = 12.22$). Given the discrepancies between G^2 and X^2, and the reported p-value for G^2, it is not so clear whether to accept or reject the model.

Perhaps a more parsimonious no-three-variable interaction model might give clearer results. Remembering from Section 4.2.1 that the growth curve for the use of marijuana expressed in terms of log odds appeared to be curvilinear, model $\{TI,IR,TR\}$ for table ITR was fitted with linear and also curvilinear restrictions on the growth curve, i.e., on the parameters λ^{ITR}_{*jk}. It turned out that parabolic growth curve models for marijuana and for alcohol fit the data significantly better than linear growth curve models, but also the parabolic growth curve model in model $\{TI,IR,TR\}$ has to be rejected. Some variant of model $\{ITR\}$ is needed. Given previous results, a restricted version of model $\{ITR\}$ for table ITR was applied in which (a) R is treated as an interval-level variable concerning the relationship with T, (b) the loglinear growth curves for both marijuana and alcohol are assumed to be curvilinear (parabolic), and (c) the differences between the loglinear growth curves of marijuana and alcohol increase linearly over time. All this implies the following restrictions on the two-variable parameters λ_{*jk}^{ITR} and the three-variable parameter parameters λ_{ijk}^{ITR} in loglinear model $\{ITR\}$ for table ITR:

$$\lambda_{*jk}^{ITR} = \alpha T_j R_k + \beta (T_j^2 - 2) R_k$$

(4.11)

$$\lambda_{ijk}^{ITR} = v_i^I T_j R_k.$$

To ensure the usual effect coding restrictions, the scores for R_k in Eq. (4.11) are $-1, 0, 1$, and for T_j they are $-2, -1, 0, 1, 2$. The quadratic component of λ_{*jk}^{ITR} as a function of T was represented by the contrast $(T^2 - 2)$ (i.e., $\{2, -1, -2, -1, 2\}$) to ensure that $\sum_k \lambda_{*jk}^{ITR} = 0$; finally, $\sum_i v_i^I = 0$. The test results are $G^2 = 19.83$, $df = 13$ ($p = .099$, $X^2 = 15.33$). The model in Eq. 4.11 can be accepted. Its relevant parameter estimates $\hat{\lambda}_{jki}^{TR|I}$ for the growth curves of marijuana and alcohol and $\hat{\lambda}_{ikj}^{IR|T}$ for the differences in use of alcohol and marijuana at each point in time are reported in Table 4.18.

The estimates in Table 4.18 contain a lot of information on the overall increase of alcohol and marijuana use. Just a few highlights will be presented here, concentrating on the (log) odds of being a regular user ($R = 3$) rather than a never user ($R = 1$), and starting with $\hat{\lambda}_{jki}^{TR|I}$ representing the form of the growth curves. Looking first at the growth curve of the use of marijuana ($I = 1$), from $\hat{\lambda}_{131}^{TR|I}$ and $\hat{\lambda}_{111}^{TR|I}$ it can be derived that at age 13 the log odds of regularly rather than never using marijuana are $= -.814 - .814 = -1.628$ less than the average log odds of regularly versus never using. At age 17, on the other hand, it can be seen from $\hat{\lambda}_{531}^{TR|I}$ and $\hat{\lambda}_{511}^{TR|I}$ that the corresponding log odds 'regular' versus 'never' are 1.145 ($= .572 - (-.572)$) higher than the corresponding average odds. In other words, the odds of regularly using rather than never using increase from age 13 to age 17 by a factor $\exp(1.145 - (-1.628)) = \exp(2.773) = 16.003$. Similar calculations for the use of alcohol lead to the conclusion that the odds of regularly using alcohol increase from age 13 to age 17 by a factor of 49.897. From age 13 to age 17, the increase in the odds of regularly using rather than never is $49.897/16.003 = 3.118$ times larger for alcohol than for marijuana. The growth curve of alcohol is steeper than for marijuana. The changes of the odds in favor of regular use are incremental, and for each next age category more and more people start using alcohol or marijuana. However, the growth curves are not linear, but parabolic. Although there is always an increase over time in the odds of regularly rather than never using for both curves, that increase in the odds diminishes for each next period by a factor of 1.273, as can be derived from the parameter estimates $\hat{\lambda}_{jki}^{TR|I}$. Due to the model specifications in Eq. 4.11, this diminishingly increasing factor is the same for alcohol and marijuana.

The growth curves can also be considered from the viewpoint of the differences in use between alcohol and marijuana at each age category. Due to the linear restrictions on the three-variable interaction term in Eq. 4.11, and the resulting estimates, the differences between the two growth curves of marijuana and alcohol become

Table 4.18. Conditional parameters $\hat{\lambda}_{jk\,i}^{TR|I}$ and $\hat{\lambda}_{ik\,j}^{IR|T}$ for *Marijuana* and *Alcohol use*; data based on model in Eq. 4.11; see text

$\hat{\lambda}_{jk\,i}^{TR|I}$		$T(ime)$ $(= Age)$			
	13	14	15	16	17
[-6pt] *R Marijuana* ($I = 1$)					
1. Never	.814	.286	−.121	−.407	−.572
2. Once a month	0	0	0	0	0
3. More than once a month	−.814	−.286	.121	.407	.572
R Alcohol ($I = 2$)					
1. Never	1.098	.428	−.121	−.549	−.857
2. Once a month	0	0	0	0	0
3. More than once a month	−1.098	−.428	.121	.549	.857

$\hat{\lambda}_{ik\,j}^{IR|T}$		$T(ime)$ $(= Age)$			
	13	14	15	16	17
R Marijuana ($I = 1$)					
1. Never	.366	.437	.508	.579	.650
2. Once a month	−.374	−.374	−.374	−.374	−.374
3. More than once a month	.008	−.063	−.134	−.205	−.276
R Alcohol ($I = 2$)					
1. Never	−.366	−.437	−.508	−.579	−.650
2. Once a month	.374	.374	.374	.374	.374
3. More than once a month	−.008	.063	.134	.205	.276

linearly larger and larger. The relevant parameter estimates are the $\hat{\lambda}_{ik\,j}^{IR|T}$ parameters Again the odds of regularly using rather than never using are discussed. Now, from $((\hat{\lambda}_{23\,1}^{IR|T} - \hat{\lambda}_{21\,1}^{IR|T}) - (\hat{\lambda}_{13\,1}^{IR|T} - \hat{\lambda}_{11\,1}^{IR|T}))$, it can be derived that at age 13 the odds of regular rather than never use are $\exp((-.008 - (-.366) - (.008 - .366)) = \exp(.716) = 2.046$ times larger for alcohol than for marijuana. At age 17, this difference between alcohol and marijuana use has become $\exp(.276 - (-.650)) - (-.276 - .650)) = \exp(1.854) = 6.383$, which is $6.383/2.046 = 3.118$ times larger than at age 13 (as found above from a different point of view). Given Eq. 4.11, the odds ratio with the odds of regular rather than never using alcohol in the numerator and the corresponding odds for marijuana in the denominator become larger with a factor of 1.329 from one age category to the next.

These kinds of analyses are very much in line with the analyses in Section 4.1 on parallel changes over time in two characteristics. The special point here in this section is that the two characteristics were not only measured on the same people, but also at the same time points. The general marginal-modeling approach takes these 'double dependencies' automatically and in a straightforward way into account.

Simultaneous Changes in Location

For the analysis of the mean uses of alcohol and marijuana, R is regarded as an interval-level variable with category scores 1, 2 and 3. The ten observed means $\hat{\mu}_{ij}^{IT}$ (I referring to *Item* with responses 1 = marijuana, 2 = alcohol and T to the five points in time) were depicted in Fig. 4.1. Models for the mean growth curves can be defined starting from the saturated model

$$\mu_{ij}^{IT} = \alpha_{**}^{IT} + \alpha_{i*}^{IT} + \alpha_{*j}^{IT} + \alpha_{ij}^{IT} \tag{4.12}$$

where the parameters on the right-hand side sum to zero over any subscript. The model in which all ten means are set equal to each other ($\mu_{ij}^{IT} = \alpha_{**}^{IT}$) did not fit the data at all ($p = .000$). A similar bad fit was obtained for the model in which the means are supposed to be stable over time, but with a (constant) difference between alcohol and marijuana ($\mu_{ij}^{IT} = \alpha_{**}^{IT} + \alpha_{i*}^{IT}$), and for the model in which there are differences in means over time, but not between the use of marijuana and alcohol ($\mu_{ij}^{IT} = \alpha_{**}^{IT} + \alpha_{*j}^{IT}$). Model

$$\mu_{ij}^{IT} = \alpha_{**}^{IT} + \alpha_{i*}^{IT} + \alpha_{*j}^{IT} \ ,$$

in which it is assumed that there are both item differences but constant over time, as well as time differences, but identical for both items (parallel growth curves) does not fit the data: $G^2 = 28.34$, $df = 4$ ($p = .000$, $X^2 = 18.64$). Obviously, some form of dissimilarity of the growth curves of marijuana and alcohol, expressed by α_{ij}^{IT} in Eq. 4.12 is needed. A very parsimonious model in this respect would be obtained by replacing α_{ij}^{IT} with a linear-by-linear term containing a single unknown parameter ϕ

$$\mu_{ij}^{IT} = \alpha_{**}^{IT} + \alpha_{i*}^{IT} + \alpha_{*j}^{IT} + \phi I_i T_j \ , \tag{4.13}$$

with I_i having scores $-1, 1$ and T_j having scores $-2, -1, 0, 1, 2$. This restricted model fits the data: $G^2 = 5.84$, $df = 3$ ($p = .120$, $X^2 = 4.88$). Making the interaction terms α_{ij}^{IT} parabolic in T does not improve the fit of the model at all. In this model, no restrictions on the shape of the growth curves of marijuana and alcohol are imposed, except that the distance between the two mean curves increases linearly over time. However, the growth curve of the means for marijuana use was shown before to be linear in form. Assuming both growth curves to be linear, and (linearly) different from each other, leads to model

$$\mu_{ij}^{IT} = \alpha_{**}^{IT} + \alpha_{i*}^{IT} + \gamma T_j + \phi I_i T_j \ .$$

This linear growth curve model need not be rejected: $G^2 = 8.97$, $df = 6$ ($p = .175$, $X^2 = 8.19$. Neither is the less restricted model defined by Eq. 4.13 to be preferred above this one: $G^2 = 8.97 - 5.84 = 3.13$, $df = 6 - 3 = 3$ ($p = .372$). Parabolic growth

Table 4.19. Estimated means for data on *Marijuana* and *Alcohol use* based on Eq. 4.13; see text.

	$T(ime)$ $(= Age)$				
	13	14	15	16	17
$\hat{\mu}$ (Marijuana)	1.107	1.237	1.367	1.497	1.627
$\hat{\mu}$ (Alcohol)	1.339	1.561	1.782	2.003	2.224

curves for either marijuana or alcohol or both do not improve the fit of the model at all.

The means for the linear growth curve model are reported in Table 4.19. It turns out that the means of marijuana use and alcohol significantly increases from one age category to the next with slopes of .130 and .221, respectively. The statistically significant difference in slopes for marijuana and alcohol use is equal to $.130 - .221 = -.091$. Formulated differently, the difference between the means of alcohol and marijuana use increases by .091 for each successive age, ultimately from $.232 (= 1.339 - 1.107)$ at age 13 to $.597 (= 2.224 - 1.627)$ at age 17.

Just as for the net changes in one characteristic, subgroup comparisons of the net changes in two or more characteristics can easily be carried out by extending the models from this subsection with a stable characteristic such as *Gender* (*G*) to compare boys and girls regarding their growth curves for marijuana and alcohol. Viewed from another angle, models for subgroups can be seen as models for the changes in marginal association between a stable characteristic (*G*) and *Marijuana use*, and between *G* and *Alcohol use*.

4.4.2 Changes in Association Between Two Changing Characteristics

So far, the marginal distributions of the two characteristics under study have been treated separately in the sense of being considered as two one-way marginals (although involving dependent observations). But the two characteristics also have a joint distribution and it may be interesting to see how this joint distribution changes over time. The relevant marginal tables for marijuana and alcohol are reported in Table 4.20. This table will be indicated as if it were a standard *TMA* table, where *T* refers to *Time*, (i.e., to the five age categories), *M* refers to the frequency of marijuana use and *A* refers to the frequency of alcohol use. The focus in this subsection will be on the joint distribution of *A* and *M*, i.e., on the association of the uses of alcohol and marijuana. Very restrictive (loglinear) models for table *TMA* in this respect are model {*T,MA*}, in which it is assumed that all five marginal *MA* tables are exactly the same at each point in time, both regarding the marginal distributions and the odds ratios; and model {*TM,TA*} which represents the hypothesis that at each of the five time points alcohol and marijuana use are independent of each other. Not surprisingly, both very restrictive models have to be rejected ($p = .000$). It then becomes interesting to see how the association between *M* and *A* changes over time. It was expected that marijuana and alcohol use are positively correlated and that this positive

Table 4.20. *Alcohol* by *Marijuana use* over time (N=208). Source: U.S. National Youth Study; see text

T	M	A 1	2	3
13	1	143	48	2
	2	2	10	0
	3	0	2	1
14	1	92	69	11
	2	1	19	3
	3	0	10	3
15	1	75	67	7
	2	2	28	7
	3	1	6	15
16	1	52	74	13
	2	2	19	19
	3	0	5	24
17	1	36	56	28
	2	1	27	20
	3	1	4	35

association increases with age: the frequencies of alcohol and marijuana use will go together more and more when the respondents grow older. To investigate these hypotheses, one of the many available measures of association must be selected. Below, by way of example, the odds ratio, gamma and the correlation coefficient will be applied.

Within the loglinear framework, association is expressed in terms of odds ratios. The null hypothesis that the corresponding odds ratios are the same in the five marginal *MA* tables, and that there is no increase in association in terms of odds ratios, can be tested by applying (by means of marginal-modeling) model $\{TM,TA,MA\}$ to *TMA* Table 4.20. This turns out to be a viable hypothesis (contrary to the initial expectations): $G^2 = 17.91$, $df = 16$ ($p = .329$, $X^2 = 14.65$). To investigate whether a more restricted and perhaps more powerful model might be able to detect existing differences in association in the population over time, a model with a linear increase in the log odds ratios over time was tested, i.e., model $\{TMA\}$ with restrictions

$$\lambda_{t\,i\,j}^{TMA} = \mu_{i\,j}^{MA} T_t$$

where T_t has scores $-2,-1,0,1,2$ and $\mu_{i\,j}^{MA}$ sums to 0 over any subscript. This restricted variant of model $\{TMA\}$ fits the data too: $G^2 = 14.46$, $df = 12$ ($p = .272$, $X^2 = 11.51$). However, on the basis of the conditional test, the constant odds ratio model $\{TM,TA,MA\}$ has to be preferred: $G^2 = 17.91 - 14.46 = 3.45$, $df = 16 - 12 = 4$ ($p = .486$). The parameter estimates $\hat{\lambda}_{*i\,j}^{TMA}$ from model $\{TM,TA,MA\}$ are reported in the upper panel of Table 4.21.

According to the $\hat{\lambda}$ estimates and the estimated local odds ratios in the lower panel of Table 4.21, there exists a strong, statistically significant relationship between

Table 4.21. *Upper panel*: estimated values and associated standard errors (in parentheses) of $\hat{\lambda}_{*i\ j}^{TMA}$ for model $\{TM, TA, MA\}$; *Lower panel*: log odds ratios and odds ratios (in parentheses); see text

			Alcohol	
Marijuana	1	2	3
1	1.437 (0.175)	-0.202 (0.106)	-1.235 (0.118)
2	-0.704 (0.120)	0.524 (0.091)	0.180 (0.109)
3	-0.733 (0.133)	-0.322 (0.113)	1.055 (0.117)
	1/2	2/3	1/3
1/2	2.867 (17.588)	.688 (1.991)	3.556 (35.012)
2/3	-.818 (.441)	1.721 (5.590)	.903 (2.468)
1/3	2.050 (7.764)	2.409 (11.128)	4.459 (86.399)

marijuana and alcohol use that is the same at all ages. The extreme log odds ratio involving cells $(1,3)$ of both M and A equals 4.459, and the odds ratio is 86.399: the odds that one becomes a regular (more than once a month) marijuana user rather than a never user of marijuana are more than 86 times larger for regular alcohol users than for never users of alcohol (and the same is true, of course, if one interchanges the roles of marijuana and alcohol). In terms of local odds ratios, the relationship is not consistently monotonic. As can be seen in Table 4.21, there is one negative local log odds ratio.

When M and A are considered as interval-level variables and a linear relationship is assumed, the product moment correlation ρ (see Chapter 1) is an often-used measure of association. The sample values r_t for the five age categories are as follows:

t	r_t
13	.363
14	.359
15	.551
16	.583
17	.495

All sample correlations are highly significantly different from 0 ($p = .000$). The hypothesis that all five correlations have the same value in the population can be tested by means of the marginal-modeling procedures described in Chapter 3, although the necessary matrices defined in the generalized exp-log notation are not very simple (but see the book's website). Despite the considerable variation among the sample values, this hypothesis that they are all the same in the population

$$\rho_t = \rho_* , \qquad \text{for all } t$$

cannot be clearly rejected: $G^2 = 8.88$, $df = 4$ ($p = .064$, $X^2 = 5.849$). The constant correlation is estimated as $r = 0.469$ ($s.e. = .031$). A similar borderline result is obtained for the hypothesis of a linear increase in correlation

$$\rho_t = \rho_* + t\alpha.$$

The test results are $G^2 = 7.33$, $df = 3$ ($p = .062$, $X^2 = 4.479$). The estimated increase in the correlation from one time point to another equals $\hat{\alpha} = 0.03$. Comparing the two models to each other, preference should be given to the constant ρ model: $G^2 = 8.88 - 7.33 = 1.55$, $df = 4 - 3 = 1$ ($p = .214$). One should therefore decide either in favor of the constant ρ model or take the sample values as they are. In the latter case, it is concluded that the correlation is the same at ages 13 and 14 and then increases to a higher more or less stable level for ages 15 through 17. And, of course, suspending judgment until further evidence has been gathered should also be a serious option. Finally, it should be kept in mind that it was seen above that the marginal distributions of M and A change over time and not exactly in the same way. This also may influence the sizes of the correlations (unlike the odds ratios).

When marijuana and alcohol use are seen as ordinal-level variables, having a monotonically increasing or decreasing relationship, Goodman and Kruskal's γ (gamma, see Chapter 1) is a reasonable choice as a measure of association. The sample values of γ are as follows:

t	$\hat{\gamma}_t$
13	.879
14	.792
15	.839
16	.850
17	.721

They are all statistically significantly different from 0 ($p = .000$). The hypothesis that all five γ's are equal at the five age categories can be accepted: $G^2 = 5.66$, $df = 4$ ($p = .226$, $X^2 = 3.84$). The common value equals $\hat{\gamma} = .781$ ($s.e. = .033$). The outcomes of a linear trend model for the γ's provide evidence against the increasing association hypothesis, because according to that model's estimates, the $\hat{\gamma}$'s linearly *decrease* for each next age category with $-.027$. However, the linear trend model should be rejected compared to the constant γ hypothesis: $G^2 = 5.66 - 4.49 = 1.17$, $df = 4 - 3 = 1$ ($p = .279$). Accepting the equal γ hypothesis is here the best choice.

The use of γ leads to the same conclusion as for the odds ratio, in the sense that the strength of the positive association remains the same over time. However, the conclusions concerning the correlation coefficient were a bit different, although uncertain. This illustrates once more the simple, but often forgotten, fact that conclusions about interactions, i.e., about changing or differing relationships between two variables within categories of other variables, depend on the kinds of measures that are used to measure association.

Table 4.22. Turnover tables for *Marijuana* and *Alcohol Use*. Source: U.S. National Youth Study

					R (Time t)						
			1			2			3		
	S (Time $t+1$)	1	2	3	1	2	3	1	2	3	
I (Item)	P (Period)										Total
Marijuana	13-14	166	21	6	6	1	5	0	1	2	208
	14-15	143	24	5	5	10	8	1	3	9	208
	15-16	128	14	7	9	20	8	2	6	14	208
	16-17	106	25	8	14	21	5	0	2	27	208
Alcohol	13-14	89	49	7	4	48	8	0	1	2	208
	14-15	63	27	3	13	68	17	2	6	9	208
	15-16	46	28	4	7	65	29	1	5	23	208
	16-17	27	24	3	9	56	33	2	7	47	208

4.4.3 Gross Changes in Two Characteristics

When two characteristics are considered rather than one, and both are measured repeatedly over time, one cannot only look, as in the previous sections, at the simultaneous net changes or at the changing nature of their associations over time, but also must investigate whether the gross changes for successive time points t and $t+1$ are similar for both characteristics. Analogous to Section 4.3.1, the questions on 'being the same or not' over time or also over items will take three basic forms. First, are the corresponding cell entries $\pi_{ij}^{t,t+1}$ in the successive turnover tables the same? Second, are the corresponding transition probabilities $\pi_j^{t+1|j}{}_i$ identical? And, third, are the corresponding odds ratios the same?

The relevant successive marginal turnover tables for marijuana and alcohol use are presented in Table 4.22, in which each row represents a particular marginal turnover table. Table 4.22 will be referred to as if it were a standard *IPRS* table, where I indicates whether the frequencies pertain to marijuana ($I = 1$) or alcohol ($I = 2$), where P is the indicator variable with four categories for the four turnover tables between adjacent time periods (age categories) t and $(t + 1)$, and where the frequency of use at time t is denoted by R, and at time $t + 1$ by S. The cell probabilities in a particular marginal turnover table RS for a particular item i and a particular period p will be denoted as $\pi_{rs\ ip}^{RS|IP}$.

The relevant loglinear models for Table 4.22 can best be seen as logit models, in which items I and P are regarded as the independent variables and the 'combined variable' RS as the dependent variable. Such logit models indicate how the cell entries of the turnover tables RS change over time (P) and how different they are for the two items (I).

The most parsimonious model in this respect is model $\{IP,RS\}$, which states that all eight turnover tables are completely identical, i.e., the entries $\pi_{rs\ ip}^{RS|IP}$ do not vary over the periods or over the items:

$$\pi_{r\,s\ ip}^{RS|IP} = \pi_{r\,s}^{RS}.$$

However, when applying our marginal-modeling approach in a direct, simple way, i.e., just trying to obtain estimates for the joint frequencies in the complete table under the marginal restriction that model $\{IP,RS\}$ is valid for Table 4.22, the algorithm did not converge and the usual number of degrees of freedom for model $\{IP,RS\}$ for table $IPRS$ ($df = 56$) was wrong. This was as expected, given the experiences with the comparisons of successive turnover tables with overlapping variables for one characteristic (Section 4.3.1), and is a rather sure sign that there exists an incompatibility or redundancy among the constraints implied by the model. A minimally specified model (Lang & Agresti, 1994) must be found, in which all constraints are functionally independent. In Chapter 2, this problem was discussed in more general terms, and in the last section of this chapter, this general treatment will be specified for the case of comparing turnover tables and overlapping marginals. For now, in much the same vein as in Section 4.3.1, the problem will be clarified in the concrete terms of this example. The treatment here certainly overlaps partly with the discussions in Section 4.3.1 (at some places using a somewhat different notation and emphasis), but given the complexity and importance of finding the appropriate set of independent restrictions we regard this as a virtue rather than a vice.

Table 4.22 consists of eight 3×3 turnover tables RS, one for each value of I and P. Given that the proportions in each turnover table sum to one, each turnover table RS contains eight proportions to be independently estimated. In model $\{IP,RS\}$, it is assumed that all corresponding entries in the eight turnover tables are the same. Therefore, there seem to be $7 \times 8 = 56$ independent restrictions and a corresponding number of degrees of freedom. This reasoning is correct and the algorithm will not give problems if the successive turnover tables do not have overlapping marginals, e.g., when comparing a set of turnover tables such as $t_1 - t_2$, $t_3 - t_4$, $t_5 - t_6$, and $t_7 - t_8$. However, the turnover tables in Table 4.22 with successive turnover tables $t_1 - t_2$, $t_2 - t_3$, etc. do have overlapping marginals. For each item, the marginal distribution of S at $P = p$ is, by definition, the same as the marginal distribution of R at $P = p + 1$. For some particular models (here for model $\{IP,RS\}$), this makes for fewer independent proportions to be estimated and for fewer independent restrictions on the cell proportions than seemed necessary in the naive direct approach. Or, stating the same thing differently, the direct application of model $\{IP,RS\}$ and imposing all $7 \times 8 = 56$ restrictions seemingly implied by $\pi_{r\,s\ ip}^{RS|IP} = \pi_{r\,s}^{RS}$ means ending up with redundant restrictions on the cell proportions. To find an appropriate set of nonredundant restrictions that is necessary and sufficient to fit model $\{IP,RS\}$, it is helpful to formulate the model restrictions in somewhat different ways.

Model $\{IP,RS\}$ first of all implies that the corresponding odds ratios in the eight turnover tables are the same. In terms of the relevant loglinear parameters, this means that $\lambda_{r\,s\ i\ p}^{RS|IP}$ does not vary over I and P and is the same for all i and p. In other words,

$$\lambda_{r\,s\ ip}^{RS|IP} = \lambda_{**rs}^{IPRS},$$

where λ_{**rs}^{IPRS} is a two-variable effect parameter in model $\{IP,RS\}$ for table $IPRS$. Because each of the eight ($= 2 \times 4$) turnover tables (corresponding to pairs of values (i,p) for IP) contains in principle four independent parameters $\lambda_{rs\,ip}^{RS|IP}$ or odds ratios, this restriction yields $7 \times 4 = 28$ degrees of freedom. Then, to achieve the equality of the corresponding cell probabilities in the eight turnover tables, as required by model $\{IP,RS\}$, the corresponding one-variable loglinear parameters should be the same: $\lambda_{r*ip}^{RS|IP}$ and $\lambda_{*s\,ip}^{RS|IP}$ should not vary over i and p. In other words,

$$\lambda_{r*ip}^{RS|IP} = \lambda_{**r*}^{IPRS}$$
$$\lambda_{*s\,ip}^{RS|IP} = \lambda_{***s}^{IPRS},$$

where λ_{**r*}^{IPRS} and λ_{***s}^{IPRS} are one-variable effect parameters in model $\{IP,RS\}$ for table $IPRS$. Alternatively, instead of imposing the additional restrictions on the one-variable loglinear parameters, one might require that the corresponding marginal distributions should be the same: $\pi_{r+\,ip}^{RS\,|IP} = \pi_r^R$ and $\pi_{+s\,ip}^{RS|IP} = \pi_s^S$. The latter restriction on the one-way marginal probabilities gives the same results as the former restriction on the one-variable loglinear parameters, because of the already imposed equality restrictions on the two-variable parameters. These restrictions on the one-variable parameters seem to imply 28 independent restrictions: $7 \times 2 = 14$ independent restrictions for the row marginals and 14 independent restrictions for the column marginals. However, as explained in Section 4.3.1, because of the overlapping marginals in the successive turnover tables, the row and the column marginals in model $\{IP,RS\}$ all become equal to each other and the restrictions actually become:

$$\lambda_{r*ip}^{RS|IP} = \lambda_{*r\,ip}^{RS|IP} = \lambda_{**r*}^{IPRS} = \lambda_{***r}^{IPRS},$$

or

$$\pi_{r+\,ip}^{RS\,|IP} = \pi_{+r\,ip}^{RS|IP} = \pi_r^R = \pi_r^S.$$

For the two trichotomous items and the five age groups, only 10 independent, nonoverlapping one-way marginal distributions or one-variable parameters have to be considered. The equality of these ten trichotomous one-way distributions yields $9 \times 2 = 18$ rather than $14 + 14 = 28$ independent restrictions. So, in total, summing the restrictions for the two-variable and the one-variable parameters, there are only $28 + 18 = 46$ degrees of freedom rather than the earlier-obtained 56. The algorithm works well when only the independent restrictions for the two-way and one-way parameters are imposed: it converges and the correct number of degrees of freedom is reported.

The test results for model $\{IP,RS\}$ are $G^2 = 239.01$, $df = 46$ ($p = .000$, $X^2 = 197.52$) and it is concluded that the model has to be rejected. It is possible, however, that the turnover tables are different for marijuana and alcohol, but the same for each item over the periods: maybe model $\{IP,IRS\}$ with the implied restrictions

$$\pi_{rs\,ip}^{RS|IP} = \pi_{rs\,i}^{RS|I}$$

is true in the population. When calculated in the naive way, not taking into account the overlapping marginals, model $\{IP,IRS\}$ would have had 48 degrees of freedom, seemingly eight less than the naive count of 56 for previous model $\{IP,RS\}$. Both counts (48 and 8) are wrong because of the overlapping marginals. To see this, remember that model $\{IP,IRS\}$ for Table 4.22 implies identical turnover tables for marijuana and identical turnover tables for alcohol use. First focus on marijuana use. There are four successive transition tables in which the $\lambda_{rs\,1p}^{RS|IP}$'s should be equal to each other for the different values of p. Given the four two-variable parameters to be estimated in each table, this gives $3 \times 4 = 12$ degrees of freedom. Then, all five marginal distributions in the four turnover tables should be the same (or the corresponding five one-variable parameters), yielding another $4 \times 2 = 8$ degrees of freedom. Therefore, the restrictions on the turnover tables for marijuana use yield $12 + 8 = 20$ degrees of freedom. Because the same is true for alcohol use, model $\{IP,IRS\}$ implies 40 degrees of freedom (or independent restrictions), only six rather than eight less than the correct degrees of freedom for model $\{IP,RS\}$. But also model $\{IP,IRS\}$ has to be rejected: $G^2 = 158.71$, $df = 40$ ($p = .000$, $X^2 = 121.13$).

The other natural relaxation of model $\{IP,RS\}$ is model $\{IP,PRS\}$, in which the turnover tables may be different for the periods, but identical for the two drugs at each period:

$$\pi_{rs\,ip}^{RS|IP} = \pi_{rs\,p}^{RS|P} \, .$$

Model $\{IP,PRS\}$ does not fit the data either: $G^2 = 129.98$, $df = 26$ ($p = .000$, $X^2 = 87.73$). The number of degrees of freedom for model $\{IP,PRS\}$ with nonoverlapping marginals would have been 32. The correct number of degrees of freedom is 26. First, because the four turnover tables of alcohol and marijuana use are the same for each period, the two-variable parameters $\lambda_{rs\,1p}^{RS|IP}$ and $\lambda_{rs\,2p}^{RS|IP}$ should be the same, giving rise to $4 \times 4 = 16$ degrees of freedom. Then, for each turnover table at $P = p$, the row marginals should be equal to each other, as well as the column marginals. But because of the overlapping marginals, the restrictions overlap: once the column marginals (S) for the turnover tables for alcohol and marijuana use are the same for $P = 1$, the row marginals (R) for these two turnover tables at $P = 2$ are also the same. Given the five time points (age categories), there are only $5 \times 2 = 10$ independent restrictions on the one-way marginals (or the one-variable parameters). All together there are $16 + 10 = 26$ degrees of freedom to test model $\{IP,PRS\}$.

The negative outcomes for models $\{IP,RS\}$, $\{IP,IRS\}$, and $\{IP,PRS\}$ naturally lead to model $\{IP,PRS,IRS\}$, which takes possible variations in the turnover tables both over the successive periods and between the two items into account, but without extra period-item interaction effects. The period differences among the turnover tables are the same for both items, and the item differences among the turnover tables are the same for all successive periods. Model $\{IP,PRS,IRS\}$ can no longer be defined in simple, direct equality restrictions on the cell probabilities, but it also no longer implies simple, direct restrictions on marginal distributions. The restrictions in model $\{IP,PRS,IRS\}$ are defined in terms of restrictions on the odds ratios. Because of this, overlapping marginals no longer pose a problem and applying our marginal-modeling algorithm in the simple, direct way to implement the restrictions in model $\{IP,IRS,PRS\}$ does not meet any special difficulties. Model $\{IP,PRS,IPS\}$ fits the data well: $G^2 = 26.99$, $df = 24$ ($p = .303$, $X^2 = 18.53$). However, its results will not be discussed, as some other more parsimonious and well-fitting models will be presented below, that have a more easy and interesting interpretation.

So far, the starting point has been models with more or less similar cell probabilities. A different and often theoretically interesting starting point is to compare turnover tables from the viewpoint of more or less similar transition probabilities. The most restrictive model from this point of view is

$$\pi_{s\ rip}^{S|RIP} = \pi_{s\ r}^{S|R} ,$$

in which it is assumed that the transition probabilities are the same in all four successive turnover tables for both marijuana and alcohol use. As discussed in Section 4.3.1, this restriction can be rendered in the form of loglinear model $\{IPR,RS\}$ for Table 4.22. And as before, defining restricted models for transition probabilities does not imply, in general, imposing redundant restrictions, even when the turnover tables have overlapping marginals. Therefore, the straightforward application of model $\{IPR,RS\}$ is not problematic. However, this very restrictive model $\{IPR,RS\}$, with identical transition probabilities for all periods and both items did not fit the data ($p = .000$). Allowing for differences in transition probabilities for the periods, but not for the items (model $\{IPR,PRS\}$), did not improve the fit ($p = .000$). However, model $\{IPR,IRS\}$ for Table 4.22 fit well: $G^2 = 40.11$, $df = 36$ ($p = .293$, $X^2 = 29.70$). From this, it is concluded that the transition tables for alcohol and marijuana are definitely different from each other, but remain stable over time for each item. This stability over time agrees with earlier conclusions about the gross changes in marijuana alone (Section 4.3.1). A variant of this model will be discussed below as the final model, but after a discussion of models starting from the point of view of restricted odds ratios.

Model $\{IPR,IPS,RS\}$ represents the strictest form of the similar odds ratios model: there is no variation in the corresponding local odds ratios among the turnover tables, among the successive periods, nor between the two items. As for the transition probabilities, restrictions on the odds ratios can be formulated and applied directly in terms of loglinear models for the pertinent tables because they do not involve direct restrictions on (overlapping) marginals. Model $\{IPR,IPS,RS\}$ fits the data very well: $G^2 = 26.27$, $df = 28$ ($p = .558$, $X^2 = 18.04$). Had model $\{IPR,IPS,RS\}$ not fit the data, then natural candidates would have been model $\{IPR,IPS,PRS\}$, to take possible changes in odds ratio over time into account, model $\{IPR,IPS,IRS\}$ to account for possible item differences, or model $\{IPR,IPS,PRS,IRS\}$, to account for both types of variation in the odds ratios. Ultimately, saturated model $\{IPRS\}$ might have been considered.

Whether the equal odds ratio model $\{IPR,IPS,RS\}$ fits the data better than the previously discussed transition model $\{IPR,IRS\}$ cannot be determined by means of conditional tests: the two models are not hierarchically nested. However, given that the equal odds ratio model $\{IPR,IPS,RS\}$ fits the data so well, perhaps the more parsimonious stable transition model $\{IPR,IRS\}$ might be further simplified by assuming

Table 4.23. Estimated stable transition probabilities for model $\{IPR,IS,RS\}$; see text. Source: U.S. National Youth Study; see text

	Marijuana			
	S (time $t+1$)			
R (time t)	1	2	3	Total
1	.824	.142	.033	1.000
2	.264	.467	.268	1.000
3	.071	.160	.769	1.000

	Alcohol			
	S (time $t+1$)			
R (time t)	1	2	3	Total
1	.612	.335	.053	1.000
2	.114	.638	.248	1.000
3	.032	.227	.740	1.000

equal odds ratios for both items and in all successive transition tables. Another indication that this extra restriction might be valid is the fact that in model $\{IPR,IRS\}$, almost none of the parameter estimates $\hat{\lambda}_{ijk}^{IRS}$ (not reported here) are significant and that the few significant parameters are just on the borderline. To restrict model $\{IPR,IRS\}$ in this way, model element IRS in model $\{IPR,IRS\}$ must be replaced by model element IR,IS,RS leading to model $\{IPR,IS,RS\}$. Model $\{IPR,IS,RS\}$ fits the data well with $G^2 = 43.80$, $df = 40$ ($p = .313$, $X^2 = 34.24$) and not significantly worse than the original transition model $\{IPR,IRS\}$ with $G^2 = 43.80 - 40.11 = 3.69$, $df = 40 - 36 = 4$ ($p = .450$). Model $\{IPR,IS,RS\}$ is therefore accepted as the final model for the analyses of the changes over time in the uses of alcohol and marijuana. It is concluded that the successive transition tables for marijuana for all time points t to $t+1$ are identical, and that the same applies for alcohol transition tables. However, the transitions for marijuana use are not identical to the ones for alcohol use, although the corresponding odds ratios in the two sets of transition tables are the same again. The estimated transition tables are reported in Table 4.23, the loglinear parameters $\hat{\lambda}_{**ij}^{IPRS}$ in Table 4.24.

From the parameter estimates in Table 4.24, it can be inferred that there is a strong association between a drug's use at time t and $t+1$, for all t regardless whether it concerns the use of marijuana or alcohol. The extreme log odds ratio in the transition tables involving the four extreme cells (for RS, cells: $11, 13, 31, 33$) equals 5.595 and the odds ratio itself equals 269.078. Further, the association is of a strict ordinal-by-ordinal type: all local log odds ratios are positive, which means that the more one uses a drug at time t, the more one uses the drug at time $t+1$. But although the association in the two transition tables in terms of odds ratios is the same for both drugs, the in- and outflow probabilities themselves are different for marijuana and

Table 4.24. Estimated parameters $\hat{\lambda}_{**ij}^{IPRS}$ and their standard errors for model $\{IPR,IS,RS\}$; see text. Source: U.S. National Youth Study (see text)

R	S		
	1	2	3
1	1.524 (.087)	-.106 (.063)	-1.418 (.088)
2	-.328 (.079)	.367 (.060)	-.039 (.064)
3	-1.196 (.131)	-.261 (.086)	1.457 (.097)

alcohol (see Table 4.23). For example, the probability of remaining a nonuser of a drug from one age category to the next is higher for marijuana (.824) than for alcohol (.612). Further, never users of alcohol switch more easily to drinking once a month (.335) than people who never use marijuana switch to using once a month (.142). Finally, the once-a-month users of alcohol are more inclined than the once-a-month users of marijuana to keep using the drug at that level (.638 vs. .467), especially because the once a month marijuana users comparatively more often switch back to nonuse.

Such interpretations of the transition probabilities may be theoretically and practically very important. But precisely because of that, it is also emphasized again here that the warnings issued at the end of Section 4.3.1 also apply here: the differences among the transition probabilities may also be caused by just a little bit of unreliability of the measurements in combination with different sizes of the categories (see the references mentioned there).

The above expositions have explained the principles involved in comparing gross changes in two characteristics. Extensions of this basic approach to more complicated situations follow easily from these principles. It poses no special difficulties to apply the above models for gross change comparisons for one group to several subgroups and investigate whether and in what respects the gross changes in the subgroups differ from each other. Also, models for the comparisons of gross changes in more than two characteristics are simple extensions of the above principles. It is also not necessary that the turnover tables to be compared be successive or of equal length periods. Researchers might be interested in the study of turnover tables (t,t') such as $(1,2)$, $(1,3)$, and $(2,5)$. Further, higher-way tables for gross change may be the focus of research, such as the comparison in a four wave panel of three-way turnover tables, like $(1,2,3)$ and $(2,3,4)$. An important *caveat* of such comparisons remains, of course, that redundant restrictions must be avoided. In sum, the basics outlined in this subsection can be applied to also answer even more complicated, but very analogous, research questions. One useful extension not mentioned so far will be explicitly discussed in the last section of this chapter, viz. combining hypotheses about net and gross changes.

4.4.4 Combining Hypotheses about Net and Gross Changes

Research hypotheses may not only concern either gross or net changes, but often involve simultaneously both kinds of changes. Social mobility research with its interests in free and forced mobility offers a prime example of this (see Section 4.3.3). An illustration of such a combined approach will be given here, using the drug use data and the models discussed above. From the loglinear analyses of the net changes in drug use (see Eq. 4.11 and Table 4.18), it was concluded first, that both alcohol and marijuana use increased over the years in a parabolic way, second, that the growth curve was steeper for alcohol than for marijuana, and third, that the distance between the two curves increased linearly over time. For the loglinear analyses of the gross changes (see Table 4.23), it was found that model $\{IPR,IS,RS\}$ fitted the data well. In this model, the transition tables for the successive points in time t and $t+1$ are stable over time, but different for marijuana and alcohol use, although the odds ratios in both stable transition tables are the same for both drugs.

Now, imagine a researcher who, from theoretical insights into the overall net growth in drugs use during adolescence and from ideas about the transitions from one age category to the next, had formulated exactly these two models as research hypotheses in the beginning of an investigation. Marginal modeling can then be used to simultaneously test the hypotheses implied by both models for the net and gross changes. A necessary requirement for the estimation procedure to work is, of course, that the two sets of restrictions do not involve redundant or incompatible restrictions (Lang & Agresti, 1994, Bergsma & Rudas, 2002a; see also Section 2.3.3). At first sight, perhaps somewhat naively, we did not expect any troubles in these respects. However, when the restrictions in Eq. 4.11 and model $\{IPR,IS,RS\}$ were implemented simultaneously, the algorithm failed to converge. The source of the problem turned out to be incompatibility between the two sets of restrictions. Once, in model $\{IPR,IS,RS\}$ for the gross changes, the marginal distribution of drug use at time 1 is fixed in combination with the stable transition table, all successive marginals are fixed, but in a way that is incompatible with the model for the net changes in Eq. 4.11. More technically, the Jacobian of the constraint function implied by the model is not of full rank. In Section 4.5, an explanation that this can be expected to happen for this model (and not for the one in the next paragraph, in which some restrictions are omitted), is given using the framework of Bergsma and Rudas (2002a).

Such a problem was not encountered when the well-fitting model for the odds ratios $\{IPR,IPS,RS\}$ discussed above was used in combination with the same model for the net changes as before. The values that an odds ratio can assume are functionally independent of the marginals and, in general, it is not problematic to simultaneously test hypotheses about odds ratios and marginal distributions (even for turnover tables with overlapping marginals). When model $\{IPR,IPS,RS\}$, with identical odds ratios over time and for both items, is combined with the preferred model for the net change in Eq. 4.11, no convergence problems were encountered. The test results are $G^2 = 36.61, df = 41$ ($p = .666, X^2 = 26.471$), and both sets of restrictions can be accepted. The estimates of the (statistically significant) coefficients for the parameters λ_{*jk}^{ITR} and λ_{ijk}^{ITR} in the net change model in Eq. 4.11, their estimated standard errors, and the corresponding z-values are:

	Estimate	S.E.	z
$\hat{\alpha}$	.386	.026	14.85
$\hat{\beta}$	−.042	.014	−2.99
$\hat{\upsilon}_i^I$	−.061	.019	−3.20

From these coefficients, the values of the estimated conditional parameters $\hat{\lambda}_{jki}^{TR|I}$ and $\hat{\lambda}_{ikj}^{IR|T}$, indicating the nature of the net change can be computed. They are given in Table 4.25. Comparisons of these estimates with those obtained when testing the model defined by Eq. 4.11 in isolation (Table 4.18) reveal small differences, but none of them are substantively important or lead to different substantive interpretations. But of course, this need not always be true. Anyhow, the net change estimates obtained in the simultaneous analyses form the best estimates given that all restrictions are true in the population.

Regarding the gross changes, Table 4.26 contains the estimates of relevant log-linear parameters λ_{**jk}^{IPRS} (and their standard errors between parentheses) in model $\{IPR, IPS, RS\}$ for Table $IPRS$ under the joint analysis. These results indicate a very strong association between the responses at two consecutive time points, that is the same for all successive turnover tables and the same for alcohol and marijuana use. The extreme odds ratio for the extreme cells in the table is $e^{3.53} = 34.12$.

In a way, combining restrictions about joint and marginal distributions or lower- and higher-way marginals is straightforward, were it not for redundancy and incompatibility issues. Because of the importance of these problems for many important research questions involving the comparison of turnover tables, these issues will be treated again for successive turnover tables involving overlapping marginals in a more general way in the last section of this chapter.

4.5 Minimally Specified Models for Comparing Tables with Overlapping Marginals; Detection of Problematic Models

The problem discussed in this section is how to derive a minimally specified marginal model for testing whether several bivariate turnover matrices with overlapping univariate marginal distribution are equal. To attack this problem, the results of Bergsma and Rudas (2002a), which were summarized in Section 2.3.3, will be used. The basic principles of the approach followed here will first be discussed by means of an example in which it is assumed that a random variable X with K categories is measured at four time points. Then, its generalization to an arbitrary number of repeated measurements will be stated.

When a variable X is measured at four time points, represented by A, B, C, and D, the probabilities of the joint distribution of the four measurements is represented

Table 4.25. Conditional parameters $\hat{\lambda}_{jk\,i}^{TR|I}$ and $\hat{\lambda}_{ik\,j}^{IR|T}$ for net changes in *Marijuana* and *Alcohol use* data based on the conjunction of model in Eq. 4.11 for Table *ITR* and model {*IPR,IPS,RS*} for Table *IPRS*. See text

$\hat{\lambda}_{jk\,i}^{TR	I}$	\multicolumn{5}{c}{$T(ime)\ (=Age)$}
	13	14	15	16	17
R Marijuana					
1. Never	.734	.282	-.084	-.367	-.565
2. Once a month	0	0	0	0	0
3. More than once a month	-.734	-.282	.084	.367	.565
R Alcohol					
1. Never	.978	.405	-.084	-.489	-.810
2. Once a month	0	0	0	0	0
3. More than once a month	-.978	-.405	.084	.489	.810

$\hat{\lambda}_{ik\,j}^{IR	T}$	\multicolumn{5}{c}{$T(ime)\ (=Age)$}
	13	14	15	16	17
R Marijuana					
1. Never	.355	.417	.478	.539	.600
2. Once a month	-.343	-.343	-.343	-.343	-.343
3. More than once a month	-.012	-.074	-.135	-.196	-.257
R Alcohol					
1. Never	-.355	-.417	-.478	-.538	-.600
2. Once a month	.343	.343	.343	.343	.343
3. More than once a month	.012	.074	.135	.196	.257

Table 4.26. Loglinear parameters λ_{**jk}^{IPRS} (and their standard errors between parentheses) in model {*IPR,IPS,RS*} for Table *IPRS* under the joint analysis. See text

	\multicolumn{3}{c}{S}
	1	2	3
1	.727	.291	-1.018
	(.055)	(.048)	(.080)
R 2	-.121	.269	-.148
	(.075)	(.055)	(.062)
3	-.607	-.559	1.166
	(.092)	(.098)	(.110)

by $\pi_{i\,j\,k\,l}^{ABCD}$. The turnover probabilities for the transition of the first to the second measurement are then defined by $\pi_{i\,j++}^{ABCD}$. The model for equality of bivariate turnover tables is formulated as

$$\pi_{i\,j++}^{ABCD} = \pi_{+i\,j+}^{ABCD} = \pi_{++i\,j}^{ABCD}$$

or, more concisely, as

$$\pi_{i\,j}^{AB} = \pi_{i\,j}^{BC} = \pi_{i\,j}^{CD} \tag{4.14}$$

for $i, j = 1, \ldots, K$. The model for the equality of transition (conditional turnover) tables is

$$\pi_{j\,i}^{B|A} = \pi_{j\,i}^{C|B} = \pi_{j\,i}^{D|C} \tag{4.15}$$

for $i, j = 1, \ldots, K$ which can equivalently be written as

$$\frac{\pi_{i\,j}^{AB}}{\pi_i^A} = \frac{\pi_{i\,j}^{BC}}{\pi_i^B} = \frac{\pi_{i\,j}^{CD}}{\pi_i^C} \, .$$

It can be shown that Eq. (4.14) is equivalent to the Eq. (4.15) combined with the constraint

$$\pi_i^A = \pi_i^B \, .$$

This is perhaps surprising. One might think that in order to show that all three turnover tables are equal, it is necessary to assume that, in addition to equal transitions, all univariate marginals are equal. However, this is not true. The restrictions on the one-way marginals for C and D are redundant. This can be proven in a very elementary way. From

$$\frac{\pi_{i\,j}^{AB}}{\pi_i^A} = \frac{\pi_{i\,j}^{BC}}{\pi_i^B}$$

and

$$\pi_i^A = \pi_i^B \, ,$$

it follows that

$$\pi_{i\,j}^{AB} = \pi_{i\,j}^{BC} \, .$$

Summing over i yields

$$\pi_{+j}^{AB} = \pi_{+j}^{BC} \, ,$$

or

$$\pi_j^B = \pi_j^C \, .$$

In a similar way, one can prove that $\pi_j^C = \pi_j^D$.

Given the size of each turnover table, the number of constraints that are involved in Eq. (4.14) is $2K^2$, but clearly some of these have to be redundant, besides the obvious restriction that the sum of the probabilities in each turnover table equals one (which would make the naive number of degrees of freedom equal to $2(K^2 - 1)$). The question is what the minimal number of independent constraints needs to be to specify the models defined by Eqs. (4.14) and (4.15). This is important for the ML estimation procedure of this book, and for determining the correct number of degrees of freedom of asymptotic chi-square tests.

To determine the minimal set of constraints, the constraints will be written in terms of the relevant marginal loglinear parameters. Note that in what follows, the marginal notation introduced in Chapter 2 is used, so that λ_{ij}^{AB} refers to a two-variable interaction term in the loglinear model for marginal table AB, and λ_i^A to a main effect in the loglinear model for the univariate table A. The model (4.14) can then be specified as follows. Since equality of the turnover tables implies equality of the corresponding odds ratios in these turnover tables, the following restrictions apply

$$\lambda_{ij}^{AB} = \lambda_{ij}^{BC} = \lambda_{ij}^{CD}$$

for $i, j = 1, \ldots, K$. Further, Eq. (4.14) also implies equality of one-way marginals of the turnover tables:

$$\pi_i^A = \pi_i^B = \pi_i^C = \pi_i^D.$$

These restrictions on the one-way marginals can also be written in terms of appropriate restrictions on the one-variable effects within the one-dimensional marginals

$$\lambda_i^A = \lambda_i^B = \lambda_i^C = \lambda_i^D$$

for $i = 1, \ldots, K$. Given the usual identifying restrictions on the loglinear parameters, the number of independent constraints for the loglinear association parameters is $2(K-1)^2$ and $3(K-1)$ for the one-way parameters. In total, there are then $(K-1)(2K+1)$ independent constraints or degrees of freedom left, which is less than $2K^2$ and then $2(K^2-1)$ obtained previously. To be precise, $[2(K^2-1)] - [(K-1)(2K+1)] = (K-1)$, which is, in this case, for four waves and three consecutive turnover tables, exactly equal to the number of redundant independent restrictions on one of the marginals (as shown above). Following a similar reasoning in terms of two- and one-variable parameters, these results for comparing and restricting the cell probabilities in successive turnover tables in the indicated way can be generalized in a straightforward manner to an arbitrary number of time points: for T time points, there are $(T-2)(K-1)^2 + (T-1)(K-1)$ independent constraints.

Now consider the restrictions on the transition probabilities in Eq. (4.15). Calculated in the normal, standard way, given that each transition table has $K(K-1)$ independent transitions to be estimated, Eq. (4.15) implies $2K(K-1)$ independent restrictions or degrees of freedom. Going to the loglinear representation, this number of degrees of freedom is confirmed. The model in Eq. (4.15) can be obtained by restricting the loglinear two-variable parameters:

$$\lambda_{ij}^{AB} = \lambda_{ij}^{BC} = \lambda_{ij}^{CD},$$

in combination with the equality of relevant one-way effects in the two-way tables:

$$\lambda_{*j}^{AB} = \lambda_{*j}^{BC} = \lambda_{*j}^{CD}$$

This yields $(K-1)^2$ independent restrictions for the first constraint and $2(K-1)$ for the second, giving (again) a total of $2(K-1)K$ independent constraints. Its generalization to T measurement occasions is straightforward: $(T-2)(K-1)K$. In models like this, restrictions on transition probabilities do not suffer from incompatibilities.

The results of Bergsma and Rudas (2002a) guaranteeing the absence of problems only apply if the same loglinear effect is not restricted in two different marginal tables (see Section 2.3.3). Thus, they do not apply to the specification of Eq. (4.14) where we set both $\lambda_i^A = \lambda_i^B$ and $\lambda_{*j}^{AB} = \lambda_{*j}^{BC}$, i.e., the B effect is restricted twice, in both tables B and BC. Thus, to apply the result of Bergsma and Rudas, assuring minimal specification and no further problems with the model, the first specification above is needed. Alternatively, it can be shown that the model is equivalent to the equality of bivariate associations as well as

$$\lambda_{j*}^{AB} = \lambda_{*j}^{AB} = \lambda_{*j}^{BC} = \lambda_{*j}^{CD},$$

which also satisfies the Bergsma-Rudas regularity conditions.

Minimal Specifications for Some Models from Chapter 4

In Section 4.4.3, some models were discussed for the data in Table 4.22, which was taken as a normal table *IPRS*. Variable I represents *Item* with two values, marijuana and alcohol. Variable P represents *Period* with four values: transitions from time 1 to 2, from time 2 to 3, from time 3 to 4, and from time 4 to 5. Finally, R and S represent the response variable at the first and second measurement occasion for each transition, respectively.

A first model that was considered is model $\{IP, RS\}$ that assumes that all eight turnover tables are completely identical,

$$\pi_{1i\ j}^{IT_1T_2} = \pi_{1i\ j}^{IT_2T_3} = \pi_{1i\ j}^{IT_3T_4} = \pi_{1i\ j}^{IT_4T_5} = \pi_{2i\ j}^{IT_1T_2} = \pi_{2i\ j}^{IT_2T_3} = \pi_{2i\ j}^{IT_3T_4} = \pi_{2i\ j}^{IT_4T_5},$$

where T_k represents the measurement at time point k. Using a notation comparable with that for the transition probabilities, a minimal specification for this model is obtained by first, imposing equality of the conditional marginal association parameters:

$$\lambda_{i\ j\ 1}^{T_1T_2|I} = \lambda_{i\ j\ 1}^{T_2T_3|I} = \lambda_{i\ j\ 1}^{T_3T_4|I} = \lambda_{i\ j\ 1}^{T_4T_5|I} = \lambda_{i\ j\ 2}^{T_1T_2|I} = \lambda_{i\ j\ 2}^{T_2T_3|I} = \lambda_{i\ j\ 2}^{T_3T_4|I} = \lambda_{i\ j\ 2}^{T_4T_5|I},$$

and second, by constraining the following univariate marginals:

$$\lambda_{i\ 1}^{T_1|I} = \lambda_{i\ 1}^{T_2|I} = \lambda_{i\ 1}^{T_3|I} = \lambda_{i\ 1}^{T_4|I} = \lambda_{i\ 1}^{T_5|I} = \lambda_{i\ 2}^{T_1|I} = \lambda_{i\ 2}^{T_2|I} = \lambda_{i\ 2}^{T_3|I} = \lambda_{i\ 2}^{T_4|I} = \lambda_{i\ 2}^{T_5|I}.$$

The number of independent constraints in this minimal specification is $(K-1)(7K+2)$.

The second model considered was model $\{IP, IRS\}$, which assumes that the turnover tables are different for the two items but are identical over time for each item. A minimal specification for this model is obtained by imposing

$$\lambda_{ij1}^{T_1T_2|I} = \lambda_{ij1}^{T_2T_3|I} = \lambda_{ij1}^{T_3T_4|I} = \lambda_{ij1}^{T_4T_5|I},$$

$$\lambda_{ij2}^{T_1T_2|I} = \lambda_{ij2}^{T_2T_3|I} = \lambda_{ij2}^{T_3T_4|I} = \lambda_{ij2}^{T_4T_5|I},$$

and

$$\lambda_{i1}^{T_1|I} = \lambda_{i1}^{T_2|I} = \lambda_{i1}^{T_3|I} = \lambda_{i1}^{T_4|I} = \lambda_{i1}^{T_5|I},$$

$$\lambda_{i2}^{T_1|I} = \lambda_{i2}^{T_2|I} = \lambda_{i2}^{T_3|I} = \lambda_{i2}^{T_4|I} = \lambda_{i2}^{T_5|I}.$$

The number of independent constraints in this minimal specification is $(K-1)$ $(6K+2)$.

Finally, the third model $\{IP, PRS\}$ assumes that the turnover tables change over period, but are the same for both items at each period. A minimal specification of this model is given by

$$\lambda_{ij1}^{T_1T_2|I} = \lambda_{ij2}^{T_1T_2|I}$$

$$\lambda_{ij1}^{T_2T_3|I} = \lambda_{ij2}^{T_2T_3|I}$$

$$\lambda_{ij1}^{T_3T_4|I} = \lambda_{ij2}^{T_3T_4|I}$$

$$\lambda_{ij1}^{T_4T_5|I} = \lambda_{ij2}^{T_4T_5|I},$$

and second, by constraining the following univariate marginal parameters:

$$\lambda_{i1}^{T_1|I} = \lambda_{i2}^{T_1|I}$$

$$\lambda_{i1}^{T_2|I} = \lambda_{i2}^{T_2|I}$$

$$\lambda_{i1}^{T_3|I} = \lambda_{i2}^{T_3|I}$$

$$\lambda_{i1}^{T_4|I} = \lambda_{i2}^{T_4|I}$$

$$\lambda_{i1}^{T_5|I} = \lambda_{i2}^{T_5|I}.$$

The number of independent constraints in this minimal specification is $(K-1)$ $(4K+1)$.

Detection of Compatibility Problems for Models in Section 4.4.4

In Section 4.4.4, we found that Eq. 4.11 and model $\{IPR, IS, RS\}$ were incompatible (I is item, P is period, R and S are the responses at first and second time points, resp.), in the sense that the Jacobian of the combined constraints was not of full rank. On the other hand, Eq. 4.11 and the model $\{IPR, IPS, RS\}$ *were* compatible. We found out about the problem by trying to fit the former combined model and running into convergence problems. However, it is possible to use the Bergsma and Rudas (2002a) framework to detect the possibility of a problem occurring before actually fitting the model. To use this, we need to formulate the model constraints using loglinear parameters for the original variables to find out if the same effect is restricted twice.

The original variables are $M_1, \ldots, M_5$ (marijuana use at time points 1 to 5) and $A_1, \ldots, A_5$ (alcohol use at time points 1 to 5). Regarding Eq. 4.11, we have the equality

$$\lambda_{1jk}^{ITR} = -\lambda_{2jk}^{ITR} = \frac{1}{2}\left(\lambda_k^{M_j} - \lambda_k^{A_j}\right)$$

That is, Eq. 4.11 restricts the M_j effect in marginal table M_j of the original table, and the A_j effect in marginal table A_j ($j = 1, \ldots, 5$). Model $\{IPR, IS, RS\}$ is equivalent to model $\{IPR, IPS, RS\}$ with the added restriction that

$$\lambda_{*j*l}^{IPRS} = 0$$
$$\lambda_{ij*l}^{IPRS} = 0 .$$

But since

$$\lambda_{*j*l}^{IPRS} = \frac{1}{2}\left(\lambda_{*\ l}^{M_jM_{j+1}} + \lambda_{*\ l}^{A_jA_{j+1}}\right)$$
$$\lambda_{1j*l}^{IPRS} = -\lambda_{2j*l}^{IPRS} = \frac{1}{2}\left(\lambda_{*\ l}^{M_jM_{j+1}} - \lambda_{*\ l}^{A_jA_{j+1}}\right) ,$$

this is easily seen to be equivalent to the following marginal loglinear restrictions for the original table

$$\lambda_{*\ l}^{M_jM_{j+1}} = \lambda_{*\ l}^{A_jA_{j+1}} = 0 ,$$

which restricts the M_{j+1} and A_{j+1} effects in marginal tables M_jM_{j+1} and A_jA_{j+1}, respectively ($j = 1, \ldots, 4$).

Thus, we see that in the combined model of Eq. 4.11 and $\{IPR, IS, RS\}$, the M_{j+1} (A_{j+1}) effect is restricted twice ($j = 1, \ldots, 4$) — once in marginal table M_{j+1} (A_{j+1}), and once in marginal table M_jM_{j+1} (A_jA_{j+1}) — and this is the reason that the compatibility problems occur. On the other hand, it can be verified that no effects are restricted twice in the combined model of Eq. 4.11 and $\{IPR, IPS, RS\}$, guaranteeing the absence of the possible problems described in Section 2.3.3.

Note that, although there is no guarantee that problems will occur if the same effect is restricted in different marginal tables, in practice we always encountered problems.

5

Causal Analyses: Structural Equation Models and (Quasi-)Experimental Designs

Marginal modeling and causal analysis are often seen as two opposite and mutually exclusive approaches. This is understandable because marginal modeling is frequently defined in contrast to conditional analyses, and the outcomes of conditional analyses are typically interpreted in causal terms. However, this opposition of marginal and causal analyses is wrong or at least way too simple. For instance, most, if not all, marginal models discussed in the previous chapters might have been derived from causal theories. The tests on the validity of these marginal models could then be regarded as tests on the validity of these underlying causal theories. Further, marginal modeling is often an indispensable part of approaches that are explicitly intended to investigate causal relationships, such as (quasi-)experimental designs and causal models in the form of structural equation models (SEMs). In this chapter, the focus will be on the role of marginal modeling in these last two approaches.

Although this chapter is about 'causal modeling,' the term 'causality' itself will not be used very often (unless it leads to cumbersome and awkward phrases). Unfortunately, it is still common social science practice to use 'causality' in a rather weak and general sense, more pointing towards some assumed asymmetry in the relationships rather than necessarily implying proper causal connections. The disadvantages and dangers of this careless use of the term are first of all, that when drawing the final conclusions researchers tend to forget the *caveats* surrounding their causal analyses, even with research designs and research questions that hardly permit inferences about causal connections; and second, that researchers do not think hard enough before the start of their investigations about the appropriate research design for answering their causal questions. Therefore, even when the term causality is used below, the reader should keep these reservations in mind. For just a few treatments of causality with different emphases, see Rubin (1974), Cartwright (1989), Sobel (1995), McKim and Turner (1997), Sobel (2000), Pearl (2000), Snijders and Hagenaars (2001), Winship and Sobel (2004), Heckman (2005) (with discussion), and Morgan and Winship (2007), and the references cited in these publications.

W. Bergsma et al., *Marginal Models: For Dependent, Clustered, and Longitudinal Categorical Data*, Statistics for Social and Behavioral Sciences,
DOI: 10.1007/978-0-387-09610-0_5, © Springer Science+Business Media, LLC 2009

5.1 SEMs - Structural Equation Models

Causal modeling is often equated with structural equation modeling (SEM), especially in the form of a recursive system of linear regression equations. The set of linear regression equations represents the assumed asymmetric (causal) relationships among a set of variables that are supposedly normally distributed and measured at interval level. The interval-level variables may be directly or indirectly observed — that is, they may be manifest or latent. In the latter case, usually some of the manifest variables act as indicators of the latent variables. The 'causal' relationships between a latent variable and its indicators are of a special nature, as the relationships between a hypothetical construct and observed variables. For an overview of this standard SEM and many extensions, see Bollen (1989); for treatments of the nature of the relationships between a latent variable and its indicators and causality involving latent variables, see Sobel (1994), Sobel (1997), Borsboom (2005), and Hand (2004).

Somewhat less well known among social science researchers, although already developed in the seventies of the previous century are SEMs for categorical variables in which loglinear equations essentially take the place of the linear regression equations (Goodman, 1973). These categorical SEMs are in many important respects similar to SEMs for continuous data.

Over the last decades, an enormous amount of publications have appeared in which these two frameworks of SEM (viz. for continuous and for categorical data) have been extended into several directions, and at the same time have been integrated into one approach. User-friendly computer programs are now available to estimate the basic models and many of their extensions. To mention just a very few instances from this huge amount of publications and programs: Pearl (2000), Stolzenberg (2003), Skrondal and Rabe-Hesketh (2004), Jöreskog and Sörbom (1996), Vermunt (1997b), Vermunt and Magidson (2004), and Muthén and Muthén (2006).

The emphasis in this chapter is on extending SEMs for categorical data by combining marginal modeling and traditional categorical data SEM (Agresti, 2002; Lang & Agresti, 1994; Croon et al., 2000). Especially for those readers who are not familiar with SEMs in the form of loglinear models, a very concise outline of the basic principles of categorical structural equation modeling will be presented in the next two subsections, including an extended empirical example.

5.1.1 SEMs for Categorical Data

Goodman introduced what he called 'modified path models' for the analysis of categorical data 'when some variables are posterior to others' (Goodman, 1973). Imagine a set of five manifest categorical variables A through E. The causal order among these five variables is denoted by means of the symbol $\Longrightarrow$:

$$A \Longrightarrow B \Longrightarrow C \Longrightarrow D \Longrightarrow E.$$

According to this causal order, A is an exogenous variable not influenced by the other variables. An exogenous variable such as A can be a joint variable, e.g., (A_1, A_2, A_3), consisting of all combinations of the categories of A_1, A_2, and A_3. The joint distribution of the separate elements A_i may be unrestricted or restricted, e.g., by imposing independence relations among the elements. But whether restricted or not, the relationships among the separate elements A_i are always treated as symmetrical and no assumptions are made about their causal nature.

The other variables B through E are endogenous variables. In principle, each endogenous variable can be directly causally influenced by the variables that precede it in the causal order but not by variables that appear later. For example, D may be determined by A, B, and C but not by E.

With categorical variables, the data can be summarized in the form of a multidimensional contingency table $ABCDE$ with cell entries $\pi_{i\,j\,k\,\ell\,m}^{ABCDE}$. Goodman's approach starts from the decomposition of the joint probability distribution of all variables into a product of successive (conditional) probability distributions referring to successively higher way (marginal) distributions following the causal order of the variables:

$$\pi_{i\,j\,k\,\ell\,m}^{ABCDE} = \pi_i^A\,\pi_{j\,i}^{B|A}\,\pi_{k\,i\,j}^{C|AB}\,\pi_{\ell\,i\,jk}^{D|ABC}\,\pi_{m\,i\,jk\ell}^{E|ABCD} \qquad (5.1)$$

The first right-hand side element in Eq. 5.1 (π_i^A) is an unconditional probability pertaining to the marginal distribution of the exogenous variable; the other right-hand side elements are all conditional probabilities.

In a way, this decomposition of the joint probability distribution of the five variables is trivial since it represents a mathematical identity or tautology, and not an empirically testable theory. On the other hand, many other tautological decompositions are possible, e.g., starting with the distribution of E (π_m^E) and working back, against the causal order, to $\pi_{j\,k\ell m}^{B|CDE}$ and, finally, to $\pi_{i\,jk\ell m}^{A|BCDE}$. However, the decomposition in Eq. 5.1 is special because it is the only tautological decomposition that is in complete agreement with the assumed causal order. Following the adage that what comes causally later cannot influence what is causally prior, the distribution of A should be investigated in marginal table A with entries π_i^A; the effects of A on B should be determined in marginal table AB using $\pi_{j\,i}^{B|A}$, not conditioning on C, D, and E; the effects on C must be estimated in marginal table ABC with $\pi_{k\,i\,j}^{C|AB}$ not controlling for D and E, etc.

The second step in Goodman's modified path approach consists of the formation of an appropriate set of (non)saturated loglinear or logit submodels, one separate submodel for each of the right-hand side elements of Eq. 5.1 An ordinary loglinear submodel is applied to the first right-hand side element referring to the (joint) distribution of the exogenous variable(s) (A). This loglinear submodel provides the estimates $\hat{\pi}_i^A$ for right-hand side element π_i^A. For the other right-hand side elements, which are conditional probabilities (e.g., $\pi_{k\,i\,j}^{C|AB}$), logit submodels are defined, in which the odds of belonging to a particular category (k) of the dependent variable (C) rather than to another (k') are modeled conditional on the independent variables (A,B). Or, what amounts to exactly the same, ordinary loglinear submodels are defined (in this example for table ABC with entries $\pi_{i\,j\,k}^{ABC}$) that contain the parameters referring to the joint distribution of the independent variables (AB) plus all assumed direct effects of the independent variables on the dependent variable (C). The resulting estimates of the joint probabilities (e.g., $\hat{\pi}_{i\,jk}^{ABC}$) are then used to compute the estimates of the conditional probabilities: $\hat{\pi}_{k\,i\,j}^{C|AB} = \hat{\pi}_{i\,jk}^{ABC}/\hat{\pi}_{i\,j+}^{ABC}$. For each of the conditional probabilities on the right-hand side of Eq. 5.1, an appropriate logit or loglinear (sub)model is defined. In all these submodels, the restrictions on the parameters only pertain to the effects of the particular independent variables on the pertinent dependent variable, while the distribution of the particular independent variables is treated as given. This can be immediately seen by way of example from the following equations for $\pi_{k\,i\,j}^{C|AB}$, in which ultimately only the parameters having C as superscript appear. Starting from the saturated loglinear submodel (in its multiplicative form) for $\pi_{i\,jk}^{ABC}$:

$$\pi_{k\,i\,j}^{C|AB} = \pi_{i\,jk}^{ABC}/\pi_{i\,j+}^{ABC}$$

$$= \tau\tau_i^A\tau_j^B\tau_k^C\tau_{i\,j}^{AB}\tau_{i\,k}^{AC}\tau_{jk}^{BC}\tau_{i\,jk}^{ABC}\Big/\sum_k \tau\tau_i^A\tau_j^B\tau_k^C\tau_{i\,j}^{AB}\tau_{i\,k}^{AC}\tau_{jk}^{BC}\tau_{i\,jk}^{ABC}$$

$$= \tau\tau_i^A\tau_j^B\tau_k^C\tau_{i\,j}^{AB}\tau_{i\,k}^{AC}\tau_{jk}^{BC}\tau_{i\,jk}^{ABC}\Big/[(\tau\tau_i^A\tau_j^B\tau_{i\,j}^{AB})\sum_k \tau_k^C\tau_{i\,k}^{AC}\tau_{jk}^{BC}\tau_{i\,jk}^{ABC}]$$

$$= \tau_k^C\tau_{i\,k}^{AC}\tau_{jk}^{BC}\tau_{i\,jk}^{ABC}\Big/\sum_k \tau_k^C\tau_{i\,k}^{AC}\tau_{jk}^{BC}\tau_{i\,jk}^{ABC}$$

If one or more submodels for the right-hand side elements of Eq. 5.1 are non-saturated rather than all saturated, the decomposition in this equation is no longer tautological and the model becomes empirically testable with respect to the imposed restrictions. Each submodel can be tested in the usual (chi-square) way by comparing its estimates $N\hat{\pi}$ with the observed frequencies n (N being the sample size).

A test of the empirical validity of the complete model — that is, a test that all restrictions in all submodels are simultaneously true — can be obtained by calculating the estimates for the cell entries $\pi_{i\,jk\ell m}^{ABCDE}$ at the left-hand side using the obtained estimates for the right-hand side elements in Eq. 5.1 The resulting estimates of $\pi_{i\,jk\ell m}^{ABCDE}$ under the condition that all postulated submodels are simultaneously true in the population will be indicated by $\hat{\pi}^{*ABCDE}_{i\,jk\ell m}$. The symbol $\hat{\pi}^*$ is used here rather than the usual $\hat{\pi}$, to be able to distinguish between estimates of the left-hand side element $\pi_{i\,jk\ell m}^{ABCDE}$ ($\hat{\pi}^{*ABCDE}_{i\,jk\ell m}$ under the complete model) and estimates of the right-hand side element $\pi_{m\,i\,jk\ell}^{E|ABCD}$ (calculated from $\hat{\pi}_{i\,jk\ell m}^{ABCDE}$ under the logit submodel for the effects of A, B, C, and D on E). The structural equation model as a whole — that is, the set of all submodels — can then be tested by comparing $N\hat{\pi}^{*ABCDE}_{i\,jk\ell m}$ with the observed cell entries $n_{i\,jk\ell m}^{ABCDE}$ in the usual way. Alternatively, the likelihood ratio chi-squares G^2 obtained for each of the right-hand side submodels in Eq. 5.1 can be summed, as can their degrees of freedom df to obtain an overall likelihood ratio test of the complete model, which is identical to the likelihood ratio chi-square directly involving the comparison of $N\hat{\pi}^{*ABCDE}_{i\,jk\ell m}$ and $n_{i\,jk\ell m}^{ABCDE}$. Note that the equality of the two ways of computing G^2 does not apply in general to Pearson's X^2.

In a way, modified path analysis can be looked upon as a form of marginal modeling, because analyses are carried out simultaneously on several marginal distributions coming from the same overall joint distribution. Not surprisingly then, the marginal-modeling algorithm explained in Chapters 2 and 3 can be used to estimate and test categorical SEMs. However, as Goodman has shown, in the simple recursive structural equation model discussed above, in which no reciprocal relations or causal loops occur and in which no restrictions are imposed that simultaneously involve two or more elements from the right-hand side of Eq. 5.1, the appropriate loglinear or logit submodels can be applied separately to each of the probabilities on the right-hand side. This is because of the simple form of Eq. 5.1, in which the ML estimates of the right-hand side elements are independent of each other, and for which the maximum likelihood function can be factorized into factors that each only pertain to one of the right-hand side elements of Eq. 5.1 (see also Fienberg, 1980, Chapter 7). In this step-by-step way, the ML effect parameter estimates and estimated standard errors are obtained, which will be identical to the outcomes of the alternative simultaneous marginal-modeling approach. This is analogous to the standard linear recursive systems without latent variables, in which the parameters and their standard errors in each of the regression equations can be estimated separately by means of ordinary multiple regression equations.

Goodman's basic framework has been elaborated in several ways, two of which will be further discussed below. First, as with linear SEMs, it is possible to introduce latent variables into the categorical structural equation model. Combining Goodman's modified path analysis with Lazarsfeld, Goodman and Haberman's work on latent class analysis (Goodman, 1974a; Goodman, 1974b; Haberman, 1977; Haberman, 1979; Lazarsfeld, 1950; Lazarsfeld, 1959; Lazarsfeld & Henry, 1968), categorical latent variables can be introduced into categorical SEMs to take care of many kinds of measurement errors and selection problems, very much analogous to the role continuous latent variables play in linear SEMs (Bassi et al., 2000; Hagenaars, 1990, 1993, 1998, 2002; Vermunt, 1997b). Latent variable models will be further discussed in Chapter 6.

Second, insight in Goodman's modified path model approach (including latent variables) has been enhanced by viewing categorical SEMs as DAG's, or Directed Acyclical Graphs (Lauritzen, 1996; Whittaker, 1990.) An immediate advantage of looking at a SEM as a DAG is that the very general mathematical and statistical results from the theory of graphical modeling can directly be applied to causal modeling. More specifically, Yule, Lazarsfeld and Goodman's fundamental insights into the equivalence of causal modeling for categorical and continuous data are much deepened, generalized, and formalized by treating both categorical and continuous data SEMs as graphs, as explicitly shown by Kiiveri and Speed (1982) and Cox and Wermuth (1996) among others (see also Pearl, 2000; Lazarsfeld, 1955; Goodman & Kruskal, 1979).

To understand another advantage of 'SEM as DAG', it is necessary to remember that conditional statistical independence among variables is the defining core of a graphical model. These graphical independence properties can be used to simplify the decomposition of the joint probability that corresponds to the assumed structural equation model. Such simplifications often make estimation procedures feasible that would otherwise not be possible, especially for forbiddingly large (sub)tables. This can be illustrated by means of a (too) simple example in terms of the five variables model in Eq. 5.1. Assume that for $\pi_{\ell\,i\,jk}^{D|ABC}$, variable D is not directly influenced by A, but only by B and C. In other words, D is conditionally independent of A, given B and C: in symbols, $A \perp\!\!\!\perp D|BC$. Because D does not vary with A, given the scores on B and C, the equality $\pi_{\ell\,i\,jk}^{D|ABC} = \pi_{\ell\,jk}^{D|BC}$ holds. This is easily seen as follows. By definition, conditional independence restriction $A \perp\!\!\!\perp D|BC$ implies

$$\pi_{i\,\ell\,jk}^{AD|BC} = \pi_{i\,jk}^{A|BC}\,\pi_{\ell\,jk}^{D|BC} .$$

Therefore,

$$\pi_{\ell\,jk}^{D|BC} = \pi_{i\,\ell\,jk}^{AD|BC} \Big/ \pi_{i\,jk}^{A|BC}$$

$$= \frac{\pi_{i\,jk\ell}^{ABCD}}{\pi_{jk}^{BC}} \Big/ \frac{\pi_{i\,jk}^{ABC}}{\pi_{jk}^{BC}}$$

$$= \pi_{i\,jk\ell}^{ABCD} \Big/ \pi_{i\,jk}^{ABC}$$

$$= \pi_{\ell\,i\,jk}^{D|ABC} .$$

Consequently, given that A has no direct effect on D, $\pi_{\ell\,i\,jk}^{D|ABC}$ can be replaced by $\pi_{\ell\,jk}^{D|BC}$ and Eq. 5.1 can be rendered in a simplified form as:

$$\pi_{i\,jk\ell\,m}^{ABCDE} = \pi_i^A\,\pi_{j\,i}^{B|A}\,\pi_{k\,i\,j}^{C|AB}\,\pi_{\ell\,jk}^{D|BC}\,\pi_{m\,i\,jk\ell}^{E|ABCD} . \tag{5.2}$$

Especially in large tables, e.g., for very large markov chains with 15 or more time points, such graphical simplifications are very useful for making ML estimation feasible.

The graphical simplification from Eq. 5.1 to Eq. 5.2 is completely equivalent to applying nonsaturated logit model $\{ABC, BCD\}$ to $\pi_{\ell\,i\,jk}^{D|ABC}$ in Eq. 5.1. Moreover, it is possible to further restrict graphical (sub)models by means of loglinear restrictions and define, for example, a nonsaturated submodel for $\pi_{\ell\,jk}^{D|BC}$ in Eq. 5.2, e.g., the no-three-variable interaction model $\{BC, BD, CD\}$. Because the absence of the three-variable interaction in model $\{BC, BD, CD\}$ cannot be represented in terms of conditional independence relationships, the model in Eq. 5.2 is no longer a graphical model in the strict sense when this particular restriction is applied.

Note that even if saturated loglinear models are defined for each of the right-hand side elements in Eq. 5.2, this equation is no longer a tautological equation, but is only true if A has no direct effect on D. In line with this, after the graphical simplification, it is no longer true that the G^2 test statistics and their degrees of freedom for the separate submodels can be summed to obtain the correct overall test of the whole model. Overall tests of these simplified graphical models should be based on $\hat{\pi}_{i\,jk\ell\,m}^{*ABCDE}$ discussed above. To see this immediately, imagine having defined saturated submodels for all right-hand side elements of Eq. 5.2. All G^2's would then be 0 while still dealing with a restricted, non-tautological model.

5.1.2 An Example: Women's Role

The above principles of categorical causal modeling will be illustrated by means of an example regarding peoples views on women's lives and roles. The data come from the survey *European Values Study* (EVS) (web page: *www.europeanvalues.nl*). Only the (unweighted) Dutch data from the 1999/2000 EVS wave will be considered, excluding missing data on the pertinent variables ($N = 960$). The particular aspect of the view on women's role taken here is the (dis)agreement with the statement 'A job is all right but what women really want is a home and children'. This variable, labeled here as W, originally had four categories, but has been dichotomized for illustrative purposes here: 1 = agree (33%), 2 = disagree (67%). In the present analysis, the connections between W and a very limited set of explanatory variables will be studied. Because of the possible presence in this simple example of some plausible alternative explanations for the relationships found, one must be very careful with any causal claims. A first natural explanatory variable to consider in this context is *Sex* (S): 1 = men (48%) and 2 = women (52%), along with *Age* (A), which has been trichotomized: 1 = born before 1945 (27%), 2 = between 1945-1963 (42%), 3 = after 1963 (31%). A and S are treated as exogenous variables. *Sex* and *Age* may be correlated because of differential mortality between men and women or because of differential nonresponse. Further, *Education* (E) may be of direct importance for the agreement with the statement on women's purposes in life (W), and can itself be determined by S and A. Also, level of *Education E* has been trichotomized: 1 = lower level (31%), 2 = intermediate (35%), 3 = higher level (34%). Finally, *Religiosity* (R) may directly influence W and may be influenced itself by S, A, and E. Because we were interested in the effects of 'personal, subjective religiosity,' we took the answer to the question: 'Independently whether you go to church or not, would you say you are a 1 = religious person (61%), 2 = not a religious person (32%), or 3 = a convinced atheist (7%).' The observed cross-classification of these variables is presented in Table 5.1.

The above implies the following causal order among the variables,

$$(S,A) \Longrightarrow E \Longrightarrow R \Longrightarrow W$$

corresponding with the causal decomposition

$$\pi_{sae\,rw}^{SAERW} = \pi_{sa}^{SA}\pi_{e\ sa}^{E|SA}\pi_{r\ sae}^{R|SAE}\pi_{w\ sae\,r}^{W|SAER} . \tag{5.3}$$

It was expected that each variable had a direct influence on each later variable in the causal chain, and that higher-order interaction effects involving S would be present. The more specific hypotheses (not reported here) were (conditionally) tested by means of the appropriate submodels for each of the right-hand side elements. The test results for the final (sub) models are reported in Table 5.2.

Perhaps because of the compensating influences of differential mortality and differential participation in the EVS survey, the hypothesis that *Age* (A) and *Sex* (S) are independent of each other in marginal table SA need not be rejected (see the test outcome for model $\{S,A\}$ in Table 5.2). Consequently, in graphical modeling terms, right-hand side element $\pi_{s\,a}^{SA}$ in Eq. 5.3 can be replaced by $\pi_s^S\pi_a^A$.

For the effects of S and A on E, the choice was between saturated model $\{SAE\}$ for table SAE or the no three-factor interaction model $\{SA, SE, AE\}$. Models in which one or both direct main effects on E were deleted did not fit at all. Model

Table 5.1. Observed frequencies for women's role data; see text for value labels of variables

			Religiosity R					
			1		2		3	
			Women's Role W					
			1	2	1	2	1	2
Sex S	Age A	Education E						
1. Male	1	1	24	13	10	3	1	1
		2	13	10	7	3	2	0
		3	19	15	4	8	2	2
	2	1	12	16	9	7	1	0
		2	16	25	6	18	1	4
		3	6	31	4	23	2	9
	3	1	5	8	4	7	0	1
		2	9	11	3	29	1	6
		3	2	17	4	20	2	3
2. Female	1	1	41	13	6	6	1	2
		2	11	16	2	1	0	1
		3	7	9	1	1	0	1
	2	1	20	26	16	10	0	1
		2	5	41	2	16	1	3
		3	5	47	1	19	0	5
	3	1	5	11	2	10	0	3
		2	15	34	4	19	0	3
		3	4	28	2	19	0	6

Table 5.2. Test results of loglinear SEM analyses on Table 5.1

Table	Model	G^2	df	p
SA	$\{S,A\}$	4.818	2	.090
SAE	$\{SAE\}$	0	0	1
SAER	$\{SAE,SR,AR\}$	22.846	28	.741
SAERW	$\{SAER,SW,AW,EW,RW\}$	58.315	46	.105
Total		85.979	76	.203

$\{SA, SE, AE\}$ yielded a borderline test results ($p = .066$). Despite a general preference for parsimonious models, the saturated model was chosen: from other sources it is well-established that there are differences between men and women regarding their educational opportunities, and that these differences have diminished over generations. According to the parameter estimates in saturated model $\{SAE\}$, the direct partial effects of age are rather large, indicating that especially the oldest generation has a much lower educational level (e.g., $\hat{\lambda}_{*11}^{SAE} = .500$). The partial main effects of sex on education are much smaller, but significant, pointing toward somewhat lower opportunities for women, e.g., $\hat{\lambda}_{1*1}^{SAE} = -.146$. The three-variable interaction terms, together with the partial effects of sex, indicate that the educational sex differences are more or less linearly increasing with age, meaning that in the oldest generation the educational sex differences are the largest, smaller in the middle generation, and absent in the youngest generation (e.g., $\hat{\lambda}_{11\,1}^{SE|A} = -.276$, but $\hat{\lambda}_{11\,3}^{SE|A} = .000$).

S, A, and E were assumed to have a direct influence on R. The choice was between models $\{SAE, SR, AR, ER\}$ and $\{SAE, SR, AR\}$, the latter without a main effect of education on religiosity. Models with higher-order effects on R did not improve the fit significantly. A conditional test of restricted model $\{SAE, SR, AR\}$ against the less-restricted model $\{SAE, SR, AR, ER\}$ turned out in favor of the more restricted model ($p = .132$; see also the test outcome for model $\{SAE, SR, AR\}$ in Table 5.2). It was therefore concluded that education has no direct effect on religiosity. According to the parameter estimates for the direct partial effects in model $\{SAE, SR, AR\}$, women are somewhat more religious than men (e.g., $\hat{\lambda}_{1**1}^{SAER} = -.222$) and the older generations or age groups are more religious than the younger ones (e.g., $\hat{\lambda}_{*1*1}^{SAER} = .375$). In terms of graphical simplification, the acceptance of model $\{SAE, SR, AR\}$ means that $\pi_{r\,sae}^{R|SAE}$ in Eq. 5.3 can be replaced by $\pi_{r\,sa}^{R|SA}$, but also that the appropriate estimates $\hat{\pi}_{r\,sa}^{R|SA}$ must be obtained by next applying model $\{SA, SR, AR\}$ to $\pi_{r\,sa}^{R|SA}$ to get rid of the three-variable interaction SAR.

The final variable in this analysis is W (*Women's role*), and the effects of S, A, E, and R on W in table $SAERW$. The starting point was the main-effects model $\{SAER, SW, AW, EW, RW\}$. Although even simpler models were acceptable and, of course, more complicated models also fitted well, this main-effects model was chosen as the final model on the basis of conditional tests (see Table 5.2). Looking at the parameter estimates, age and education have the biggest direct partial effects on W: the younger one is, and the higher one's educational level, the more one disagrees with the statement about women and the more one objects to confining the women's role to the home (e.g., $\hat{\lambda}_{*1**1}^{SAERW} = .443$ and $\hat{\lambda}_{**1*1}^{SAERW} = .373$). Somewhat surprisingly, the direct effects of sex are not very large, although they are significant: men agree with the statement somewhat more than women ($\hat{\lambda}_{1***1}^{SAERW} = .107$). The direct effects of religiosity are also small and borderline significant: the more religious one is, the more one agrees with the statement about women (e.g., $\hat{\lambda}_{***11}^{SAERW} = .120$).

The model as a whole can certainly be accepted (see Table 5.2). Perhaps the most interesting thing about these outcomes are the findings contrary to our expectations. Much larger effects of sex on W had been expected, as well as interaction effects of sex on W with age or education: although different from each other, men and

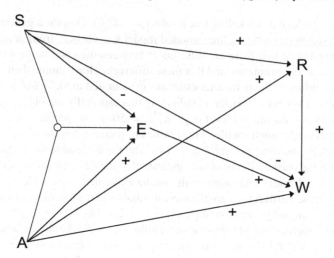

Fig. 5.1. Graphical representation of the revised modified path model for explaining *Women's role*

women are more alike in this context than assumed before. Further, religiosity played a somewhat less important role than expected in the Dutch situation. It might be that a more formal definition of religiosity by means of belonging to particular religious denominations would have behaved somewhat differently.

As usual in causal modeling, the SEM outcomes can be depicted in a path diagram, including the positive or negative signs of the direct relations (for ordered data). The diagram is given in Fig. 5.1. Note the special way, following Goodman, in which higher-order interactions are depicted by means of knots connecting the variables that have a higher-order interaction. This is different from strict graphical representations where the knots are not drawn and higher-order interactions are automatically assumed among directly connected variables.

The path diagram in Fig. 5.1 has, in most important respects, the same interpretations as in a standard linear SEM (Pearl, 2000; Hagenaars, 1993). Discussions have arisen about how to define and compute indirect and total effects (certainly not by simply multiplying the loglinear coefficients found for each direct relationship). This topic will not be dealt with here, but see Hagenaars (1993, pp. 49-50), Pearl (2001), and Robins (2003).

5.1.3 Marginal Modeling and Categorical SEM

Although the marginal-modeling approach advocated in this book can handle categorical SEMs, for strictly recursive models, the more simple and straightforward Goodman (or graphical) approach is usually preferred. Adaptations of the basic Goodman algorithm have been developed to handle restrictions that concern two or more right-hand side elements of the causal decomposition simultaneously. For example, one may additionally assume in the previous example that the direct effect of *Sex* (*S*) on *Religiosity* (*R*) is exactly as large as the direct effect of *S* on the *Opinion about women's role* (*W*). Or, the structural equation model may take the form of a homogenous first-order markov chain in which the (employment) status at time t is only directly influenced by the status at the immediately preceding time $t - 1$, with the additional restriction that all transition tables from t to $t + 1$ are exactly the same. For many of these kinds of restrictions, the required simultaneous estimation procedures have been developed and can be routinely imposed — for example, by means of Vermunt's LEM (Vermunt, 1997b; see also Bassi et al., 2000). As a matter of fact, such restrictions can also be handled by means of the advocated marginal-modeling approach.

In general, marginal modeling provides the means to estimate SEMs for categorical data under a very large range of additional marginal restrictions, and for dependent marginal tables without having to derive the appropriate estimation procedures for each (class of) model(s) separately. Two examples to illustrate this will be given below — a more specific and a more generally applicable example (for some other applications, see also Luijkx (1994), Agresti (2002), and Croon et al. (2000)). The specific example pertains to the situation that the researcher wants to test a particular structural equation model, but in combination with the hypothesis that the (positive) direct and (negative) indirect effects of a particular independent variable on the ultimate dependent variable cancel each other out, resulting in statistical independence in the appropriate marginal table. The more general example has to do with structural equation modeling when the data are not independently observed. A prime example might be that one wants to compare a causal model at t_1 with a causal model involving the same characteristics at t_2, but using panel data with the same respondents at both points in time. One cannot simply use *Time* as just another independent variable to investigate in what respects the two SEMs differ from each other. Because such marginal analyses of longitudinal data have been extensively discussed in the previous chapter, a different kind of dependent data will be used where observations are dependent because they are paired within families.

Marginal Restrictions in SEMs; Zero Total Effects

The illustrative application in this subsection has to do with the fact that variables often have both direct and indirect (or spurious) effects on the dependent variable, but are opposite in sign. An interesting hypothesis may then be that these compensating influences cancel each other out, resulting in an overall zero marginal relationship. This phenomenon is mainly known under the names of 'suppressor variables' or

Table 5.3. Joint marginal frequency distribution of *Sex* and *Women's role*; data Table 5.1

	Statement on women's Role (W)		
Sex (S)	1. Agree	2. Disagree	Total
1. Male	169 (37%)	260 (63%)	459 (100%)
2. Female	151 (30%)	350 (70%)	501 (100%)

Simpson's paradox (Simpson, 1951). A less well-known (among statisticians) but earlier instance — in 1897, way before 1951 — is provided by Durkheim's famous analysis of the relationship between marital status and suicide, with age acting as a suppressor variable (Durkheim, 1930).

Something like this occurred in the example of the previous subsection on women's roles (W). Sex (S) had a statistically significant, small positive direct partial effect on W: men are more traditional in this respect than women. But at the same time, there was a negative indirect effect of S on W through *Education* (E): men have higher educational level than women and more highly educated people are less traditional. There is also a negative indirect effect through *Religiosity* (R): men are less religious than women and less religious people have less traditional opinions. Do these opposite direct and indirect effects cancel each other out? The observed marginal relationship between S and W is shown in Table 5.3 and reveals a small positive relationship between S and W ($\hat{\lambda}_{11}^{SW} = .075$). The independence model $\{S, W\}$ for marginal table SW yields a p-value of .028 ($G^2 = 4.808$, $df = 1$), significant at the .01 but not at the .05 level. Marginal modeling provides the opportunity to estimate the previous structural equation model but now under the condition of zero total effect of S on W in the population, i.e., independency in marginal table SW, and test the marginal independence restriction for table SW simultaneously with the structural equation modeling restraints. This cannot be done by simply applying Goodman's algorithm because of the dependencies among the right-hand side elements of the basic equation (such as Eq. 5.3) that originate from this extra marginal restriction and application of marginal modeling is required.

For investigating the influence of the additional restriction that for marginal table SW model $\{S, W\}$ holds in the population, the structural equation model has been restricted as above (Table 5.2). It is also possible to impose the extra marginal restriction $\{S, W\}$ for table SW and, given this restriction, start looking for a well-fitting model. Results might then differ somewhat from the finally chosen structural equation model above. Adding the marginal restriction $\{S, W\}$ for table SW to the structural equation model reported in Table 5.2 by means of the marginal-modeling algorithm described in Chapter 2, yields the following test results: $G^2 = 88.35$, $df = 77$, $p = .177$ ($X^2 = 85.98$). The conditional test of marginal restriction $\{S, W\}$ in table SW, conditional on the assumed structural equation model without this marginal restriction yields: $G^2 = 88.35 - 85.98 = 2.37$, $df = 77 - 76 = 1$, $p = .124$. There is no

Table 5.4. Estimates loglinear parameters with and without marginal restriction on Table *SW*

	With Marginal Restrictions		Without Marginal Restrictions	
	Estimate	s.e.	Estimate	s.e.
$\hat{\lambda}^{SAE}_{*11}$	.503	.071	.500	.071
$\hat{\lambda}^{SAE}_{1*1}$	−.163	.049	−.146	.051
$\hat{\lambda}^{SAE}_{111}$	−.132	.071	−.130	.071
$\hat{\lambda}^{SAER}_{1**1}$	−.226	.055	−.222	.055
$\hat{\lambda}^{SAER}_{*1*1}$	.375	.088	.375	.088
$\hat{\lambda}^{SAERW}_{*1**1}$	.448	.055	.443	.055
$\hat{\lambda}^{SAERW}_{**1*1}$	.373	.053	.373	.053
$\hat{\lambda}^{SAERW}_{***11}$	.121	.066	.120	.068

reason to reject the hypothesis that the total effect of *Sex* (*S*) on *Women's role* (*W*) is zero, and that therefore the direct effect is equal in strength but of opposite sign to the result of the two indirect effects through education and religiosity.

Note that this conditional test outcome is different from the 'stand alone' independence test reported above for marginal table *SW* ($G^2 = 4.81$). Or stated differently, the simultaneous test result $G^2 = 88.35$ for the structural equation model plus the extra marginal restriction is smaller than the sum of the G^2's, obtained above for the structural equation model as such and separately for model $\{S,W\}$ for table *SW*: $85.98 + 4.81 = 93.79$. Actually, the G^2s cannot be summed in this way because the extra marginal restriction is not independent of the remainder of the structural equation model.

Regarding the parameter estimates of the causal model and comparing the structural equation modeling outcomes with and without the extra marginal restriction, the biggest difference pertains to the direct partial effect of *S* on *W*. With the extra marginal independence restriction, the positive direct effect of *S* on *W* is still significant and positive, but much smaller than without: $\hat{\lambda}^{SAERW}_{1***1} = .051, \hat{\sigma}_\lambda = .013$ (and before: $\hat{\lambda}^{SAERW}_{1***1} = .107, \hat{\sigma}_\lambda = .039$). The observed small, positive marginal association SW is made into an expected zero association mainly by diminishing the direct positive effect of *S* on *W*. For all other variables, their effects become just a little bit stronger and their standard errors just a little bit smaller, although not always noticeable when rounding off to three decimals. Table 5.4 illustrate this for the coefficients of the structural equation model discussed before.

Categorical SEMs with Dependent Observations

Dependencies among and within the (marginal) tables from the categorical SEM may not only have their origins in the extra (marginal) restrictions, as illustrated above, but also in the research design. When the resulting dependencies among the observations are not the prime interest of the researcher, and are just considered a nuisance, marginal-modeling methods can be used to answer the research questions. For example, in a panel study with three waves involving the same respondents over time, a researcher may be interested in estimating at each time point a structural equation model among a particular set of variables and in comparing the results of these three SEMs. This is actually a form of multiple group comparison, but with dependent observations among the three groups (time points). Dependent multiple group comparisons will be discussed explicitly at the end of this subsection. But first, a general illustration of structural equation modeling with dependent observations will be presented.

In Section 2.2.3, a module from the Netherlands Kinship Panel Survey (NKPS) has been introduced in which each respondent is coupled with one of its parents. The data from this module will be used again, but it is used now to estimate a SEM in which the information about the child-parent dyad is regarded as a nuisance. The variables in this SEM are similar to the ones used in the EVS example above about the opinion on women's roles. But with the NKPS data, some difficulties arise when trying to estimate a similar SEM. For example, if the effect of sex on the opinion about women's roles is studied, and men and women are being compared regarding their opinions, dependencies among the observations in the two categories arise since many male respondents are coupled to female respondents as part of a child-parent dyad. Of course the same is true for many female respondents being coupled with men in the child-parent dyad. At the same time, male (female) children are coupled with a male (female respectively) parent. For the researcher, these dependencies may not be substantively important given the research questions, but they nevertheless have to be taken into account when estimating and testing the structural equation model.

The symbols used for the variables concerned are similar to the ones used in the EVS example above. The ultimately dependent variable *Attitude towards women's roles*, denoted as W, is now an index with three categories — the same as the one used before in Section 2.2.3, but with reversed order to make the comparison with the EVS analyses easier: 1 = traditional, 2 = moderately traditional, and 3 = nontraditional. *Sex S* has two categories (1 = men, 2 = women); *Age* (A) has three categories (1 = old ($\geq$ 55), 2 = middle (38-54), and 3 = young ($\leq$ 37)) and also *Education* (E) has three categories: 1 = low, 2 = intermediate, and 3 = high. Finally, an operationalization different from EVS had to be used for *Religiosity* (R), viz. *Church attendance* with three categories: 1 = often/regular, 2 = sometimes, 3 = never/rarely. Therefore, when comparing the NKPS results with the EVS outcomes above, one has to keep in mind that religiosity has been measured differently, that W is now a more reliable index than the single EVS item, and that in NKPS there are 3,594 respondents (and 1,797 dyads) compared to $N = 960$ in EVS. The full data set, which includes also the variable 'family,' is too large to represent here, but can be found on the book's website. As before, the usual shorthand notation for indicating loglinear models will be used, treating table *SAERW* as if it were a normal table.

The causal order of the variables and the corresponding causal decomposition is supposed to be the same as above in the EVS example (Eq. 5.3). However, the analysis is now much more complex and the marginal-modeling approach of Chapter 2

Table 5.5. Test results of loglinear SEM analyses NKPS data; see text; † for comparison, the sum of separate G^2's is 123.89

Table	Model	separate models G^2	df	p
SA	$\{S,A\}$	5.88	2	.053
SAE	$\{SAE\}$	0.	0	1
SAER	$\{SAE,AR,ER\}$	27.52	26	.482
SAERW	$\{SAER,SEW,ARW\}$	90.48	80	.198
Overall	all submodels	$121.52^{\dagger}$	108	.177

is needed. In the first place, the pairwise dependency of the data has to be taken into account when estimating the parameters of a particular loglinear or logit sub-model for a particular right-hand side element in Eq. 5.3. Moreover, because of the presence of dependent observations, it cannot be taken for granted any longer that the right-hand side elements and their submodels are asymptotically independent of each other. Therefore, all parameters and their standard errors for all submodels have to be estimated simultaneously. In line with this, it is no longer necessarily true that the G^2's for the submodels, even when for each submodel the dependencies in the data have separately been taken into account by means of separate marginal submodels, sum to the appropriate overall G^2 (see Table 5.5).

Nevertheless, for exploratory purposes, first separate appropriate marginal models were investigated for each right-hand side element in Eq. 5.3 using marginal modeling. The final submodels and test results from these separate analyses are reported in Table 5.5. Thereafter, the selected models were all estimated simultaneously, yielding a well-fitting model (last row, Table 5.5). It was further investigated whether additional modifications were called for, given the parameter estimates and their estimated standard errors. No indications for such modifications were found here. This need not be the case in general. As stated above, it must be kept in mind that the successive submodels are not independent of each other, and that the likelihood function cannot be separated into independent parts. Consequences hereof are that the estimates for the different submodels and subtables may be correlated, that the separate submodel G^2's no longer sum to the appropriate overall G^2, but also that the usual sufficient statistics present in models for independent observations no longer function as expected. Or rather, the usual sufficient statistics of ordinary loglinear models are no longer sufficient statistics in the loglinear marginal analyses. For example, when one uses the final estimated (marginal) probabilities $\hat{\pi}^{*SAERW}_{saerw}$ to compute the estimated marginal 'submodel' probabilities $\hat{\pi}^{*SAERW}_{sae++}$, then the estimated marginal frequencies $N\hat{\pi}^{*SAERW}_{sae++}$ for table SAE are not the same as the observed marginal frequencies n^{SAERW}_{sae++}, in spite of the fact that the saturated model $\{SAE\}$ is specified for subtable SAE.

The parameter estimates based on the final model for the NKPS data partly lead to the same conclusions as the above final model for the EVS data and partly to different ones. In both cases, it was concluded that age and sex are independent of each other. Further, it was even much more clear in the NKPS than in the EVS data that a three-variable interaction was needed to obtain a well-fitting model for table SAE. The main partial effects of age are very strong — stronger than found in EVS but leading to the same conclusion that older generations having lower educational levels (e.g., $\hat{\lambda}_{*11}^{SAE} = .922$). Further, as in EVS, men have on average a somewhat higher educational level than women (e.g., $\hat{\lambda}_{1*1}^{SAE} = -.189$). Finally, given the nature of the significant three-variable interaction effects, the educational sex differences are more or less linearly increasing with age (e.g., $\hat{\lambda}_{11\ 1}^{SE|A} = -.387$, $\hat{\lambda}_{11\ 2}^{SE|A} = -.189$ and $\hat{\lambda}_{11\ 3}^{SE|A} = .008$).

An interesting difference between the EVS and the NKPS data concerns the effects on religiosity (personal, subjective religiosity in EVS, but church attendance in NKPS). With the NKPS data, it was estimated that Sex (S) did not have a direct effect on religiosity, but age and education did, while in EVS, religiosity was only directly influenced by sex and age. According to the NKPS data and the estimates in submodel $\{SAE, AR, ER\}$, age and educational level have more or less linear direct relationships with religiosity: the older one is, and the lower one's educational level, the less religious (e.g., $\hat{\lambda}_{*1*1}^{SAER} = .373$ and $\hat{\lambda}_{**11}^{SAER} = .260$). It might be interesting to further investigate whether these EVS-NKPS differences are indeed due to the different operationalization of religiosity.

For the final effects on the attitude towards women's roles, model $\{SAER, SEW, ARW\}$ was chosen for the NKPS data, while above for the EVS data, the main-effects-only model was selected without any three-variable interactions. It might be that, given the greater reliability of the measurement of W and the larger sample size in NKPS (persons and dyads) than in EVS, the tests are now more powerful and better able to detect three-variable interactions in the population. In terms of the direct partial effects, men have definitely more traditional opinions about women's roles than women (e.g., $\hat{\lambda}_{1***1}^{SAERW} = .299$), the oldest age group is more traditional than the two younger ones (e.g., $\hat{\lambda}_{*1**1}^{SAERW} = .290$), the lower one's educational level the more traditional one's opinions are about women's roles ($\hat{\lambda}_{**1*1}^{SAERW} = .627$), and the more religious, the more traditional one is (e.g., $\hat{\lambda}_{***11}^{SAERW} = .589$). All these partial direct effects are much stronger than the ones found in EVS, which is, as suggested before, probably due to the greater reliability of W. There are two significant three-variable interactions terms: the effects of S on W vary somewhat with E (or, what amounts to the same, the effects of E on W vary with S), and the effects of R on W vary somewhat with A (or the effects of A on W with R). The men-women differences regarding the attitude towards women's roles decrease as education increases, e.g., $\hat{\lambda}_{11\ *1*}^{SW|AER} = .447$, $\hat{\lambda}_{11\ *2*}^{SW|AER} = .299$, and $\hat{\lambda}_{11\ *3*}^{SW|AER} = .142$ (or viewed from the other way: the educational effects on W are somewhat larger among women than among men, although in the same direction). Further, in all age groups, religious people are more traditional regarding women's roles than nonreligious people, but these differences are much larger among the younger than among the older respondents (e.g.,

$\hat{\lambda}_{11\ *1*}^{RW|SAE} = .361$, $\hat{\lambda}_{11\ *2*}^{RW|SAE} = .470$, and $\hat{\lambda}_{11\ *3*}^{RW|SAE} = .937$), or from the other point of view, age has not much effect on W among the most religious people, but is important for the other two religious categories.

As usual, not only the data but also the information about the model specification for this extended SEM are reported on the book's website.

Comparative Analyses with Dependent Observations

This very general approach towards categorical structural equation modeling in the presence of dependent observations can easily be extended towards, or interpreted in terms of, *comparative analyses* applying *multiple group comparisons* when the observations from the different groups are not independent of each other.

Let us start with the simple situation in which the multiple groups to be compared all have independent observations. Multiple group comparison then amounts to ordinary (causal or structural equation) modeling, once it is recognized that the grouping variable can be treated as any other normal variable. This grouping variable G may refer to observations grouped at particular points in time or to different cultures, to different regions, or to gender, birth cohorts, etc. Assume that the researcher's interest lies in comparing the relationships among variables A, B, and C — more specifically, in the effects of A and B on C among the several groups. The model of interest is the no-three-factor interaction model $\{AB, AC, BC\}$. Within each group (at each point in time, within each nation, birth cohort, etc.), an independent sample has been drawn. If this researcher would fit model $\{AB, AC, BC\}$ separately to each of the groups, this would be exactly the same as fitting model $\{GAB, GAC, GBC\}$ in the standard way to table $GABC$. If the researcher then wants to constrain particular effects to be the same in all groups, e.g., the effect of A on C (but not the effect of B on C), this researcher must test model $\{GAB, AC, GBC\}$ in the usual standard way for table $GABC$.

The same kind of logic can be applied to comparative analyses involving dependent observations. Imagine a three-wave panel study with measurements for characteristics A, B, and C at each point in time. The time of the measurement of the characteristics will be denoted by means of subscripts, e.g., A_i is the measurement of characteristic A at time i. All these measurements give rise to the high-dimensional frequency table $A_1B_1C_1A_2B_2C_2A_3B_3C_3$. Now the focus of this study is not on the effects of previous measurements on later ones, but on the question of whether or not the effects of A and B on C remained the same at the three points in time. The necessary models and restrictions can best be understood by setting up a marginal table $GABC$, where $G = 1$ refers to the wave one measurements (marginal table $A_1B_1C_1$), $G = 2$ to marginal table $A_2B_2C_2$, and $G = 3$ to marginal table $A_3B_3C_3$. Estimating the joint frequencies in the original high-dimensional table under restriction $\{GABC\}$ (the saturated model) for table $GABC$ by means of the marginal-modeling approach amounts to allowing the (lower and higher-order) effects of A and B on C to be completely different for the three-waves. Model $\{GAB, GAC, GBC\}$ for table $GABC$ means that there are no three-variable interaction effects on C at none of the three time points, but otherwise the effects of A and B on C may be completely different at each point in time. Finally, model $\{GAB, AC, GAB\}$ would introduce the extra restriction that the direct partial effects of A on C are the same in all three-waves.

Also, the results of the final SEM for the NKPS data (Table 5.5) can be interpreted in terms of group comparisons involving dependent observations. The purpose of the investigation might be the comparison of (the partially matched) men and women regarding the relationships among variables A, E, R, and W, with S as the grouping variable,

- regarding the distribution of age: model $\{S, A\}$ for table SA implies that the age distribution is the same for men and women
- regarding the effects of age on education: model $\{SAE\}$ for table SAE implies that the effects of age on education are different for men and women
- regarding the effects of age and education on religiosity: model $\{SAE, SR, AR, ER\}$ for table $SAER$ (different from Table 5.5) implies that for men nor for women are there three-variable effects of age and education on religiosity. Further, the direct effects of age and education on religiosity are the same for men and women. Model $\{SAE, SR, AR, ER\}$ for table $SAER$ differs from model $\{SAE, AR, ER\}$, chosen as the final model for table $SAER$ in Table 5.5. The latter model not only investigates the comparability of the effects of age and education on religiosity for men and women, but moreover implies that the gender differences in religiosity disappear when age and education differences are controlled for, when age and education are held constant
- regarding the effects of age, education, and religiosity on the attitude towards women's roles: model $\{SAER, SEW, ARW\}$ for table $SAERW$ implies that the effects of education on women's roles are different for men and women and that age and religiosity interact in their effects on women's roles, but in the same way for men and women

Each of these above comparative statements about men-women differences can be investigated and tested separately one by one by means of the marginal-modeling approach taking the clustered observations for men and women into account. It is also possible, as essentially done above in the previous subsection, to investigate all comparisons at the same time.

5.2 Analysis of (Quasi-)Experimental Data

Next to structural equation modeling, the other main tradition in causal analysis within the social, behavioral, and life sciences is the (quasi-) experimental approach. Although the two traditions overlap in many respects, in the SEM tradition, the emphasis is on the data analysis phase — how to control for confounding factors in the (causal) analyses. In the quasi-experimental approach, the main focus is on the design of a causal study — how to set up an investigation to minimize the influences of confounding factors. Especially Campbell and his associates have made unique and decisive contributions to the latter tradition, evaluating a large variety of research designs in terms of their internal and external validity (for overviews, see Campbell and Stanley (1963) and Shadish, Cook, and Campbell (2002)). Three designs will be discussed below — the one-group pretest-posttest design, the nonequivalent control group design, and the randomized experimental design. Given the purposes of this book, the focus of the next discussions will not be on the internal validity aspects of these designs, but on the analyses of (quasi-)experimental data and especially the role played by marginal modeling.

5.2.1 The One-group Pretest-Posttest Design

To represent a particular research design, Campbell has developed what has become, with some minor variations, the standard notation in the methodological literature for representing (quasi-)experimental designs. Rows indicate research groups (experimental and control groups) and columns the time order. Further, O_i is a statistic (which could be the entire distribution) representing the value of the dependent variable at a particular time and in a particular group (a mean, a proportion, etc.) and X_e symbolizes the experimental intervention, i.e., a category of the independent variable X. The one-group, pretest-posttest design is represented as

$$O_1 \quad X_e \quad O_2.$$

In this diagram, the symbol O_1 refers to the distribution of pretest and O_2 to the distribution of the posttest for the same group, with X_e occurring in between pre- and posttest. The purpose of the pretest-posttest design is to evaluate the influence of X_e on the dependent variable by analyzing the difference between O_1 and O_2. The fact that the pre- and posttest are conducted on the same people must be taken into account. In uncontrolled circumstances, the biggest threat to the internal validity of this design is 'history.' Between pre- and posttest, besides X_e many other things may occur that can cause a difference between O_1 and O_2. The effects of these intervening events may be confounded with the effects of X_e. Another possible threat to the internal validity of this design is 'testing': the pretest can have a direct confounding influence on the posttest that can be mistaken for the effect of X_e.

To illustrate the analysis of this design, the data from Table 5.6 on the opinion about U.S. Supreme Court nominee Clarence Thomas will be used (CBS News and the New York Times, 2001). These data were previously analyzed by Bergsma and Croon (2005). Clarence Thomas was nominated in 1991 as member of the U.S. Supreme Court by President George H.W. Bush. The nomination provoked some public debate because of Clarence Thomas' race (black) and because of his allegedly extremely conservative social and political views. A panel of U.S. citizens was interviewed regarding their opinion on Clarence Thomas' candidacy during September 3–5 (A) and on October 9 (B). After the first wave, more precisely on September 25, a charge of sexual harassment was brought against Clarence Thomas by his former aide, Anita Hill. The data in Table 5.6 provide the information about how this charge and the enormous media attention it got, symbolized by X_e, influenced opinion about Clarence Thomas. More specifically, the differences between the marginal distributions of A and B, indicated in Table 5.6 as O_1 and O_2, respectively, will be used

Table 5.6. Opinions about Clarence Thomas in the United States in September and October 1991. Source: CBS and the New York Times (2001)

September (A)	October (B)					All Categories		Without Cat. 4	
	1	2	3	4	Total	O_1	O_2	P_1	P_2
1. Favorable	199	15	51	11	276	27.9	44.3	45.5	49.8
2. Not favorable	25	90	21	3	139	14.0	20.8	22.9	23.3
3. Undecided	69	44	58	21	192	19.4	23.9	31.6	26.9
4. Haven't heard enough	146	57	107	74	384	38.7	11.0	–	–
Total	439	206	237	109	991	100	100	100	100

Table 5.7. Adjusted residuals for the MH model applied to Table 5.6

A	B				O_1
	1	2	3	4	
1	0	.28	−3.00	−13.61	−9.37
2	−.28	0	−2.64	−10.29	−4.97
3	3.00	2.64	0	−9.98	−2.44
4	13.61	10.29	9.98	0	14.78
O_2	8.77	5.57	2.66	−14.81	0

for measuring the effect of X_e. In terms of the previous chapter, the nature of the net changes must be investigated. Because O_1 and O_2 pertain to the same people, marginal-modeling methods are required. Ignoring possible alternative explanations in terms of history and testing, the null hypothesis of no effect of X_e amounts to $O_1 = O_2$ — in other words, to MH model

$$\pi_{i+}^{AB} = \pi_{+i}^{AB}$$

for all i. This MH model has to be rejected for the data in Table 5.6 ($G^2 = 269.47$, $df = 3$, $p = .000$, $X^2 = 230.88$), and it is concluded that there are significant net changes between September and October. These significant changes have mostly to do with increasing interest. As the O_1 and O_2 columns in Table 5.6 show, the number of people in Category 4 who had not heard enough about the case to decide dropped seriously from 39% in September to 11% in October. The adjusted residuals under the MH model for both the relevant marginal frequencies O_1 and O_2, and the joint frequencies in table AB, confirm this conclusion about the main source of the significant net changes (Table 5.7). From inspection of the marginal residuals in Table 5.7, it is seen that the most marked changes occurred regarding Category 4 and also, to a somewhat lesser extent, regarding Category 1. The interior cell residuals reveal that the last row and column contain most of the bad fit — there is a consistent movement away from Category 4 towards the first three categories. Note that the zero residuals on the main diagonal in Table 5.7 are a necessary consequence of the maximum likelihood estimation method for the MH model (see Bishop et al., 1975, p. 295).

It is clear that the Anita Hill charge and the subsequent media attention steered the public's opinions away from 'not knowing enough' about the nomination. But did this result in a more favorable or a more unfavorable opinion? Leaving out the last Category 4, a comparison can be made among those who said the first time that they had enough information at their disposal to form an opinion and those who said so the second time. The two relevant response distributions are presented in the last two columns of Table 5.6. Response distribution P_1 is based on the 607 respondents that belonged to Categories 1, 2, or 3 of A and response distribution P_2 on the 882 respondents in categories 1, 2, or 3 of B. The pre- and posttest distributions P_1 and P_2 now overlap only partially: 572 respondents had heard enough to decide both in September and October, but this partial dependency also must and can be taken into account. The no-effects MH hypothesis $P_1 = P_2$ can be accepted: $G^2 = 5.06$, $df = 2$, ($p = .080$, $X^2 = 5.07$). Although there is a very slight observed tendency towards 'favorable' (Category 1) and away from 'undecided' (Category 3), as shown in the last two columns of Table 5.6, these net changes are not statistically significant. The adjusted residuals for the marginal frequencies (not shown here) are on the borderline of significance (around ± 2.07) for categories 1 and 3. Given all these outcomes, it is probably wisest to accept $P_1 = P_2$ or postpone judgment till further evidence is available.

Given the nature of the charges and the variables involved, it may well be that women reacted to these charges differently from men or Republicans differently from Democrats. To investigate these possible subgroup differences, one might set up a (causal) model for the relations among the relevant variables as explained in the previous section, or apply the kinds of net change models for subgroups explained extensively in the previous chapter.

5.2.2 The Nonequivalent Control Group Design

From the viewpoint of threats to the internal validity, one of the strongest quasi-experimental designs Campbell et al. discussed is the nonequivalent control group design, usually represented as

$$O_1 \ X_e \ O_2$$
$$- - - - -$$
$$O_3 \ X_c \ O_4$$

This design has (at least) two groups — an experimental and a control group with pre- and posttests for both groups; X_e and X_c represent the experimental and control intervention respectively. Because the respondents are not randomly allotted to the two groups, the potentially nonequivalent groups are separated from each other in the diagram by a dotted line. This design is much stronger than the one-group pretest-posttest design because of the presence of a control group. Where in the one-group pretest-posttest design, conclusions about the experimental group undergoing X_e potentially suffered from history and testing among other confounding factors, the same disturbing factors are in the nonequivalent control group design also operating in the control group undergoing X_c. They can therefore, in principle, be controlled by comparing $(O_2 - O_1)$ with $(O_4 - O_3)$. The nonequivalent control group design is also much stronger than the 'after only design' without pretests. In the nonrandomized after only design, the posttest difference $(O_2 - O_4)$ may be caused not only by the fact that one group received X_e and the other group not (X_c), but also by selection. Perhaps the posttest differences were already present before the experimental group experienced the experimental stimulus. By means of the pretest difference $(O_1 - O_3)$, this possibility can be investigated in the nonequivalent control group design and therefore selection can in principle be controlled. This does not mean that there are no problems left. After all, this is not a randomized design and some kinds of confounding factors remain a possible threat, especially in the form of interaction effects between selection and other disturbing factors such as history. Nevertheless, the nonequivalent control group design has, in principle, a higher internal validity than most other quasi-experimental designs.

How to control exactly for confounding factors such as history and selection is a much debated issue. There are several ways in which the net changes in the experimental group $(O_2 - O_1)$ can be compared with and corrected by the net changes $(O_4 - O_3)$ in the control group. Or, stating the same thing differently, there are several ways in which the posttest differences $(O_2 - O_4)$ can be corrected for by the pretest differences $(O_1 - O_3)$. The different approaches that have been suggested in the literature involve different assumptions and may lead to different conclusions about the effect of X. The partially overlapping issues that play a role in this debate are: whether to use fixed or random-effect models or mixtures thereof; whether to apply conditional or unconditional (marginal) models; what assumptions to make about the (self)selection mechanism; about the nature of the growth curves; about the causal lag; what to do about autocorrelated error terms; what about measurement error; how to deal with regression artifacts (Allison, 1990, 1994; Bryk & Weisberg, 1977; Campbell & Kenny, 1999; Duncan, 1972; Hagenaars, 1990, 2005; Hsiao, 2003; Judd & Kenny, 1981; Kenny & Cohen, 1980; Lord, 1960, 1963; Halaby, 2004; Shadish et al., 2002; Wainer, 1991).

Because this section is certainly not intended as a complete treatment of the nonequivalent control group design, most of these issues will be neglected. The focus will be on how to carry out unconditional marginal analysis with a brief comparison between the marginal and the conditional fixed-effects approach. In the context of regression analysis, Lord (1967) has pointed out most clearly why different results may be obtained with marginal and conditional fixed-effect analysis. Hagenaars (1990) showed that Lord's paradox also occurs with categorical data and loglinear models.

By way of illustration, a classical data set will be introduced that has been used several times in the past, but not analyzed by means of the methods advocated in this book (Glock, 1955; Campbell & Clayton, 1961; Hagenaars, 1990, pp. 215-233, and Hagenaars, 1990, Section 5.3).

The data are from a panel study among 503 white Christians living in and around Baltimore. The study's purpose was to determine the effect of seeing the film 'Gentleman's Agreement' on reducing the level of antisemitism (Glock, 1955, p. 243). *Antisemitism* was measured in November 1947 (variable A) prior to the movie being locally shown and consisted of three categories: 1 = high, 2 = moderate, and 3

Table 5.8. Pre- and posttest antisemitism in relation to seeing film. Source: Glock, 1955. (T_1 is November 1947, T_2 is May 1948)

$X = 1$ (Seen film)		B. Antisemitism T_2			
		1. High	2. Moderate	3. Low	Total
A. Antisemitism T_1	1. High	20	6	6	32
	2. Moderate	4	12	10	26
	3. Low	3	5	49	57
	Total	27	23	65	115
$X = 2$ (Not seen film)		B. Antisemitism T_2			
		1. High	2. Moderate	3. Low	Total
A. Antisemitism T_1	1. High	92	20	20	132
	2. Moderate	34	15	27	76
	3. Low	24	28	121	173
	Total	150	63	168	381

= low. *Antisemitism* was measured again in May 1948 (variable *B*). In addition, the respondents were asked whether or not they had (voluntary) seen the movie, which had been shown in Baltimore theaters during the period between the two interviews (variable *X*). The experimental group (with $X = 1$ or X_e) consisted of those respondents who saw the movie; the control group (with $X = 2$ or X_c) consisted of those who did not. Table 5.8 contains the observed frequencies.

Table 5.9 summarizes the data in the form of the marginal distributions for pre- and posttest measurements in both groups in correspondence with the above diagram of the nonequivalent control-group design. Table 5.9 will be referred to below as an *XTR* table in which *R* is the response variable on *Antisemitism* measured with the three categories mentioned above. *T* refers to the time of measurement and has two categories (pretest: $T = 1$ and posttest: $T = 2$), and *X* is the dichotomous independent variable with the two categories indicated above.

First, the *unconditional* or *marginal approach* will be applied that focuses on the analysis of the marginal distributions labeled O_i in Table 5.9. Note that the comparison O_1 with O_2 involves the same respondents, as does the comparison O_3 with O_4, but with the response observations for those who did see the movie compared to those who did not pertain to different people. In a purely descriptive sense, the percentages in Table 5.9 indicate that there is a slight net tendency over time towards less antisemitism in the experimental group and a very weak tendency towards more antisemitism in the control group. Viewed from the other perspective, the experimental group is a bit less antisemitic than the control group, but this difference is a bit larger for the posttest than for the pretest. As it turns out, the posttest differences are also statistically significant: hypothesis

Table 5.9. Marginal distributions for pre- and posttest measurements in experimental and control group. Source: see Table 5.8

Antisemitism	$X = 1$ (X_e Seen film) Pretest (O_1)	Posttest (O_2)	$X = 2$ (X_c Not seen film) Pretest (O_3)	Posttest (O_4)
1. High	32 (28%)	27 (23%)	132 (35%)	150 (39%)
2. Moderate	26 (23%)	23 (20%)	76 (20%)	63 (17%)
3. Low	57 (50%)	65 (57%)	173 (45%)	168 (44%)
Total	115 (100%)	115 (100%)	381 (100%)	381 (100%)

$$\pi_k^{R|XT}{}_{1\,2} = \pi_k^{R|XT}{}_{2\,2}$$

has to be rejected, using a straightforward nonmarginal test: $G^2 = 10.24$, $df = 2$ ($p = .006$, $X^2 = 9.767$). The pretest differences on the other hand are not significant: hypothesis

$$\pi_k^{R|XT}{}_{1\,1} = \pi_k^{R|XT}{}_{2\,1}$$

can be accepted, again on the basis of a straightforward nonmarginal test: $G^2 = 1.91$, $df = 2$ ($p = .384$, $X^2 = 1.88$).

All these outcomes point towards a small effect of seeing the movie on antisemitism in the expected decreasing direction. However, the unique advantages of the nonequivalent control-group design consists in the simultaneous comparison of the net changes in the experimental and control group or, what amounts to the same, in the simultaneous analyses of the posttest and pretest differences, and the previous separate tests are not the optimal way to proceed.

A good starting point is a marginal test of the most restrictive assumption regarding a complete absence of any effect of the movie — viz., the hypothesis that all four percentage distributions O_i in Table 5.9 are the same

$$\pi_k^{R|XT}{}_{1\,1} = \pi_k^{R|XT}{}_{1\,2} = \pi_k^{R|XT}{}_{2\,1} = \pi_k^{R|XT}{}_{2\,2}$$

for all values of k. In other words, logit model $\{XT,R\}$ is valid for the data in Table 5.9. The test outcomes are $G^2 = 13.00$, $df = 6$ ($p = .043$, $X^2 = 12.72$) which is an ambiguous result. Acceptance of model $\{XT,R\}$ means that one accepts the null hypothesis that the movie did not influence the degree of antisemitism. But rejection of model $\{XT,R\}$ does not automatically imply that there is an effect of the movie, because less restrictive models may be valid that still do not imply an effect of the movie. Given the borderline test outcome for model $\{XT,R\}$, the adjusted residuals for the four distributions O_i under model $\{XT,R\}$ were inspected to see how the model could be improved. The adjusted residuals (not reported here) point out that the degree of antisemitism is significantly overestimated for O_2 (the posttest for the experimental group), and significantly underestimated for O_4 (the posttest scores in the control group). Perhaps, as the descriptive analysis of the percentages also suggested, some changes took place in the expected direction.

To compare the net changes in the experimental and control group, first the no net change hypothesis in the experimental group is investigated. The MH hypothesis for the experimental group

$$\pi_k^{R|XT} \underset{1\,1}{} = \pi_k^{R|XT} \underset{1\,2}{}$$

can be accepted: $G^2 = 2.96$, $df = 2$ ($p = .227$, $X^2 = 2.89$). The test statistics for the no net change MH hypothesis for the control group

$$\pi_k^{R|XT} \underset{2\,1}{} = \pi_k^{R|XT} \underset{2\,2}{}$$

are $G^2 = 3.45$, $df = 2$ ($p = .178$, $X^2 = 3.43$), and this hypothesis can also be accepted. Because the experimental and the control group have been independently observed, the test statistics for the simultaneous hypothesis of no net change in either experimental or control group

$$\pi_k^{R|XT} \underset{1\,1}{} = \pi_k^{R|XT} \underset{1\,2}{} \text{ and } \pi_k^{R|XT} \underset{2\,1}{} = \pi_k^{R|XT} \underset{2\,2}{} \,,$$

which is equivalent to logit model $\{XT,XR\}$ for table XTR, can be obtained by summing the separate results for the experimental and control group: $G^2 = 2.96 + 3.45 = 6.42$, $df = 2 + 2 = 4$ ($p = .170$). There seems to be no reason to assume that there are any net changes in either the experimental or the control group, and the hypothesis that intervening event X_e did not affect the degree of antisemitism can be accepted.

At first sight, it may seem strange that accepting the no net change model $\{XT,XR\}$, in which there is an effect of X on R, leads to the conclusion that the movie did not have an effect. However, although model $\{XT,XR\}$ allows for differences in response between the experimental and control groups reflected in the effect of X on R, model $\{XT,XR\}$ implies most importantly that these group differences are the same at both points in time. And, as the logic underlying the nonequivalent control-group design dictates, equal posttest and pretest differences lead necessarily to the conclusion that the intervening event X_e does not have an influence. In terms of possible underlying causal mechanisms, less prejudiced people are more inclined to go and see the movie than the more prejudiced respondents and less prejudiced people just remain less prejudiced, as do the more prejudiced ones.

If the no net change model $\{XT,XR\}$ had been rejected, an alternative model to be considered might have been model $\{XT,TR\}$ for table XTR, in which there are no pretest and no posttest differences between the experimental and the control group, but net changes are allowed, although they are identical for both groups. Therefore, also in model $\{XT,XR\}$, no effect of the movie (X_e) is present. Above it was seen that the posttest differences between experimental and control groups are significant, while the pretest differences are not. These two separate test results cannot be added because the pre- and posttests involve the same respondents. Using the marginal-modeling approach, the test results for the simultaneous hypotheses

$$\pi_k^{R|XT} \underset{1\,2}{} = \pi_k^{R|XT} \underset{2\,2}{} \text{ and } \pi_k^{R|XT} \underset{1\,1}{} = \pi_k^{R|XT} \underset{2\,1}{}$$

(or logit model $\{XT,TR\}$) are: $G^2 = 11.13$, $df = 4$ ($p = .025$, $X^2 = 10.61$). This is a significant result at the .05 but not at the .01 level. If model $\{XT,TR\}$ is accepted, the hypothesis of no effect from seeing the movie must be accepted.

If both model $\{XT,XR\}$ and model $\{XT,TR\}$ had been rejected, this still would not imply that there is an effect of the movie on the degree of antisemitism. Model $\{XT,XR,TR\}$ for table XTR would provide the final test. Logit model $\{XT,XR,TR\}$ allows for group differences in antisemitism, with the restriction that they are the same before and after the introduction of X_e. An alternative interpretation of logit model $\{XT,XR,TR\}$ is that it allows for net changes that are the same in experimental and control group. Therefore, acceptance of model $\{XT,XR,TR\}$ implies the accepting of the hypothesis of no effect of seeing the movie, rejection of this model would lead to acceptance of an effect of the movie. Because of the dependencies involved, marginal modeling is required. Not surprisingly, given the earlier test outcomes, model $\{XT,XR,TR\}$ can be accepted here: $G^2 = 4.17$, $df = 2$ ($p = .124$, $X^2 = 4.07$).

If model $\{XT,XR,TR\}$ had been rejected, the conclusion would have been drawn that the movie changed the level of antisemitism and the three-variable parameter estimates $\hat{\lambda}_{i\,j\,k}^{TXR}$ in saturated model $\{TXR\}$ would then have indicated how the movie affects the level of antisemitism, because the three-variable effects $\hat{\lambda}_{i\,j\,k}^{TXR}$ show how different the posttest differences are from the pretest differences, or equivalently, how different the net changes in the experimental group are from the net changes in the control group. Just by way of example, the small and nonsignificant parameter estimates $\hat{\lambda}_{i\,j\,k}^{TXR}$ in model $\{TXR\}$ for Table 5.9 are estimated as

$$\hat{\lambda}_{1\,1\,1}^{TXR} = .068, \hat{\lambda}_{1\,1\,2}^{TXR} = -.022, \hat{\lambda}_{1\,1\,3}^{TXR} = -.046,$$

indicating that there is a very small (nonsignificant) effect of the movie into the direction of lowering the degree of antisemitism.

Despite the suggestions derived from just looking at the percentages in Table 5.9, which pointed towards a small effect of X_e in lowering the degree of antisemitism, these effects (if present) cannot be distinguished from sampling fluctuations. Sometimes more power can be achieved by not comparing whole response distributions but instead, particular characteristics of these distributions, such as means.

Because the response variable is ordinal here, comparison of the ordinal locations appears to be the most appropriate. The multiplicative measure L'_{ij} will be used for the comparison of distribution O_i with distribution O_j. L'_{12} then indicates the ordinal location shift (or net change) in the experimental group and L'_{34} is the corresponding shift in the control group. The estimated difference between $\hat{L}'_{12}$ and $\hat{L}'_{34}$ shows the difference between the net changes in location for experimental and control groups and therefore can be used as a measure of the effect of the movie:

$$\hat{L}'_{12} - \hat{L}'_{34} = .238 - (-.109) = .347.$$

The value of $\hat{L}'_{12}$ shows that in the experimental group, the ordinal location of the posttest scores is higher (less antisemitic) than the location of the pretest scores, while from $\hat{L}'_{34}$ it is concluded that this is the other way around in the control group. However, these $\hat{L}'$ values are rather small and their difference is not very big. To see that this difference is small, note that L'_{12} is a log odds coefficient (see Chapter 1), as is L'_{34}. Their difference is then a log odds ratio. In that perspective, the value of .347 corresponds to a value of $\lambda = .347/4 = .087$, being the (effect coded) loglinear association parameter in a 2×2 with a log odds ratio equal to .347. This would be regarded as a very modest association. A formal test that the L''s are equal ($L'_{12} = L'_{34}$) yields $G^2 = 4.57$, $df = 1$ ($p = .032$, $X^2 = 4.46$), which is significant at the .05, but not at the .01 level.

Given the way L' is computed, and the ways ties are dealt with, similar but not identical results are obtained when comparing the differences in ordinal locations of the pre- and posttest scores — that is, comparing $\hat{L}'_{13}$ with $\hat{L}'_{24}$. This is in contrast with the use of the arithmetic means of the response distributions as a location measure; in that case, the difference between the mean net changes is mathematically equivalent to the difference between the mean pretest and posttest differences. The $\hat{L}'$ values are

$$\hat{L}'_{13} - \hat{L}'_{24} = -.204 - (-.538) = .334.$$

At both points in time, the experimental group is less antisemitic than the control group, but the posttest differences are larger than the pretest differences. The test of $L'_{13} = L'_{24}$ now yields $G^2 = 4.53$, $df = 1$ ($p = .037$, $X^2 = 4.25$), again a borderline result.

Altogether, the conclusions from these marginal, unconditional analysis are that there is not much reason to believe that the movie did really diminish the degree of antisemitism. The results of the logit analyses were unambiguously negative and they are not very clear for L'. At the very best, the analyses by means of L' suggest the possibility of a small effect on the ordinal locations.

To determine the effect of the movie on the antisemitic attitude, the data in Table 5.8 can also be analyzed by means of a conditional approach. This will be done below, mainly to illustrate the above mentioned controversies about how to analyse the data from a nonequivalent control-group design. In the *conditional analysis* of the nonequivalent control-group design, the effects of the movie X on the posttest scores B are investigated given (conditional upon) one's initial score on the pretest A. Table 5.10 derived from Table 5.8 contains the pertinent conditional probabilities $\pi^{B|AX}_{j\,i\,k}$ of B given A and X.

The hypothesis that there is no effect of X now amounts to

$$\pi^{B|AX}_{j\,i\,1} = \pi^{B|AX}_{j\,i\,2} \tag{5.4}$$

for all values of i and j. Marginal modeling methods can be used to investigate this no-effects hypothesis by estimating the entries in Table 5.8 under the restrictions in Eq. 5.4. However, no dependent observations are now involved and a model with these restrictions is identical to a standard loglinear model $\{AX, AB\}$ for the joint frequencies in the normal table XAB. Therefore, the validity of Eq. 5.4 can be tested by means of an ordinary loglinear analysis in which model $\{AX, AB\}$ is applied to

Table 5.10. Conditional response distributions for posttest scores given pretest scores and group membership; see Table 5.8

$$\pi_{j\,i\,k}^{B|AX}$$

A X		1	2	3	Total
1	1	.625	.188	.188	1
	2	.697	.152	.152	1
2	1	.154	.462	.385	1
	2	.447	.197	.355	1
3	1	.053	.088	.860	1
	2	.139	.162	.699	1

Table 5.8. The test results are $G^2 = 17.024$, $df = 6$ ($p = .009$, $X^2 = 16.088$), and model $\{AX,AB\}$ has to be rejected. The logical next step is to formulate a logit model, including an effect of X on B: model $\{AX,XB,AB\}$. In this loglinear 'covariance model,' X is allowed to have a direct effect on B, controlling for A. This model can be accepted with $G^2 = 7.905$, $df = 4$ ($p = .095$, $X^2 = 7.822$). It appears not to be necessary to include a three-variable interaction term and estimate model $\{AXB\}$. The effects of X on B in model $\{AX,XB,AB\}$ are estimated as

$$\hat{\lambda}_{*11}^{AXB} = -.266, \quad \hat{\lambda}_{*12}^{AXB} = .101, \quad \hat{\lambda}_{*13}^{AXB} = .165.$$

These estimates show that there is a small, but significant effect of seeing the movie towards diminishing the degree of antisemitism.

The outcomes of this conditional analysis are different from the unconditional analysis in the sense that now there is more clear evidence of a small effect from seeing the movie. The literature cited above deals extensively with these and other differences between the conditional and unconditional approach. Formulated in a very general way, in the marginal approach no effect means equal net changes in the marginal distributions for the experimental and the control group or, in other words, equal pre- and posttest differences between the experimental and the control group. In the conditional approach, no effect means equal conditional distributions on the posttest for experimental and control group, conditional on the pretest scores.

5.2.3 A Truly Experimental Design

Not only the analyses of quasi-experimental designs benefit from the development and application of marginal models, but also with truly experimental data, marginal modeling offers many advantages and possibilities. The data that will be used in this section to illustrate this are from an experiment designed for the investigation of the effectiveness of a particular expert system intervention to convince people to quit smoking. The data were collected by a group of researchers at the Cancer

Prevention Research Center of the University of Rhode Island (U.S.) (Prochaska, Velicer, Fava, Rossi, & Tosh, 2001). For all 4,144 smoking subjects, information was collected on their smoking habits and their attitudes towards smoking. This was done immediately at the start of the study, providing a baseline measurement, and then at the sixth, twelfth, eighteenth, and twenty-fourth month. The sample of 4,144 smokers was randomly assigned to either the control or the experimental condition. The 1358 participants in the experimental group received a series of specific instructions on where they stood in the smoking cessation process and what they should do next to improve their chances of quitting. These instructions were sent to them by mail at the start of the study, and in the third and the sixth month. The 2,786 participants in the control condition did not get any such instructions (for more details, see Prochaska et al., 2001).

The central dependent variable in this project is a stage variable describing five different stages of the smoking cessation process (Velicer et al., 1995): 1 = pre-contemplation or smokers who are not considering quitting; 2 = contemplation or smokers who are thinking about quitting in the next six months; 3 = preparation or smokers who are planning to quit in the next 30 days and have already made a recent attempt to stop; 4 = action or people who have stopped smoking during less than six months; 5 = maintenance or people who have quit smoking for more than six months. Because at the start of the study all subjects were smoking, only the first three stages could occur at the baseline measurement.

All this gives rise to an enormous table, way too large to present here. But as usual, the models and restrictions will be denoted in terms of a marginal table (Table 5.11), which essentially contains the (dependent) marginal distributions of the stage variable at the four moments of measurement, specified according to the group one belongs to and the initial position at the beginning of the experiment. More precisely, Table 5.11 will be referred to as table $TXBR$, where T is the timing of the interviews ($T = 1$ at 6 months, $T = 2$ at 12 months, $T = 3$ at 18 months, and $T = 4$ at 24 months); X is the experimental variable ($X = 1$ for the experimental group and $X = 2$ for the control group); B represents the baseline (pretest) measurement of the stage variable ($B = 1$ for Precontemplation, $B = 2$ for Contemplation, and $B = 3$ for Preparation); and R is the (repeated) measurement of the stage variable ($R = 1$ for Precontemplation, $R = 2$ for Contemplation, $R = 3$ for Preparation, $R = 4$ for Action, and $R = 5$ for Maintenance). Table 5.11 then contains the marginal distributions of R within the combined categories of T, X, and B and, in its last column, the number of missing values on R.

The experimental intervention is supposed to bring people to stop smoking. One way to investigate such an effect is to see whether indeed the marginal distributions of R in Table 5.11 show a change in the expected direction. As a first step, it has to be investigated whether there is any (net) change at all. One has to be sure that the variations in the marginal distributions of R in Table 5.11 are not just random sampling fluctuations. The null hypothesis that all conditional marginal distributions in Table 5.11 are identical in the population, i.e.,

$$\pi_{\ell}^{R|TXB}{}_{i\,j\,k} = \pi_{\ell}^{R},$$

Table 5.11. Marginal distributions of *Stages of change R* at different time points *T* for groups *X* and baseline levels *B*. Data provided by W. F. Velicer, J. O. Prochasky and B. B. Hoeppner of the Cancer Prevention Research Centre, University of Rhode Island, Kingston, R.I.

			Stage of Change R					
Time *T*	Group *X*	Base *B*	1	2	3	4	5	*Missing*
1	1	1	301	84	22	12	11	135
		2	127	223	66	31	20	98
		3	36	56	56	20	10	50
	2	1	741	185	28	51	5	170
		2	297	462	114	70	4	156
		3	90	195	102	39	7	70
2	1	1	224	75	25	24	15	202
		2	96	155	51	45	41	177
		3	31	51	35	25	13	73
	2	1	572	159	40	63	22	324
		2	220	381	99	76	56	271
		3	93	115	84	55	26	130
3	1	1	196	74	16	20	18	241
		2	84	128	36	35	55	227
		3	26	28	40	24	22	88
	2	1	477	156	35	43	46	423
		2	217	302	87	63	70	364
		3	73	119	81	41	44	145
4	1	1	188	64	20	26	33	234
		2	86	113	41	34	66	225
		3	30	26	29	17	29	97
	2	1	436	145	43	43	55	458
		2	193	269	95	67	80	399
		3	85	85	70	51	52	160

is equivalent to the hypothesis that logit model $\{TXB,R\}$ is true in the population for table *TXBR*. Luckily, model $\{TXB,R\}$ can be rejected: $G^2 = 1533.8$, $df = 92$ ($p = .000$, $X^2 = 1425.5$). Obviously, some change is going on. Exactly what changes take place and how these are related to the experimental intervention can be investigated in several different ways, given the design of the experiment and taking the nature of the central dependent stage variable into account. Just two possible approaches will be presented here. Later on, marginal loglinear analyses will be carried out on the proportion of people that quit smoking ($R = 4, 5$). But first, marginal models will be introduced that are very similar to (M)ANOVA's for repeated measurements. In these marginal analyses, the dependent stage variable R will be treated as a categorical interval-level variable. Compared to the standard (M)ANOVA approach, the categorical marginal-modeling approach advocated here makes less stringent statistical assumptions. For example, in a repeated-measures (M)ANOVA, it is assumed that the dependent variable (R) is normally distributed and quite restrictive assumptions about the error covariance matrix must be true for the standard F-tests to be valid. In the analysis proposed here, none of these assumptions have to be satisfied. The only necessary assumption pertains to the way in which the data were collected: the sampling of the respondents has taken place according to a (product-)multinomial or Poisson sampling scheme, which is a not very demanding assumption. More about (M)ANOVA and the comparison with marginal modeling for categorical data can be found in Chapter 7.

Marginal Models for the Mean Response

Because this experiment is a truly randomized repeated measurement design, in principle, unconfounded estimates of the effects of X on R can be obtained by comparing the mean responses between the control and the experimental group at each point in time. For the computation of the means, the categories of R were given equidistant integer scores 1 through 5. Because this assignment of scores is somewhat arbitrary given the ordinal nature of R, a few other scoring patterns were tried out, such as $1, 2, 3, 5, 6$, emphasizing the difference between the first three categories (still smoking) and the last two (stopped smoking). But no serious differences with the equidistant scoring pattern were found. An alternative might have been to use the ordinal location parameters L or L', but this was not investigated.

First, the relevant means were calculated from the entries in Table 5.11, summed over B ($\pi_{\ell\,i\,j}^{R|TX} = \pi_{\ell i\,j+}^{RTXB} \big/ \pi_{+i\,j+}^{RTXB}$), because the initial baseline scores are not expected to be different for the two groups with a successful randomization. These means are reported in Table 5.12. From inspection of this table, it is seen that the means of R systematically increase over time in both groups in more or less the same way, indicating a similar tendency towards nonsmoking in the experimental and control group. Further, the means in the experimental group are about .15 higher than in the control group at all points in time. At a first descriptive sight, the expert system intervention does seem to have an impact on smoking cessation, perhaps not a dramatically big one, but from an applied perspective — convincing people to stop smoking — an important one.

A more formal analysis may start from an increase of the means over time (every six months) as a simple function of time and group. This simple model for the means only contains the additive main effects of time of observation (T) and the intervention variable X on the response stage variable R,

$$\mu_{t\,g}^{TX} = \upsilon + \alpha_t^T + \beta_g^X,$$

Table 5.12. Mean *Stage of change R* as a function of Group X and Time T. Based on Table 5.11; see also text

	Time T			
Group X	1	2	3	4
1	1.934	2.171	2.286	2.404
2	1.784	2.018	2.113	2.213

with $\mu_{t\,g}^{TX}$ denoting the mean response for $T = t$ and $X = g$, υ the overall effect, α_t^T the effect of T, and β_g^X the effect of X and with identifying restrictions $\sum_t \alpha_t^T = \sum_g \beta_g^X = 0$. When fitted, using the marginal-modeling approach described in Chapter 3, this main-effects model fits the data almost perfectly: $G^2 = 0.57, df = 3$ ($p = .904$, $X^2 = 0.57$). Therefore, there is no need to introduce interaction terms $\delta_{t\,g}^{TX}$ for the interaction effects of T and X on R: the effects of the intervention X are the same at each point in time. Further, the main effects β_g^X of X are significant. This follows from the parameter estimates and their standard errors reported below, and from the fact that the model with no effects of X, i.e., the above main-effects model but with $\beta_g^X = 0$ for $g = 1, 2$ has to be rejected: $G^2 = 21.35, df = 4$ ($p = .000$, $X^2 = 21.80$). The estimated effects of T and X on R are (with the standard errors between parentheses)

$$\hat{\alpha}_t^T = \begin{pmatrix} -.253\ (.015) \\ .018\ (.013) \\ .082\ (.014) \\ .189\ (.014) \end{pmatrix}$$

and

$$\hat{\beta}_g^X = \begin{pmatrix} .082\ (.017) \\ -.082\ (.017) \end{pmatrix} .$$

The formal analysis leads essentially to the same conclusions as the descriptive analysis based on the observed outcomes in Table 5.12. The intervention has a statistically significant effect in the intended direction causing an estimated mean difference of .164 between the experimental and the control group — a difference that is the same at each point in time. Further, there is a clear and significant change over time towards giving up smoking: either the longer participation in the experiment as such has a positive effect on smoking cessation, or $\hat{\alpha}_t^T$ reflects a societal trend towards nonsmoking during these months. The fact that the decrease in smoking is largest in the beginning of the experiment points perhaps in the direction of the first option.

There is, however, one problem with the above analyses. Although this study uses a randomized design, the missing data may have destroyed the random character of the group assignment. It seems reasonable to try to correct for the possible bias caused by the missing data by conditioning on the pretest scores — that is, on

Table 5.13. Mean *Stage of change R* as a function of Group X, Baseline B, and Time T; see Table 5.11

Group X Base B		Time T 1	2	3	4
1	1	1.48	1.71	1.73	1.95
	2	2.13	2.43	2.55	2.65
	3	2.51	2.60	2.91	2.92
2	1	1.41	1.60	1.71	1.80
	2	1.97	2.24	2.28	2.39
	3	2.26	2.48	2.62	2.70

the baseline variable B. If the missing data is systematically related to the outcome variable, conditioning on B might at least mitigate the possible biasing effects of the missing data. Moreover, it might be expected that controlling for B enhances the precision of the estimates of the effects of X on R (although this turned out hardly to be the case). Table 5.13 contains the mean scores of R for the different combinations of levels of T, X, and B, based on the original entries in Table 5.11.

The simple main-effects model for the means in Table 5.13 takes the form

$$\mu_{t\,g\,b}^{TXB} = \upsilon + \alpha_t^T + \beta_g^X + \gamma_b^B ,$$

with the additional identifying restriction $\sum_b \gamma_b^B = 0$. This main-effects model fits the data very well with $G^2 = 17.18$, $df = 17$ ($p = .442$, $X^2 = 16.94$). Introducing an extra parameter $\delta_{g\,b}^{XB}$ to capture possible interaction effects of X and T on R does not improve the fit significantly: $G^2 = 17.18 - 15.76, = 1.41$, $df = 17 - 14 = 3$ ($p = .702$). Further, all three main effects on R are significant. The hypothesis that T has no significant main effect ($\alpha_t^T = 0$ for all t) has to be rejected with $G^2 = 329.22$, $df = 20$ ($p = .000$, $X^2 = 306.17$). The hypothesis that the baseline level B has no effect ($\gamma_b^B = 0$ for all b) has also to be rejected: $G^2 = 462.43$, $df = 19$ ($p = .000$, $X^2 = 368.03$). Finally, and most importantly, the hypothesis that the experimental intervention X has no effect ($\beta_g^X = 0$ for all g) must be rejected with $G^2 = 37.64$, $df = 18$ ($p = .004$, $X^2 = 37.95$) and, conditionally, compared to the main-effects model: $G^2 = 37.64 - 17.18 = 20.46$, $df = 18 - 17 = 1$ ($p = .000$). The estimates of the model parameters with the standard errors between parentheses are

$$\hat{\alpha}_t^T = \begin{pmatrix} -.246\ (.014) \\ -.015\ (.013) \\ .075\ (.014) \\ .186\ (.016) \end{pmatrix} ,$$

$$\hat{\beta}_g^X = \begin{pmatrix} .071\ (.016) \\ -.071\ (.016) \end{pmatrix} ,$$

and

$$\hat{\gamma}_b^B = \begin{pmatrix} -.496 \ (.019) \\ .099 \ (.020) \\ .397 \ (.016) \end{pmatrix}.$$

The introduction of the baseline measurement B as a covariate does not seriously alter the conclusions regarding the effects of time ($\hat{\alpha}_t^T$) and the intervention ($\hat{\beta}_g^X$). In that sense, the missing data effects as far as those captured through B did not distort our effect estimates. Parameter estimates $\hat{\gamma}_b^B$ reflect the effect of the initial attitudes towards smoking. B has the biggest effect on R. In that sense, the respondents are remarkably consistent over time (not surprisingly).

The analogy of the above categorical marginal analyses of means with the analysis of covariance can be taken further by computing measures that are akin to proportions of unexplained and explained variance on the basis of the observed and estimated data. Although in categorical data analysis the 'error terms' are not independent of the predicted data and some of the usual 'additivity' properties may no longer be valid, meaningful analogous measures of explained variance have been developed for categorical dependent variables, especially in the context of logistic regression (Long, 1997; Hosmer & Lemeshow, 2000) This issue will not be pursued further here.

A Marginal Logit Model for the Probability of Quitting Smoking

The main aim of the intervention study was to stimulate the participants to quit smoking. Therefore, the ultimate test of the success of the intervention is the number of people that eventually stopped smoking — more precisely, the probability that a participant ends in one of the two last stages ($R = 4$ (Action) or $R = 5$ (Maintenance)) at the end of the study. Table 5.14 contains those probabilities for the different levels of X, B, and T. For example, the probability of 'stopped smoking' for a respondent in $TXB = 111$ is .053, and the probability to continue smoking is $1 - .053 = .947$; for $TXB = 423$ these probabilities are .300 and .700, respectively.

A visual inspection of Table 5.14 suggests that all explanatory variables T, X, and B have an effect on the probability of quitting smoking. Treating Table 5.14 as if it were an ordinary $TXBS$ table, where S is a dichotomous variable having stopped smoking ($S = 1$) or not ($S = 2$), the main effects of the explanatory variables on S are estimated in logit model $\{TXB, TS, XS, BS\}$. This model fits the data very well ($G^2 = 12.16, df = 17, p = .790, X^2 = 12.05$), and no further interaction effects have to be added. The estimates of the main effects (and the standard errors) are (using the notation of the previous subsection for ease of comparison rather than our standard notation (with $\hat{\alpha}_t^T$ equal to $\hat{\lambda}_{t**1}^{TXBS}$; $\hat{\beta}_g^X$ equal to $\hat{\lambda}_{*g*1}^{TXBS}$; and $\hat{\gamma}_b^B$ equal to $\hat{\lambda}_{**b1}^{TXBS}$)),

$$\hat{\alpha}_t^T = \begin{pmatrix} -.355 \ (.023) \\ .021 \ (.016) \\ .110 \ (.017) \\ .224 \ (.017) \end{pmatrix},$$

Table 5.14. Probability of being in stages Action or Maintenance ($R = 4$ or 5) as a function of Group X, Baseline B, and Time T; from Table 5.11

		Time T			
Group X	Base B	1	2	3	4
1	1	.053	.107	.117	.178
	2	.109	.222	.266	.294
	3	.169	.245	.329	.351
2	1	.055	.099	.118	.136
	2	.078	.159	.180	.209
	3	.106	.217	.237	.300

$$\hat{\beta}_g^X = \begin{pmatrix} .080\ (.021) \\ -.080\ (.021) \end{pmatrix},$$

and

$$\hat{\gamma}_b^B = \begin{pmatrix} -.260\ (.028) \\ .038\ (.026) \\ .222\ (.029) \end{pmatrix}.$$

The pattern of the estimates is much the same as found for the mean responses, although now the effect of T is as big (or a bit larger) than the effect of B. Testing the hypothesis that T has no effect on the probability of having stopped smoking yields $G^2 = 304.54$, $df = 20$ ($p = .000$, $X^2 = 285.16$). Similarly, testing that B has no effect leads to $G^2 = 105.40$, $df = 19$ ($p = .000$, $X^2 = 102.36$). However, the crucial test that X has no effect unexpectedly yields a nonsignificant result: $G^2 = 27.62$, $df = 18$ ($p = 0.068$, $X^2 = 28.57$). On the other hand, the more powerful conditional tests of X having no effect against the main-effects logit model, yields $G^2 = 27.62 - 12.16 = 15.46$, $df = 18 - 17 = 1$ ($p = .000$), which is a clearly significant result. The final conclusion must be that the intervention brought people to stop smoking, although the effect is not very large compared to the effects of T. From the estimates $\hat{\beta}_g^X$, it can be computed that the odds of stopping smoking rather than not stopping smoking are 1.38 times larger for the experimental group than for the control group. From an applied perspective, this may might well be an important finding.

To repeat the beginning of this chapter, marginal modeling is certainly not to be opposed to causal modeling. On the contrary, marginal modeling offers the possibility of utilizing normal SEMs even with data that contain dependent observations by design but where the researcher is not interested in the dependency, as such. By means of marginal modeling, the usual models occurring in comparative research can be handled even when comparing groups with matched observations. When necessary, additional marginal restrictions can be added to SEMs that are not feasible with the standard algorithms.

Regarding quasi-experimental and experimental designs with repeated measures, marginal modeling makes analyses possible for categorical dependent variables that are very similar to the ones for continuous dependent variables. Moreover, even with an interval-level dependent variable with five or more categories, regarded usually as enough to treat them as continuous, the marginal-modeling approach for categorical variables may have advantages above traditional (M)ANOVA models because of the weaker assumptions about the sampling distributions and the absence of strict assumptions about the nature of the dependencies in the data.

6

Marginal Modeling with Latent Variables

So far in this book, all analyses have been conducted at the manifest, observed data
level with the implicit assumption that the observed data resulted from perfectly
valid and reliable measurements of the intended concepts. However, observed data
are often only indirect measures of what researchers want to measure and are often
contaminated by measurement errors. Latent variable models provide very powerful
tools for discovering many kinds of systematic and unsystematic measurement errors
in the observed data, and offer many ways to correct for them. For categorical data —
where the measurement errors are usually denoted as misclassifications — the most
appropriate and flexible latent variable model is the latent class model (Lazarsfeld &
Henry, 1968; Goodman, 1974a; Goodman, 1974b; Haberman, 1979; Clogg, 1981b;
Hagenaars & McCutcheon, 2002). Combining latent class models with marginal restrictions makes it, in principle, possible to conduct the marginal analyses at the
latent level. Important pioneering work regarding maximum likelihood estimation
for marginal latent class models has been conducted by Becker and Yang (1998).

Because not all readers may be familiar with latent class models, the basic latent
class model will be introduced in the next section, mainly in the form of a loglinear
model with latent variables. Thereafter, in Section 6.2, a simple marginal homogeneity model for the joint distribution of two latent variables will be presented. Further,
it will be shown that application of the marginal-modeling approach advocated in
this book makes it possible to handle latent class models that lie outside the loglinear framework (Section 6.3). Next, two examples of marginal categorical SEMs with
latent variables will be presented in Section 6.4. The explanation of the appropriate ML estimation procedures for latent marginal modeling is presented in the final
Section 6.5.

6.1 Latent Class Models

Only the very basic principles of latent class analysis will be dealt with here, and only
from the viewpoint of discovering and correcting the unreliability of measurements.

W. Bergsma et al., *Marginal Models: For Dependent, Clustered, and Longitudinal
Categorical Data*, Statistics for Social and Behavioral Sciences,
DOI: 10.1007/978-0-387-09610-0_6, © Springer Science+Business Media, LLC 2009

Table 6.1. *Party preference* and *Candidate preference for Prime Minister*, February-March 1977. Variables A and B are *Party preference* at $T1$ and $T2$, respectively; variables C and D are *Candidate preference* at $T1$ and $T2$, respectively. All variables have categories 1 = Christian-Democrats, 2 = Left wing, and 3 = Other. Source: Hagenaars 1990, p.171

C		1			2			3			Total
D		1	2	3	1	2	3	1	2	3	
A	B										
1	1	84	9	23	6	13	7	24	8	68	242
	2	0	1	0	0	8	1	2	2	3	17
	3	3	1	2	0	2	3	2	3	9	25
2	1	1	1	0	1	2	2	1	0	1	9
	2	2	4	0	1	293	6	1	22	21	350
	3	1	0	0	1	8	7	0	0	9	26
3	1	6	1	1	4	5	0	9	1	16	43
	2	0	1	1	0	31	0	2	9	7	51
	3	14	1	15	3	48	23	12	21	200	337
Total		111	19	42	16	410	49	53	66	334	1100

For an overview of the many other applications of latent class analysis, see Hagenaars and McCutcheon (2002). The data in Table 6.1 will be used for illustration. These data come from a Dutch panel study (T_1 – February 1977, T_2 – March 1977) and concern the questions for which party the respondent intends to vote (variables A and B, respectively) and which candidate the respondent prefers to become the next Prime Minister (C and D). The data have been analyzed before (Hagenaars, 1986, 1988, 1990), and more information on the panel study and the outcomes may be obtained from these references.

If it is assumed that *Party preference* and *Candidate preference* are both indicators of the underlying concept *Political orientation*, and if it is also believed that this underlying variable is stable over time during the interview period, the relationships among the four variables A through D can be explained by the latent class model in Fig. 6.1.

In the basic latent class model in Fig. 6.1, the symbol X represents an unobserved categorical latent variable that is assumed here to have three latent categories or classes in agreement with the categories of the manifest variables. Observed variables A through D are considered to be the indicators of X. In general, they will not be completely reliable and valid indicators and the relationship between the latent variable and an indicator is not perfect. There are, in this sense, misclassifications. Due to all kinds of accidental circumstances, some people who have a truly left-wing *Political orientation* ($X = 2$) will nevertheless say they intend to vote for the Christian-Democratic Parties ($A = 1$), etc. An important characteristic of this basic latent class model is that there are no direct relations among the indicators. This is the local independence assumption: all indicators are conditionally independent of each other, given the score on X. For example, for indicators A and B,

$$\pi_{i\,j\,t}^{AB|X} = \pi_{i\;t}^{A|X}\,\pi_{j\;t}^{B|X} \; .$$

In other words, all that A through D have in common with each other is completely due to their separate, direct dependencies on X, i.e., to their being an indicator of X. This crucial local independence assumption also implies that the misclassifications of the several indicators are independent of each other.

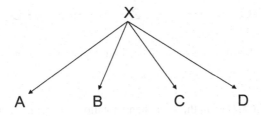

Fig. 6.1. The basic latent class model — the manifest variables A,B,C and D are independent given the value of the latent variable X

The assumed existence of the latent variable X with T latent classes means that the population can be divided into T mutually exclusive and exhaustive categories, and that

$$\sum_t \pi_{t\;i\;jkl}^{XABCD} = \pi_{i\;jkl}^{ABCD} \; .$$

Because of the local independence assumption, the joint probability for the latent and observed variables can be written as

$$\pi_{t\;i\;jkl}^{XABCD} = \pi_t^X\,\pi_{i\;t}^{A|X}\,\pi_{j\;t}^{B|X}\,\pi_{k\;t}^{C|X}\,\pi_{\ell\;t}^{D|X} \; . \tag{6.1}$$

Parameter π_t^X denotes the probability that an individual belongs to latent class $X = t$. The other parameters are conditional response probabilities, e.g., $\pi_{i\;t}^{A|X}$ represents the conditional probability that someone answers i on A, given that person is in latent class $X = t$.

The equivalent loglinear representation of this basic latent class model is (using the 'marginal notation')

$$\ln \pi_{t\;i\;jkl}^{XABCD} = \lambda_{*****}^{XABCD} + \lambda_{t\;****}^{XABCD} + \lambda_{*i\;***}^{XABCD} + \lambda_{**j**}^{XABCD} + \lambda_{***k*}^{XABCD} + \lambda_{****\ell}^{XABCD}$$
$$+ \lambda_{t\;i\;***}^{XABCD} + \lambda_{t\;*j**}^{XABCD} + \lambda_{t\;**k*}^{XABCD} + \lambda_{t\;***\ell}^{XABCD} \tag{6.2}$$

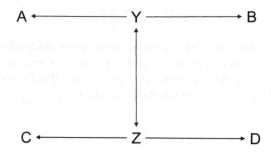

Fig. 6.2. The latent class model with two stable latent variables, *Party preference* and *Candidate preference*

or model $\{AX,BX,CX,DX\}$ in the short-hand notation for hierarchical models.

The model parameters in Eqs. 6.1 or 6.2 can be estimated in standard ways, e.g., using the EM algorithm. In general, because of the presence of the latent variable(s) in the loglinear model, identification of the parameters can be problematic (but not in this example). More on estimation procedures and identifiability of latent class models can be found in the references in the beginning of this chapter.

The model can be tested by means of chi-square tests by comparing the observed frequencies n_{ijkl}^{ABCD} with the estimated manifest frequencies $N\hat{\pi}_{+ijkl}^{XABCD}$, where N is the sample size. The number of degrees of freedom for testing identifiable models equals, as usual, the number of independent 'knowns' minus the number of nonredundant 'unknowns.' The independent 'knowns' are here the number of cells in the joint observed table $ABCD$ minus one: $3^4 - 1 = 80$; the non-edundant 'unknowns,' are the number of parameters to be estimated, taking into account that probabilities sum to 1 where appropriate or that the loglinear parameters are subjected to the usual (effect or dummy coding) identifying restrictions. Here, 26 parameters are to be estimated. Therefore, the number of degrees of freedom here is $80 - 26 = 54$. Applying this basic latent class model with one trichotomous latent variable to the data in Table 6.1 yields $G^2 = 362.71$, $df = 54$ ($p = .000$, $X^2 = 491.24$). The one latent variable model has to be rejected: it cannot be maintained that the preferences for political parties and for candidates for Prime Minister are the indicators of one underlying variable *Political orientation* that is stable over time.

Perhaps the stability restriction is the problem. It might be the case that *Party preference* and *Candidate preference* are the indicators of one underlying latent variable *Political orientation*, but that some respondents changed their underlying political position from February to March. Two trichotomous latent variables Y and Z

Table 6.2. Estimates of conditional probabilities for model $\{YZ, YA, YB, ZC, ZD\}$: Y is the latent *Party preference* (with indicators A, B), Z the latent *Candidate preference* (with indicators C, D); data in Table 6.1

| | | | $\hat{\pi}_{i\,r}^{A|Y}$ | | | $\hat{\pi}_{j\,r}^{B|Y}$ | | | $\hat{\pi}_{k\,s}^{C|Z}$ | | | $\hat{\pi}_{\ell\,s}^{D|Z}$ | | |
| | | | A | | | B | | | C | | | D | | |
Y	Z	$\hat{\pi}_{r\,s}^{YZ}$	1	2	3	1	2	3	1	2	3	1	2	3
1	1	.168	.87	.02	.11	.94	.02	.04	.68	.08	.24	.74	.08	.17
1	2	.015	.87	.02	.11	.94	.02	.04	.01	.91	.08	.00	.99	.01
1	3	.089	.87	.02	.11	.94	.02	.04	.04	.11	.85	.04	.07	.89
2	1	.003	.03	.87	.10	.00	.98	.02	.68	.08	.24	.74	.08	.17
2	2	.343	.03	.87	.10	.00	.98	.02	.01	.91	.08	.00	.99	.01
2	3	.027	.03	.87	.10	.00	.98	.02	.04	.11	.85	.04	.07	.89
3	1	.026	.04	.05	.91	.03	.03	.94	.68	.08	.24	.74	.08	.17
3	2	.054	.04	.05	.91	.03	.03	.94	.01	.91	.08	.00	.99	.01
3	3	.278	.04	.05	.91	.03	.03	.94	.04	.11	.85	.04	.07	.89

may then be assumed to exist, one for the true political orientation position at time 1 (Y), and one for the underlying position at time 2 (Z). In loglinear (shorthand) terms, the model becomes $\{YZ, YA, YC, ZB, ZD\}$. However, when applied to the data in Table 6.1, this two latent variable model does not perform remarkably better than the one latent variable model: $G^2 = 351.13$, $df = 48$ ($p = .000$, $X^2 = 501.02$). Stability of the latent variable does not seem to be the main problem.

An alternative two-latent variable model is depicted in Fig. 6.2. In this model, it is no longer assumed that *Party preference* and *Candidate preference* are the indicators of one underlying concept. Rather, a stable trichotomous latent variable *Party preference* (Y) is postulated with indicators A and B, as well as a stable trichotomous latent variable *Candidate preference* (Z) with indicators C and D. This model, represented in loglinear terms as $\{YZ, YA, YB, ZC, ZD\}$, fits the data in Table 6.1 much better than the previous two models: $G^2 = 84.74$, $df = 48$ ($p = .000$, $X^2 = 94.33$). Obviously, the fit is still not satisfactory according to a p-level of .05 or .01. Nevertheless, this model will be explained further in this section and used in the following one to illustrate latent marginal homogeneity. In Section 6.4, model $\{YZ, YA, YB, ZC, ZD\}$ will be extended to arrive at a well-fitting model. The representation of model $\{YZ, YA, YB, ZC, ZD\}$ in terms of (conditional response) probabilities is as follows:

$$\pi_{r s i j k l}^{YZABCD} = \pi_{r s}^{YZ}\,\pi_{i\,r}^{A|Y}\,\pi_{j\,r}^{B|Y}\,\pi_{k\,s}^{C|Z}\,\pi_{\ell\,s}^{D|Z}\;. \tag{6.3}$$

The estimates of the probabilities from Eq. 6.3 are presented in Table 6.2. It is left to the interested reader to study Table 6.2 in detail. Just a few striking features will be pointed out here, to give a general feeling of how to handle and interpret latent class models. First of all, as is true in all latent variable models, the latent variables get their meaning and substantive interpretations from their relationships with their indicators. Information about the association between the latent variables and the indicators is provided by the conditional response probabilities. From their patterns, it is clear that there exists here a strong positive relationship between each latent variable and its indicators, and that the meanings of the categories of the latent variables can be interpreted in agreement with the meanings of the categories of the indicators. Association measures — such as loglinear parameters — describing the relationships between the latent variable and its indicators can be seen as a kind of reliability estimate of the indicators by means of which the reliabilities of categorical indicators can be compared.

Once the meanings of the latent variables and their categories is established, the conditional response probabilities can often be interpreted in terms of misclassification probabilities — that is, in terms of 'correct' or 'incorrect' answers in agreement with the latent class someone belongs to. Striking in the outcomes in Table 6.2, and often seen in panel analyses is that the probabilities of a correct answer are always somewhat larger in the second wave than in the first wave: $\hat{\pi}_{i}^{A|Y} < \hat{\pi}_{i}^{B|Y}$ and $\hat{\pi}_{i}^{C|Z} < \hat{\pi}_{i}^{D|Z}$. Further, all conditional probabilities of a correct answer for all indicators are around .90 (or higher), except for the conditional probability of choosing the Christian-Democratic candidate ($C = 1, D = 1$), given one is a true supporter of a Christian-Democratic candidate ($Z = 1$): $\hat{\pi}_{1\,1}^{C|Z} = .68$ and $\hat{\pi}_{1\,1}^{D|Z} = .74$. This probably reflects the fact that the Christian-Democratic candidate, in contrast to the other candidates, was a newcomer as leader of the Christian-Democrats. Even his ('true') supporters were uncertain about him.

Finally, there are the parameter estimates $\hat{\pi}_{r\,s}^{YZ}$ representing the joint distribution of the two latent variables, i.e., the true *Party preference* (Y) and the true *Candidate preference* (Z). From these estimates, it is clear that the two latent variables are highly positively correlated. In terms of the loglinear representation, it turns out that the $\hat{\lambda}$-parameters on the main diagonal of table YZ have very large values: $\hat{\lambda}_{11}^{YZ} = 1.821$, $\hat{\lambda}_{22}^{YZ} = 2.204$ and $\hat{\lambda}_{33}^{YZ} = .799$. Compared to the corresponding parameter estimates on the manifest level, the latent relationship is much stronger than the manifest one. For example, for the marginal manifest relation $B - D$ (stronger than $A - C$), the corresponding loglinear parameters are: $\hat{\lambda}_{11}^{BD} = 1.223$, $\hat{\lambda}_{22}^{BD} = 1.644$, $\hat{\lambda}_{33}^{BD} = .501$. As with continuous data, the observed association is corrected for 'attenuation due to unreliability.' Although, as Alwin (2007, Chapter 11) rightfully remarked, a satisfactory operational definition of the reliability of a categorical variable that has similar properties to continuous data is hard to find, and the latent class model provides a way to correct the relationships at the latent level for unreliability in the form of independent classification errors in much the same vein as can be done for continuous data. This opens the possibility to apply marginal models adjusted for unreliability in the measurements.

6.2 Latent Marginal Homogeneity

A simple example of a research problem leading to a marginal model involving latent variables is whether or not the latent table YZ discussed at the end of the previous section has homogenous marginals. Once this simple example is understood, the application of more complicated marginal models for latent variables follows easily.

The research problem concerns the question of whether or not the overall support for a political leader in the population is the same as the support for the corresponding party. Such a question may be of theoretical interest, but is certainly of practical political importance. A first answer to this question is possible by carrying out the analysis at the manifest level only. The manifest data in Table 6.1 can be used to find the answer for these particular elections. Under the hypothesis that the overall support for parties and leaders is the same at both points in time, the cell entries $\pi_{ijk\ell}^{ABCD}$ are estimated subject to the simultaneous marginal homogeneity restrictions that the marginal distributions of A and C are homogeneous, as well as the marginal distributions of B and D:

$$\pi_{i+++}^{ABCD} = \pi_{++i+}^{ABCD}$$
$$\pi_{+i++}^{ABCD} = \pi_{+++i}^{ABCD} .$$
(6.4)

The test outcomes for the restrictions in Eq. 6.4 applied to Table 6.1 are $G^2 = 111.044$, $df = 4$ ($p = .000$, $X^2 = 94.566$). According to this manifest data analysis, it is clear that the hypothesis that *Party preference* and *Candidate preference* have the same marginal distribution, both at time 1 and at time 2, must be rejected. From the adjusted marginal residual frequencies (not reported here), it is seen that especially the Christian-Democratic candidate for Prime Minister was less popular and the left-wing candidate more popular than their respective parties at both points in time.

However, possible misclassifications in the data might influence this marginal homogeneity test on the manifest level. It is, for example, possible that there exists true (latent) marginal homogeneity between *Candidate preference* and *Party preference*, but that different patterns of misclassifications for the two characteristics destroy the marginal homogeneity at the manifest level. If model $\{YZ, YA, YB, ZC, ZD\}$, defined by Eq. 6.3, can be accepted as valid for the data in Table 6.1 (at least for the sake of illustration), then $\hat{\pi}_{rs}^{YZ}$ represents the estimated 'true' joint distribution of *Party preference* and *Candidate preference*, adjusted for independent misclassifications. The test of latent marginal homogeneity,

$$\pi_{i+}^{YZ} = \pi_{+i}^{YZ} ,$$
(6.5)

within model $\{YZ, YA, YB, ZC, ZD\}$ is then a test for the 'true' marginal homogeneity of *Party preference* and *Candidate preference*.

When carrying out this latent MH test, it is important to make sure that the corresponding categories of the latent variables have the same substantive meaning, e.g., $Y = 1$ (Christian-Democrats) should refer to the same substantive category $Z = 1$ (Christian-Democrats). This note of caution is necessary because the order of the categories of the latent variables is not fixed in the estimation process and should always be checked by means of the relationships between the latent variables and their indicators. Choosing appropriate starting values may overcome this small, but easily overlooked problem, which, if overlooked, can have serious negative consequences.

As can be inferred from the outcomes in Table 6.2 for model $\{YZ, YA, YB, ZC, ZD\}$ without the MH restriction, the estimated latent marginal distributions of Y and Z differ from each other. For the successive response categories 1, 2, and 3, the estimated marginal probabilities $\hat{\pi}_{r+}^{YZ}$ (latent *Party preference*) are $(.269, .373, .358)$. For Z, latent *Candidate preference*, the estimated marginal probabilities $\hat{\pi}_{+r}^{YZ}$ are $(.197, .411, .392.)$. The largest difference exists for the Christian-Democratic Category 1: the Christian-Democratic candidate is less preferred than his Christian-Democratic party. The question is whether this is a statistically significant difference. Hagenaars (1986) tested the latent MH hypothesis in Eq. 6.5 by means of a conditional marginal homogeneity test applying latent (quasi-)symmetry models to the same data and using the same model $\{YZ, YA, YB, ZC, ZD\}$. The test outcomes were $G^2 = 21.46$, $df = 2$, $p = .000$. The substantive conclusions were in line with the above test outcomes of the manifest MH hypothesis in Eq. 6.4, also regarding the relative stronger or weaker support of particular candidates compared to their parties.

The test outcomes of the much more flexible marginal approach advocated here are $G^2 = 106.27$, $df = 46$ ($p = .000$, $X^2 = 114.16$). Obviously, model $\{YZ, YA, YB, ZC, ZD\}$ with the latent MH restraint in Eq. 6.5 has to be rejected for the data in Table 6.1. Further, because it was decided to accept model $\{YZ, YA, YB, ZC, ZD\}$ for the time being, the conditional test on the latent MH restriction within model $\{YZ, YA, YB, ZC, ZD\}$ yields $G_{MH}^2 = 106.27 - 84.74 = 21.53$, $df_{MH} = 48 - 46 = 2$, $p = .000$. Also, this conditional test clearly leads to a rejection of the latent MH restriction in Eq. 6.5. Even taking misclassifications into account, the 'true' distributions of *Party preference* and *Candidate preference* differ from each other in the way indicated above. As a final note, the conditional test statistic $G_{MH}^2 = 21.53$ is for all practical purposes the same as the conditional test outcome based on the (quasi-)symmetry models reported above, which was $G^2 = 21.46$. This is caused by the fact that the quasi-symmetry model fits the joint latent distribution YZ almost exactly.

6.3 Loglinear and Nonloglinear Latent Class Models: Equal Reliabilities

So far, a latent marginal model was defined for the relationships among the (two) latent variables. Other marginal models involving latent variables may concern the relationships between the latent variables and the manifest variables (or in particular, the indicators of the latent variables). For this, the associations between the latent variables and their indicators have to be parameterized. For obvious reasons, these associations will often be indicated here as reliabilities, when there is a one-to-one relationship between the categories of the latent variables and the indicators. The reliability can be expressed in terms of odds ratios, percentage differences, or agreement coefficients, among other things. Restrictions on these reliability coefficients is sometimes straightforward in the sense that the standard algorithms for latent class analysis developed by Goodman and Haberman (see references in the beginning of this chapter) can be used. These algorithms essentially define the latent class model

Table 6.3. *Party preference* and *Presidential candidate preference*, Erie County, Ohio, 1940 (August (T_1) and October (T_2)). Source: Lazarsfeld 1972, p. 392

C - Party preference (T_2)		1. Democrats		2. Republicans		
D - Candidate preference $(T2)$		1. Against	2. For	1. Against	2. For	Total
A - Party pref.-(T_1)	B - Candidate pref. (T_1)					
1. Democrats	1. Against Willkie	68	2	1	1	72
	2. For Willkie	11	12	0	1	24
2. Republicans	1. Against Willkie	1	0	23	11	35
	2. For Willkie	2	1	3	129	135
Total		82	15	27	142	266

in terms of a loglinear model with latent variables. Imposing all kinds of linear restrictions on the (log)odds ratios fits nicely within their approach. However, these standard algorithms cannot be used in general to impose restrictions on reliability coefficients like percentage differences or agreement coefficients. Imposing such restrictions is not a problem in terms of our marginal-modeling approach explained in Chapter 3, and the last section of this chapter.

For illustration, use will be made of a famous and often analyzed data set that concerns variables much like the ones in the previous sections of this chapter. These data come from the first systematic panel study on voting, conducted by Lazarsfeld and his associates in Erie County, Ohio in 1940 (Lazarsfeld et al., 1948; Lazarsfeld, 1972; Wiggins, 1973; Hagenaars, 1993). The data are presented in Table 6.3 and refer to the variables A – *Party preference at time 1* – August 1940 (1. Republican 2. Democrat), B – *Presidential Candidate preference at time 1* (1. for Willkie 2. against Willkie), C – *Party preference at time 2* – October 1940, and D – *Presidential Candidate preference at time 2*. Wendell Willkie was the (defeated) 1940 Republican presidential candidate running against the Democrat Franklin D. Roosevelt.

Hagenaars (1993) fit several loglinear latent class models to the data in Table 6.3 and found model $\{YZ,YA,YC,ZB,ZC\}$, where Y and Z are dichotomous latent variables, to fit well with $G^2 = 7.32$, $df = 4$ ($p = .120$, $X^2 = 11.53$). Note that this model is very similar to the latent *Party/Candidate preference* depicted in Fig. 6.2 (with an interchange of indicator labels). The estimated model parameters for model $\{YZ,YA,YC,ZB,ZC\}$ in terms of (conditional) probabilities are presented in Table 6.4.

6.3.1 Restrictions on Conditional Response Probabilities

If there is a one-to-one relationship between the categories of the latent variable and the indicators, the conditional response probabilities represent the probabilities of misclassification and, as such, reflect the (un)reliability of the indicators. From the seminal 1974 Goodman papers on, many authors have suggested ways to formulate and test hypotheses about the reliabilities of the indicators directly in terms of

Table 6.4. Estimates of conditional) probabilities for model $\{YZ,YA,YC,ZB,ZD\}$ applied to Table 6.3. Variable Y is the latent *Party preference* (with indicators A, C), variable Z the latent *Candidate preference* (with indicators B, D)

$Y = r$	$Z = s$	$\hat{\pi}_{rs}^{YZ}$	$\hat{\pi}_{1\,r}^{A\mid Y}$	$\hat{\pi}_{2\,r}^{A\mid Y}$	$\hat{\pi}_{1\,s}^{B\mid Z}$	$\hat{\pi}_{2\,s}^{B\mid Z}$	$\hat{\pi}_{1\,r}^{C\mid Y}$	$\hat{\pi}_{2\,r}^{C\mid Y}$	$\hat{\pi}_{1\,s}^{D\mid Z}$	$\hat{\pi}_{2\,s}^{D\mid Z}$
1	1	.315	.97	.03	.85	.15	.99	.01	.99	.01
1	2	.051	.97	.03	.08	.92	.99	.01	.00	1.00
2	1	.101	.01	.99	.85	.15	.00	1.00	.99	.01
2	2	.534	.01	.99	.08	.92	.00	1.00	.00	1.00

restrictions on the conditional response probabilities (see the general latent class references cited in the beginning of this chapter). Usually restrictions are imposed on the conditional probabilities of giving a 'correct' answer in agreement with the latent class someone is in. But it may also be interesting to constrain the probabilities of particular misclassifications and of giving 'wrong' answers. Of course, for dichotomous manifest and latent variables, restricting the misclassification is identical to restricting the probabilities of a correct answer, and vice versa.

In terms of the example of this section, the hypothesis that dichotomous indicator A — the manifest variable party preference at time 1 — is a perfect indicator of the true underlying party preference Y can be tested by setting the pertinent conditional response probabilities to one:

$$\pi_{1\,1}^{A\mid Y} = \pi_{2\,2}^{A\mid Y} = 1.$$

Each answer on A is completely in agreement with the true latent class the respondent belongs to and there are no misclassifications for this item.

When the indicators have more than two categories, it is possible to exclude particular kinds of misclassifications by setting particular conditional response categories to zero without automatically setting other response probabilities equal to one. For example, it might be ruled out in this way that a truly left person who belongs to the latent-left category expresses a manifest preference for an extreme right-wing party, or, to give a completely different example, that a truly very high-status person has a manifest occupation with a very low status (Hagenaars, 1990, p. 188; Clogg, 1981a).

Besides setting conditional response probabilities equal to a constant (0, 1, .5, etc.), it is also possible to make particular response probabilities equal to each other. For example, imposing restriction

$$\pi_{1\,1}^{A\mid Y} = \pi_{2\,2}^{A\mid Y}$$

can be used to test the hypothesis that A has the same reliability for both latent classes. The conditional response probability that a correct answer is given on A is the same for $Y = 1$ and $Y = 2$. A restriction such as

$$\pi_{i\ r}^{A|Y} = \pi_{i\ r}^{B|Z}$$

expresses the hypothesis that A and B are equally reliable indicators for their respective latent variables (for one or all latent classes).

All above restrictions on the (marginal) conditional response probabilities (in marginal tables AY, BZ, etc.) can be formulated within the loglinear model, i.e., in terms of linear restrictions on the pertinent loglinear parameters λ in model $\{YZ, YA, YC, ZB, ZD\}$, or in terms of structural zeroes in the table with estimated frequencies (see Hagenaars, 1990, p. 111 and Heinen, 1996, p. 53). Therefore, imposing such restrictions on the conditional response probabilities does not make the latent class model nonloglinear, and estimation can be achieved by means of the standard efficient algorithms proposed by Goodman (1974a), Goodman (1974b), and Haberman (1979). (A slight but important extension of this standard algorithm has been made by Mooijaart and Van der Heijden (1991); see also Vermunt, 1997a).

Restricting the reliabilities of the indicators by means of the conditional response probabilities makes intuitively much sense and provides clear and simple interpretations in terms of (equal) probabilities of misclassification. However, a serious disadvantage is that the conditional response probabilities are not only a function of the association between the latent variables and their indicators, but also of the (partial) one-variable distributions of the indicators. In loglinear terms, the conditional response probabilities are a function not only of the two-variable parameters $\lambda_{i\ r}^{AY}$, $\lambda_{j\ r}^{BY}$, etc., but also of the one variable marginal parameters λ_i^A, λ_j^B. Therefore, restrictions on the conditional response probabilities usually imply restrictions not only on the loglinear two-variable parameters, but simultaneously on the one-variable parameters. If item A is very popular (a majority agreeing) and item B very unpopular (the minority agreeing), a restriction like $\pi_{i\ r}^{A|Y} = \pi_{i\ r}^{B|Y}$ will be rejected, despite the fact that the association of the latent variable Y and the indicators A and B may be the same. For many purposes, researchers want to separate strength of association (or discrimination) from popularity (or difficulty). In the next three subsections, it will be shown how this can be done using our marginal-modeling approach, using different measures of association.

6.3.2 Restrictions on Odds Ratios

Odds ratios are measures of association that have the nice property of being variation independent of the popularity, i.e., of the marginal distributions of the indicators (Rudas, 1997; Bergsma & Rudas, 2002c). Therefore, reliability restrictions can be formulated purely in terms of 'discrimination' rather than 'discrimination plus popularity' restrictions by means of restrictions on the odds ratios, i.e., by means of restricted two-variable parameters such as $\lambda_{i\ r}^{AY}$ in the loglinear representation of the latent class model. The standard Goodman and Haberman algorithms and standard software (e.g., Vermunt's LEM or Latent GOLD (Vermunt & Magidson, 2004) or Muthén's Mplus (Muthén & Muthén, 2006)) enable the researcher to implement these restrictions (and also the above restrictions on the conditional response probabilities) in a simple and straightforward manner. Although the outcomes in this

Table 6.5. Tests of fit for reliability models based on different measures of reliability for the data of Table 6.3, model $\{YZ,YA,YC,ZB,ZD\}$; for further explanations of models, see text

Reliability Model	df	G^2 (p-value) using		
		odds ratio α	epsilon ε	kappa κ
No restriction on reliabilities	4	7.32 (.120)	7.32 (.120)	7.32 (.120)
Equal reliabilities	7	25.16 (.001)	32.98 (.000)	35.51 (.000)
Equal change	5	7.64 (.177)	15.14 (.005)	15.59 (.008)
No change	6	20.01 (.003)	19.82 (.003)	20.06 (.003)
No change party pref.	5	8.31 (.140)	8.45 (.133)	8.40 (.135)
No change candidate pref.	5	19.98 (.001)	19.68 (.001)	19.98 (.001)

subsection have been obtained using our marginal-modeling algorithm, exactly the same results might have been obtained, for example, by means of LEM.

In terms of the data in Table 6.3, the estimated 'reliability log odds ratios' $\ln \hat{\alpha}$ in model $\{YZ,YA,YC,ZB,ZD\}$ without further restrictions are

$$\ln \hat{\alpha}_{11}^{YA} = 7.666; \ \ln \hat{\alpha}_{11}^{ZB} = 4.183; \ \ln \hat{\alpha}_{11}^{YC} = 10.268; \ \ln \hat{\alpha}_{11}^{ZD} = 75.593 \ .$$

For later discussions, it is important to realize that these (log) odds ratios are collapsible in model $\{YZ,YA,YC,ZB,ZD\}$, and that one gets the same results (point estimates and standard errors) whether these odds ratios are computed using the (marginal) conditional response probabilities in Table 6.4 or the complete table $ABCDYZ$.

Note that especially the log odds ratio for the relationship $D-Z$ is enormously large. This is caused by the fact that (just) one of the estimated cell entries in the 2×2 table is almost empty or, stated otherwise, one of the pertinent conditional response probabilities is for all practical purposes estimated as zero (see the outcomes in Table 6.4). This also causes the estimates of the odds ratio and of the standard error to become rather unstable.

Clearly, the estimated reliabilities of the indicators differ from each other, when measured by means of the log odds ratios $\ln \hat{\alpha}_{11}^{YA}$, etc. However, these differences might be the result of sampling fluctuations, and it might be true that in the population these reliabilities are similar to each other in some respects. The test outcomes of several potentially interesting hypotheses regarding the reliabilities in terms of odds ratios are given in Table 6.5, Column 'odds ratio α.'

The most stringent hypothesis is that all indicators have equal reliabilities, i.e., have the same odds ratios for the relationships with their latent variables:

$$\ln \alpha_{11}^{YA} = \ln \alpha_{11}^{ZB} = \ln \alpha_{11}^{YC} = \ln \alpha_{11}^{ZD} \ .$$

The common odds ratio under this equality restriction is estimated as $\ln \hat{\alpha} = 8.699$ (s.e. = .981). But as is clear from the test outcome in Table 6.5 (Row 'Equal reliabilities,' Column 'Odds ratio'), this hypothesis has to be rejected. This test outcome is the outcome for the unconditional test for model $\{YZ,YA,YC,ZB,ZD\}$ with the equal reliability hypothesis imposed on the parameters. One can also use the outcomes in Table 6.5 to carry out conditional tests comparing model $\{YZ, YA, YC, ZB, ZD\}$ with and without the reliability restrictions, leading to the same substantive conclusions in all cases in Table 6.5.

The next test (reported in Row 'Equal change' of Table 6.5) is based on the observation that in panel analyses, reliabilities tend to increase over time for the successive waves. Because of this panel effect, the reliabilities of both observed *Party preference* and *Candidate preference* are allowed to change from wave 1 to wave 2, but with an equal amount for both types of indicators:

$$\ln \alpha_{11}^{YA} - \ln \alpha_{11}^{YC} = \ln \alpha_{11}^{ZB} - \ln \alpha_{11}^{ZD}.$$

As seen in Table 6.5, the equal change model fits the data well for the odds ratios. Because some estimated response probabilities at time 2 are very close to the boundary, the estimated odds ratios (as well as their standard errors — not reported) become very large and very unstable:

$$\ln \hat{\alpha}_{11}^{YA} = 7.496; \ \ln \hat{\alpha}_{11}^{ZB} = 4.177; \ \ln \hat{\alpha}_{11}^{YC} = 30.656; \ \ln \hat{\alpha}_{11}^{ZD} = 27.337$$
$$\ln \hat{\alpha}_{11}^{YA} - \ln \hat{\alpha}_{11}^{YC} = \ln \hat{\alpha}_{11}^{ZB} - \ln \hat{\alpha}_{11}^{ZD} = -23.160$$
$$\ln \hat{\alpha}_{11}^{YA} - \ln \hat{\alpha}_{11}^{ZB} = \ln \hat{\alpha}_{11}^{YC} - \ln \hat{\alpha}_{11}^{ZD} = 3.319.$$

As expected, the reliabilities increase over time and, due to the restrictions, with the same amount (23.160) for both *Party* and *Candidate preference*. Looking at the same 'equal change restriction' from the other point of view, observed *Party preference* (A, C) is at both points in time more reliably measured than observed *Candidate preference* (B, D), and to the same extent (3.319).

Another way of trying to remedy the failure of the equal reliability model would have been to let the reliabilities of *Party preference* (A, C) and *Candidate preference* (B, D) be different from each other, but stable over time (ignoring the 'panel effect'):

$$\ln \alpha_{11}^{YA} = \ln \alpha_{11}^{YC}$$
$$\ln \alpha_{11}^{ZB} = \ln \alpha_{11}^{ZD}.$$

Not surprisingly, this 'no change' model has to be rejected (see Table 6.5). This might be due to lack of stability of the reliability of just one of the two characteristics. Looking at the 'no change in party preference only' hypothesis,

$$\ln \alpha_{11}^{YA} = \ln \alpha_{11}^{YC},$$

a well-fitting model is obtained. This is not true for the 'no change in candidate preference only' hypothesis

$$\ln \alpha_{11}^{ZB} = \ln \alpha_{11}^{ZD}.$$

From a substantive point of view, this is not a surprising result. In general, party preferences are well-established, which results in high and stable reliabilities for the indicators. But as political candidates come and go, candidate preferences are less crystallized, leading to low reliabilities of the indicators. But as candidates will be known better and better during the campaign, these reliabilities are expected to increase over time. The reliability odds ratios in the 'no change in party preference' model are all very high:

$$\ln \hat{\alpha}_{11}^{YA} = \ln \hat{\alpha}_{11}^{YC} = 8.621 \text{ (s.e. } = .836)$$
$$\ln \hat{\alpha}_{11}^{ZB} = 4.216 \text{ (s.e. } = .428)$$
$$\ln \hat{\alpha}_{11}^{ZD} = 37.731 \text{ (s.e } = .644) \, .$$

The very large value of $\ln \hat{\alpha}_{11}^{ZD}$ underlines again the fact that in very sparse tables with odds ratio estimates very close to the boundary, the estimates for the odds ratios become very unstable and the usual standard error estimates cannot be trusted or used for constructing confidence intervals.

Given the test outcomes for the reliability models using odds ratios, one may either accept the 'equal change' model or the 'no change in candidate preference' as the preferred model for the reliabilities in the measurements. Because good theoretical but different considerations may be given in favor of both acceptable models, the choice between them cannot be made on the basis of these data and knowledge.

6.3.3 Restrictions on Percentage Differences

Researchers are often inclined to interpret the patterns of LCA outcomes, such as in Table 6.4, by means of straightforward differences between conditional probabilities, rather than in terms of the odds ratios based on the same response probabilities. For example, when looking at the parameter estimates in Table 6.4, a researcher may be inclined to use percentage differences ε (*epsilon*) rather than loglinear parameters such as λ_{ir}^{AY} to describe the relations between the latent and the manifest variables. Applied to indicator A in model $\{YZ, YA, YC, ZB, ZD\}$, the reliability of A in terms of ε is

$$\varepsilon_{1|1}^{A|Y} = \pi_{1|1}^{A|Y} - \pi_{1|2}^{A|Y}.$$

At first sight it may seem strange or even wrong to describe the relations among variables, estimated by means of a loglinear model in terms of ε's, which are unstandardized regression coefficients belonging to the additive model. However, a model like $\{YZ, YA, YC, ZB, ZD\}$ without further restrictions only implies that the frequencies in table $ABCDYZ$ are estimated given the observed table $ABCD$ and a number of statistical independence restrictions — conditional independence relations among the indicators given the latent variables, conditional independence of A and C of latent variable Z given the score on latent variable Y, and conditional independence of B and D of Y. Now, statistical independence implies and is implied by the log odds ratio being zero: $\ln \alpha = 0$, but identically also by $\varepsilon = 0$. Therefore, (conditional) independence restrictions, whether formulated in terms of odds ratios or in terms of percentage differences, lead to exactly the same expected frequencies. Consequently, the maximum likelihood estimators for the reliabilities of the indicators in model $\{YZ, YA, YC, ZB, ZD\}$, but now expressed in terms of ε, can still be computed from the estimated conditional response probabilities in Table 6.4,

$$\hat{\varepsilon}_{1\,1}^{A|Y} = .952, \text{ (s.e.} = .024)$$

$$\hat{\varepsilon}_{1\,1}^{B|Z} = .772, \text{ (s.e.} = .042)$$

$$\hat{\varepsilon}_{1\,1}^{C|Y} = .987, \text{ (s.e.} = .016)$$

$$\hat{\varepsilon}_{1\,1}^{D|Z} = .986, \text{ (s.e.} = .022),$$

where the standard errors have been calculated by means of our marginal-modeling algorithm described in Section 6.5. For later discussions, remember that the conditional response probabilities and their differences ε are collapsible; for example, $\pi_{i\,r\,s\,j\,k\,l}^{A|YZBCD} = \pi_{i\,r}^{A|Y}$, given the independence restrictions in the model.

As long as no additional restrictions are imposed on the ε's or the odds ratios, the loglinear and the additive model are the same in the sense that they imply the same (conditional) independence restrictions on the data and yield the same estimated expected frequencies. However, things generally change if extra restrictions are imposed. If it is assumed that in model $\{YZ, YA, YC, ZB, ZD\}$ the reliabilities of all indicators are the same in terms of odds ratios (see above), it does not make much sense to use the outcomes of this model to express the reliabilities in terms of percentage differences. These ε's will not be the same. Further, making the reliabilities equal to each other in terms of ε leads to a different model, and especially to a nonloglinear model (with latent variables). The Goodman/Haberman algorithms, as well as the standard programs for latent class analysis, are no longer applicable (the M-step in the EM-algorithm no longer has the simple form as in latent loglinear models). In Section 6.5, it is explained how to estimate such linear, additive latent class models by means of our marginal algorithm.

For reliabilities expressed in terms of ε, the same restricted models can be formulated as for the odds ratios above. The test results are reported again in Table 6.5, more specifically, in the Column labeled 'epsilon.' These test results are obtained by estimating the joint frequencies under the (conditional) independence restrictions implied by model $\{YZ, YA, YC, ZB, ZD\}$, plus the particular reliability restrictions on the ε's. The test outcomes for the restricted reliability epsilons are, to a large extent, similar to the outcomes for the odds ratios. The models implying 'equal reliability,' 'no change,' and 'no change candidate preference' have to be rejected. But now also model 'equal change' is not acceptable, or at most, just on the borderline. Model 'no change party preference' fits the data as well as for the odds ratios. The estimates for the reliabilities in the 'no change party preference' model are

$$\hat{\varepsilon}_{1\,1}^{A|Y} = \hat{\varepsilon}_{1\,1}^{C|Y} = .012 \text{ (s.e.} = .093)$$

$$\hat{\varepsilon}_{1\,1}^{B|Z} = .774 \text{ (s.e.} = .042)$$

$$\hat{\varepsilon}_{1\,1}^{D|Z} = .981 \text{ (s.e.} = .023).$$

As a final remark on ε, note that ε is not variation independent of the marginal distributions of the latent variable and the indicator.

6.3.4 Restrictions on Agreement

Still another way of measuring reliability might be in terms of agreement between the scores on the latent and the manifest variables. This calls for κ as a reliability measure. In principle, using κ again leads to a nonloglinear model. Compared to the use of ε, an extra complication arises that gives, in a way, rise to interpretation difficulties. The agreement coefficient κ cannot be defined purely in terms of the (marginal) conditional response probabilities (as in Table 6.4). For example, to compute κ for the marginal relationship between A and Y (κ^{YA} in table AY), one must make use of the joint probabilities $\pi_{i\,r}^{AY}$ in marginal table AY, rather than just the conditional response probabilities $\pi_{i\,|\,r}^{A|Y}$. Because $\pi_{i\,r}^{AY}$ is not collapsible in model $\{YZ,YA,YC,ZB,ZD\}$, in contrast to $\pi_{i\,|\,r}^{A|Y}$, the agreement coefficient κ is not collapsible. Therefore, there are two different ways of defining κ. First, as a partial coefficient, which is defined as some kind of average of the conditional κ's within the categories of the other variables. For example, regarding the relationship between A and Y, partial κ^{AY} can be computed as a weighted average of the conditional coefficients $\kappa^{AY|ZBCD}_{\quad s\,j\,k\,l}$. Alternatively, it is possible to compute the marginal κ^{AY}, based on the marginal entries $\pi_{i\,r}^{AY}$. In general, the marginal and partial κ's will differ from each other. The answer to the question of which one must be preferred, the marginal or the partial one depends on the research question, but is hard to answer in general.

Only marginal κ's will be dealt with further here. These marginal κ's can be subjected to the same kinds of restrictions as the odds ratios and epsilons above. The test results for several restricted models can be found again in Table 6.5 in the Column labeled 'kappa.' These test results are obtained by estimating the joint frequencies under the (conditional) independence restrictions implied by model $\{YZ,YA,YC,ZB,ZD\}$ plus the particular reliability restrictions on the κ's. The (highly significant) $\hat{\kappa}$'s for the model without any extra reliability restrictions are

$$\hat{\kappa}^{YA} = .955, \; (\text{s.e.} = .022)$$
$$\hat{\kappa}^{ZB} = .775, \; (\text{s.e.} = .041)$$
$$\hat{\kappa}^{YC} = .988, \; (\text{s.e.} = .015)$$
$$\hat{\kappa}^{ZD} = .988, \; (\text{s.e.} = .019).$$

The test results for the models with restrictions on the κ-reliabilities are essentially the same as for ε. In terms of κ, a well-fitting restricted model is the 'no change in party preference.' The estimated κ's for this model are

$$\hat{\kappa}^{YA} = \hat{\kappa}^{YC} = .971 \; (\text{s.e.} = .011)$$
$$\hat{\kappa}^{ZB} = .778 \; (\text{s.e.} = .041)$$
$$\hat{\kappa}^{ZD} = .984 \; (\text{s.e.} = .020).$$

The common conclusions from all these reliability analyses based on different measures is that the indicators are very reliable, except for a somewhat lower reliability of the candidate preference in the first wave. This particular lower reliability may reflect a true tendency in the population in the sense argued above that the opinions of the people about the candidate are rather vague but become more and more crystallized during the election campaign. Alternatively, the crystallization that takes place might be an artifact of the panel study: re-interviewing people forces them to get firm opinions about matters they would ordinarily not think so strongly about. Regarding the methodologically interesting 'equal change model,' this is only a well-fitting model for reliability in terms of odds ratios. This also illustrates the well-known fact that conclusions about association (and reliability) may depend on the measure chosen.

As a final remark, our marginal-modeling approach turned out to be very well suited to estimate not only the many useful variants of the loglinear representation of the latent class model, but also a very general class of latent class models that are outside the area of loglinear modeling.

6.4 Marginal causal analyses

In the beginning of this chapter, it was suggested that, in principle, all manifest marginal analyses shown in the preceding chapters can also be carried out when latent variables are involved. Several examples thereof have been presented above. In this section, two final illustrations will be given, showing how structural equation models with latent variables can or must make use of marginal modeling. The first example concerns a structural equation model with an MH restriction on the distributions of the two latent variables. The second example is an application of a categorical SEM with latent variables to the clustered NKPS data, used several times before.

6.4.1 SEMs with latent marginal homogeneity

The standard latent variable models that have been fitted to the data in Table 6.1 on the Dutch party and candidate preferences did not fit the data very well. The relatively best model, used for illustrative purposes, turned out to be the two latent variable model $\{YZ, YA, YB, ZC, ZD\}$ where Y represents the latent party preference and Z the latent candidate preference (see also Fig. 6.2); the test results were $G^2 = 84.74$, $df = 48$ and $p = .000$. Inspection of the two-way marginal residuals of this not quite fitting model led to the suggestion that perhaps a kind of test-retest effect took place within the same interview at time 1: the answers to the (earlier) question on the respondents' *Party preference A* may have directly influenced the answers to the (later) question on the *Candidate preference C*. This extra effect, a violation of the local independence assumption (Hagenaars, 1988), is incorporated in Fig. 6.3.

The model in Fig. 6.3 can no longer be written down in the form of a single loglinear equation, but given the causal order, a categorical structural equation model with several equations must be defined. The starting point is the graphical decomposition, taking the implied conditional independence relationships into account (see Chapter 5):

$$\pi_{r\,s\,i\,j\,k\,\ell}^{YZABCD} = \pi_{rs}^{YZ} \pi_{i\ r}^{A|Y} \pi_{j\ r}^{B|Y} \pi_{k\,i\,s}^{C|AZ} \pi_{\ell\ s}^{D|Z} \,. \tag{6.6}$$

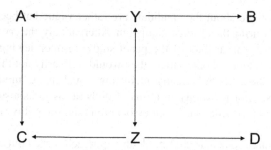

Fig. 6.3. The latent class model with two stable latent variables *Party preference* and *Candidate preference* but with a direct effect from A on C

Without further extra restrictions, the estimates for all elements at the right-hand side of Eq. 6.6, in agreement with the model in Fig. 6.3, can be obtained by specifying the appropriate saturated loglinear model for each element, except for element $\pi_{k\,i\,s}^{C|AZ}$ — or example, model $\{YZ\}$ for marginal table YZ, model $\{AY\}$ for marginal table AY, etc. However, the appropriate loglinear submodel for marginal table ACZ is not the saturated submodel, but submodel $\{AZ,AC,CZ\}$ because no three-variable interaction effect on C is assumed ($\lambda_{i\,k\,s}^{ACZ} = 0$). Application of the thus extra restricted model in Eq. 6.6 to the data in Table 6.1 yields $G^2 = 45.97$, $df = 44$ ($p = .391$, $X^2 = 44.04$). Addition of the cross-sectional test-retest effect $A - C$ in model $\{YZ,YA,YB,ZC,ZD\}$ improves the fit enormously and now leads to a well-fitting model. It is not necessary to impose a similar direct effect $B - D$ for the second wave. Again, perhaps later on in the campaign, the preferences for particular candidates have become more crystallized and are not (or much less) easily influenced by the previous party preference answer.

From the estimated, statistically significant loglinear effects (not all reported here), it can be seen that the strong direct effect of A on C is such that especially the stated preference for a Christian-Democratic or left-wing party brings the respondent to express an extra preference for the candidate from the same party ($\hat{\lambda}_{11*}^{ACZ} = .683$; $\hat{\lambda}_{22*}^{ACZ} = .744$; $\hat{\lambda}_{33*}^{ACZ} = .320$), a stated preference over and above the preferences implied by the relations between the latent variables Y and Z and the direct effect of Y on A and of Z on C. Regarding the other relationships in the model, most noticeable is that the relationship between the two latent variables $Y - Z$ is somewhat weaker, and the relation $Z - D$ much stronger, in this local dependence than in the corresponding local independence model.

Now that a well-fitting and well-interpretable model for these data has been found, again the question might be raised: is latent table YZ marginally homogeneous (Eq. 6.5). The estimates for the latent marginal distributions of Y and Z for the model in Fig. 6.3 without the MH restriction are

$$\hat{\pi}_{r+}^{YZ} : .275, .374, .351$$
$$\hat{\pi}_{+r}^{YZ} : .193, .428, .379.$$

The test outcomes for the model in Fig. 6.3, but now with the MH restriction on the latent variables (Eq. 6.5), are on the borderline of significance ($.01 < p < .05$): $G^2 = 62.33, df = 46$ ($p = .011, X^2 = 57.10$). However, the more powerful conditional test of Eq. 6.5 provides a clear result: $G^2 = 62.33 - 45.97 = 16.36, df = 48 - 46 = 2$, $p = .000$. Once more, but now in the context of a well-fitting categorical SEM, it must be concluded that the hypothesis of homogeneous distributions of party and candidate preference has to be rejected.

6.4.2 Latent Variable SEMs for Clustered Data

In Section 5.1.3, an example was presented of a manifest structural equation model using the clustered or dependent data of the NKPS study (see also Section 2.2.3). Remember that in this study, data were collected for parent-child couples, making it necessary to use marginal-modeling methods to estimate the relevant SEMs. To illustrate a SEM with latent variables, the same study and partly the same data will be used as in Section 5.1.3. The causal order of the variables is $(S,A),E,R,T,(W,M)$, where S, A, E and R stand for: S – *Sex* with categories 1 = man, 2 = woman; A – *Age* with 1 = old (≥ 55), 2 = middle (38-54), 3 = young (≤ 37); E – *Education* with three categories: 1 = low, 2 = intermediate, 3 = high; R – *Religiosity* defined as church attendance with 1 = often/regular, 2 = sometimes, 3 = never/rarely. Different from the SEM in Section 5.1.3 is the ultimately dependent variable. This is now a dichotomous latent variable T – *Traditionalism* measured by a dichotomized indicator W – *Attitude towards women's role* and a dichotomized indicator M – *Attitude towards marriage*. The categories of the indicators and the latent variable are all the same: 1 = traditional, 2 = nontraditional. Indicator W was used before, with three categories (see Sections 2.2.3 and 5.1.3). Category $W = 1$ now contains the original index scores 1 to 4, and $W = 2$ the scores lower than 1. Indicator M results from agreeing or not on a five-point scale with statements about 'children choosing their own partners,' 'no problem to live together without being formally married,' 'not having a divorce when the children are young,' etc. Category $M = 1$ consists of the index scores between 1.80 and 5; Category $M = 2$ lower than 1.80. The full data set for all variables can be found on the book's website. Adding indicator M resulted in four extra deleted respondents because of missing values compared to the analyses in the previous chapter. A possible representation of the relationships among the variables is rendered in Fig. 6.4.

If the data had come from independent observations and did not constitute clustered data, the analysis would proceed as follows. The joint distribution of all variables concerned would be decomposed, taking into account the assumed causal order

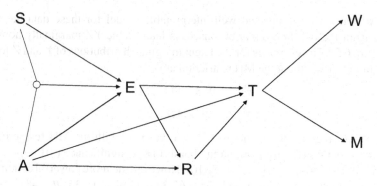

Fig. 6.4. Causal model for *Traditionalism*

among the eight variables, and arbitrarily assuming without further consequences here that W precedes M (see below):

$$\pi_{sae\,rt\,wm}^{SAERTWM} = \pi_{sa}^{SA}\,\pi_{e\ sa}^{E|SA}\,\pi_{r\ sae}^{R|SAE}\,\pi_{t\ sae\,r}^{T|SAER}\,\pi_{w\ sae\,rt}^{W|SAERT}\,\pi_{m\ sae\,rt\,w}^{M|SAERTW} \,. \qquad (6.7)$$

The last two elements at the right-hand side of Eq. 6.7 ($\pi_{w\ sae\,rt}^{W|SAERT}$ and $\pi_{m\ sae\,rt\,w}^{M|SAERTW}$) pertain to the measurement part of the SEM, whereas the first four elements at the right-hand side constitute the structural part. Each of the separate right-hand side factors would be estimated independently (for independent observations) by defining the appropriate submodels for the pertinent marginal tables. Their estimated frequencies would be used to calculate the ML estimates for the right-hand side elements and ultimately for the left-hand side joint probability $\pi_{sae\,rt\,wm}^{SAERTWM}$.

However, the NKPS data are clustered. As was explained in the discussions of the outcomes in Table 6.5, a consequence thereof is that it is not correct to estimate each of the submodels separately, not even using the marginal-modeling approach in which the dependencies in the clustered data are taken into account for each separate submodel. Even then, combining these separate 'marginal-modeling' estimates using Eq. 6.7 does not provide the correct ML estimates for the left-hand joint probability $\pi_{sae\,rt\,wm}^{SAERTWM}$. It does not even provide the correct ML estimates for each of the separate right-hand side elements under the simultaneous restrictions for the whole model. What is needed is the full marginal-modeling approach, in which all restrictions for all separate submodels are imposed simultaneously to obtain the appropriate ML estimates for the left-hand side probabilities in Eq. 6.7. This estimation procedure makes use of the complete table *SAERTWM* for the parents cross-classified with table *SAERTWM* for the kids. A practical consequence thereof, especially for computing the standard errors of the estimates, is that the ML estimation procedure requires an enormous amount of computing space. The provisional version of the program that was used for this book initially ran out of space with the original coding of the variables. That is why it was decided to dichotomize the latent variable and its indicators. Future versions of our program will be more flexible in this respect (see the book's website).

Because the same (manifest) variables are involved in the beginning of the causal chain as indicated in Table 5.5 and as represented in Fig. 6.4, the starting submodels for the relations among variables S, A, E, and R will be restricted in the same way as before: submodel $\{S,A\}$ will be assumed for marginal table SA; saturated submodel $\{SAE\}$ for marginal table SAE; and submodel $\{SAE,AR,ER\}$ for marginal table $SAER$. Regarding the effects on the latent variable T (*Traditionalism*), the starting submodel for marginal table $SAERT$ will be the model with only the two-variable effects involving T, i.e., model $\{SAER,ST,AT,ET,RT\}$. For the relations between the latent variable and its indicators, the starting submodel for the effects on W will be $\{SAERT,WT\}$ for table $SAERTW$ and for the effects on M model $\{SAERTW,MT\}$ for table $SAERTWM$. The latter restrictions regarding the effects of the latent variable T on its indicators W and M can also be estimated with the same results by replacing $\pi_{w\ sae\,r\,t}^{W|SAERT} \pi_{m\ sae\,r\,t\ w}^{M|SAERTW}$ in Eq. 6.7 by $\pi_{w\,m\ sae\,r\,t}^{WM|SAERT}$ and estimating submodel $\{SAERTW,WT,MT\}$ for table $SAERTWM$.

The test results for the complete categorical SEM with latent variables, consisting of all above indicated submodels for all subtables and applying the marginal-modeling approach described in the following section, clearly indicate that the model as a whole has to be rejected: $G^2 = 979.63$, $df = 394$, $p = .000$. A possible weakness of the complete model might be the absence of any three-variable effects involving T (see also the effects on W in the previous chapter). For example, there may be good reasons to assume an interaction effect of *Age* and *Religion* on *Traditionalism*. Maybe *Religion* only has an effect on *Traditionalism* among the younger, not among the older generations; older generations are traditional anyway, regardless of religion. Replacing model $\{SAER,ST,AT,ET,RT\}$ for subtable $SAERT$ by $\{SAER,SAT,SET,SRT,AET,ART,ERT\}$ did not improve the fit: $G^2 = 949.63$, $df = 376, p = .000$. Most three-variable interaction effects were very small, except the interaction effects of *Age* and *Religion* on *Traditionalism*; but these estimates were very unstable, depending on a few cells becoming (almost) zero.

However, that still leaves us with a SEM that has to be rejected. By inspection of the residual frequencies, more in particular by comparing the observed and the estimated odds ratios in all two-way marginal tables, it was concluded that some direct effects of the background characteristics on the indicators W and M have to be added — in particular, direct effects of *Age* (A) and *Education* (E) on *women's attitudes* (W) and of *Age* (A) and *Religiosity* (R) on *marriage attitudes* (M). The effects of A, E, and R on the indicators do not go completely through latent variable T. In other words, latent variable T does not capture all systematic variation in the indicators, and the measurement part of the SEM is not completely homogeneous or invariant (Hagenaars, 1990, Section 3.6; Millsap & Yun-Tein, 2004). To obtain estimates for $\pi_{w\,m\ sae\,r\,t}^{WM|SAERT}$ in a well-fitting model, model $\{SAERTW,AW,EW,AM,RM,WT,MT,\}$

seems to be needed rather than the earlier model $\{SAERTW, WT, MT\}$. The test results for the earlier SEM with only the main effects on T and with the specified direct effects on the indicators added led to a SEM that can be accepted: $G^2 = 396.50$, $df = 386, p = .345$. From the viewpoint of substantive interest, the outcomes of this model should only be seriously considered if a good theoretical explanation can be found for the necessity to include the direct effects of the background variables on the indicators in the final SEM. Given the purposes of this chapter, we will not discuss further (substantive) details.

From a marginal-modeling perspective, it is interesting to compare the estimated relationships among S, A, E, and R for three SEMs: the one in the previous chapter involving the same variables and data, and the present two models (i.e., the one with and the one without the extra direct effects on the indicators). If the observations would have been independently observed but not clustered, exactly the same parameter estimates and standard errors would have been found in all three cases, given the causal order of the variables in the assumed SEMs (except for the four extra missings in this chapter). With the clustered NKPS data, this is no longer true. The submodels for the various subtables are no longer separately estimated. If varying restrictions and submodels are imposed on the conditional probabilities on the right-hand side of Eq. 6.7 involving variables T, W, and M that occur later in the causal chain than S, A, E, and R, different estimates will also be obtained for the relationships among S, A, E, and R despite the fact that variables T, W, and M occur later in the causal chain.

Now looking at the outcomes, for most practical purposes the substantive conclusions regarding the relationships among S, A, E, and R are the same in all three SEMs. Nevertheless, the corresponding parameter estimates in the three models are not exactly the same and do differ from each other in absolute terms, mostly between .01 and .08. Of course, larger or smaller differences must be reckoned with depending on the nature of the clustering.

Given the purposes of this chapter, further outcomes will not be discussed: the main point of this section has been to illustrate that the SEM without latent variables for clustered data discussed in the previous chapter can be extended to SEMs with latent variables. The logic of the extensions is rather straightforward; its practical implementation may still meet some practical difficulties for tables with very large numbers of cells.

6.5 Estimation of Marginal Models with Latent Variables Using the EM Algorithm

A standard method for fitting models with latent variables is the Expectation-Maximization (EM) algorithm. In Section 6.5.1, the basic ideas behind the EM algorithm will be presented, using a simple illustration of a standard latent class model in the form of a loglinear model with a latent variable. After this exposition of basic ideas, a general EM algorithm for marginal models with latent variables will be described in Section 6.5.2. This algorithm is indeed a very general one and can, for example, be used for all latent marginal models discussed in this chapter. However, for some classes of models, it is very inefficient, sometimes to such an extent that it can hardly be implemented on a current standard computer. Therefore, in Sections 6.5.3 and 6.5.4, for particular classes of models, much more efficient algorithms will be presented. In the last section, some final issues with the EM algorithm are discussed.

6.5.1 Basic EM Algorithm

For a set of variables consisting of two subsets viz. a subset of manifest and a subset of latent variables, what will be called 'the complete table' refers to the table consisting of all (manifest and latent) variables. The EM algorithm aims to find the ML estimates of the probabilities in the population for the complete table and involves computation of estimated 'observed' probabilities for the complete table, referred to as complete data. The likelihood function for the complete table is called the complete data likelihood. Using appropriate starting estimates of the probabilities in the complete table, an 'expectation' (E) and a 'maximization' (M) step are performed iteratively. In the E-step, the expectation of the complete data likelihood given the observed data is calculated, which leads to estimated observed probabilities for the complete table. In the M-step, the estimated 'observed' proportions are treated as if they were actually observed, and the ML estimates of the probabilities in the complete table are calculated using a method for manifest variable models. The two steps are performed iteratively, until sufficiently accurate convergence is reached.

Haberman (1979, Chapter 10) described a modified Newton-Raphson (scoring) algorithm as an alternative, specifically for loglinear models with latent variables. However, the latter algorithm is more sensitive to starting values — that is, starting values need to be (much) closer to the ML estimates than for EM. To overcome this difficulty, Haberman (1988) developed a stabilized Newton-Raphson algorithm.

The basic principles of the EM algorithm can be explained easily when the model for the complete data is the simple loglinear model represented in Fig. 6.1. For this model, the EM-algorithm takes a very simple form, especially regarding the M-step. There are four manifest variables A, B, C, and D, and one latent variable X. The loglinear model is $\{AX, BX, CX, DX\}$, i.e., the manifest variables are independent given the latent variable. Given the probabilities $\pi_{i\,jkl\,t}^{ABCDX}$ for the complete table, the estimated 'observed' probabilities for the complete table, given the data $p_{i\,jkl}^{ABCD}$, are

$$\hat{p}_{i\,jkl\,t}^{ABCDX} = p_{i\,jkl}^{ABCD}\, \pi_{t\,i\,jkl}^{X|ABCD}. \tag{6.8}$$

The maximum likelihood estimates of the probabilities $\pi_{i\,jkl\,t}^{ABCDX}$ given 'observed' probabilities $\hat{p}_{i\,jkl\,t}^{ABCDX}$ are given in closed form as

$$\hat{\pi}_{i\,jkl\,t}^{ABCDX} = \hat{p}_{i\,t}^{A|X}\, \hat{p}_{j\,t}^{B|X}\, \hat{p}_{k\,t}^{C|X}\, \hat{p}_{l\,t}^{D|X}\, \hat{p}_{t}^{X}. \tag{6.9}$$

Eqs. 6.8 and 6.9 are used in the EM algorithm as follows. First, starting values $(\hat{\pi}_{i\,jkl\,t}^{ABCDX})^{(0)}$ satisfying model $\{AX, BX, CX, DX\}$ should be chosen, e.g., random or constant starting values for the parameters occurring in the model. Then for iteration $s = 0, 1, 2, \ldots$, the following steps are repeated:

E-step: $(\hat{p}_{i\,jkl\,t}^{ABCDX})^{(s)} \quad = p_{i\,jkl}^{ABCD} (\hat{\pi}_{t\,i\,jkl}^{X|ABCD})^{(s)}$

M-step: $(\hat{\pi}_{i\,jkl\,t}^{ABCDX})^{(s+1)} = (\hat{p}_{i\,t}^{A|X})^{(s)} (\hat{p}_{j\,t}^{B|X})^{(s)} (\hat{p}_{k\,t}^{C|X})^{(s)} (\hat{p}_{l\,t}^{D|X})^{(s)} (\hat{p}_{t}^{X})^{(s)}.$

The iterations are stopped when sufficient convergence has been reached, for example when successive estimates do not change very much.

More detailed descriptions of the EM algorithm for loglinear models can be found in the general latent class references in the beginning of this chapter. A general description of the EM algorithm can be found in MacLachlan and Krishnan (1997).

For decomposable loglinear models — i.e., those loglinear models with closed form ML estimates like in the above example — the M-step is particularly simple. For nondecomposable loglinear models and the marginal models to be discussed in the next subsection, the M-step is less simple and no closed-form solutions for the parameters are possible. For latent class models with marginal restrictions on particular subtables, the iterative algorithms presented in Chapters 2 and 3 will be used in the M-step. The next section describes the technical details of this nested iterative estimation method.

There are several general issues to be considered when using latent class models. First, for many latent class models and depending on the data, the likelihood function tends not to be a concave function of the parameters, and may have many local maxima. It is usually a good idea to try different (random) starting values and choose the solution with the highest likelihood value, although there is never a guarantee that the ML estimates have been found. One might perhaps think that the addition of marginal constraints increases the complexity of the fitting problem, e.g., causing there to be more local maxima. In our experience, this does not appear to be the case, however: for the models of this chapter, no multiple local maxima were encountered (except for those relating to a wrong ordering of the latent classes; see the third point below). This has perhaps to do with the fact that, in general, in our experience, restricted models are less bothered by local maxima than less restricted models, and especially with the fact that, as noted in Chapter 2, the likelihoods for many marginal models have a unique stationary point that is its global maximum (Bergsma, 1997, Section 5.2).

Second, there is the possibility of boundary estimates, i.e., some probabilities in the complete table are estimated as zero. Although this often happens for manifest variable marginal models as well, and need not be a big issue, for latent class models it can more easily lead to the Jacobian of the constraint matrix not being of full rank, which leads to difficulties in calculating the correct number of degrees of freedom of G^2.

Third, a well-known problem is that the order of the latent classes is not determined by the model; for example, with a dichotomous latent variable, latent class 1 of a particular analysis may come out as latent class 2 in another analysis. In a way, this is completely irrelevant because if the model is identified, all estimated proportions for the latent classes and all conditional response probabilities are correspondingly interchanged as well (and, of course, all test statistics are exactly the same for the two outcomes). As stated before, the interpretation and meaning of a particular latent class solution depends on the fitted values — more specifically on the relationships between the latent and the manifest variables. For example, it was found in Section 6.3.2. that the equal reliabilities model for the odds ratios produced a common estimated log odds ratio of 8.699 (and $\alpha = 5996.9$); another analysis might have produced an interchange of the latent classes and a common log odds ratio of $-.8699$ (and $\alpha = 1/5996.9$). Obviously, assigning similar category meanings to the latent variables Y and Z as to manifest variables A through D does not make much sense with the latter result of $-.8699$; this would imply not so much an extreme unreliability (statistical independence between the latent variable and the indicator) as well as a serious tendency to always and systematically choose the wrong alternative. An appropriate choice of starting values may prevent such a seemingly awkward outcome, but in this case it is even simpler to accept the strange outcome and just interchange the labels and interpretations of the latent categories. However, when additional restrictions on the latent variables are imposed, such as marginal homogeneity, such a relabeling is not always possible.

6.5.2 ***General EM for Marginal Models

With π the vector of probabilities for the complete table, the EM algorithm will now be presented for models defined in the parametric form

$$\theta(A'\pi) = X\beta \tag{6.10}$$

or, equivalently, by the constraint

$$h(A'\pi) = B'\theta(A'\pi) = 0 \tag{6.11}$$

for some possibly vector valued function h. As before, B and X are each other's orthocomplement. All models discussed in this chapter can be written in these forms (see Section 3.3 for further details). A special case is the class of loglinear marginal models, which are of the form

$$B'\log(A'\pi) = 0,$$

as discussed in Chapter 2.

The M-step of the EM algorithm consists of fitting the model in Eq. 6.11. Therefore, the algorithms discussed in Chapter 2 and 3 can be used. Below, more details will be provided.

Assume there are T_L latent variables, denoted in vector form $L = (L_1, L_2, \ldots, L_{T_L})'$, and T_M manifest variables, denoted in vector form $M = (M_1, M_2, \ldots, M_{T_M})'$. Let $\pi_{i\ t}^{ML}$ ($\hat{p}_{i\ t}^{ML}$) denote the population (estimated observed) probability in a cell of the complete table, where the index for the latent variables is $t = (t_1, t_2, \ldots, t_{T_L})'$ and the index for the manifest variables is $i = (i_1, i_2, \ldots, i_{T_M})'$. Similarly, π_t^L denotes the population probability in a cell of the table of latent variables, and π_i^M denotes the population probability in a cell of the table of manifest variables.

With notation as in Section 2.3.4, let $H = H(A'\pi)$ be the Jacobian of $h(A'\pi)$, let

$$l(\pi, \lambda | p) = \frac{p}{\pi} - 1 - H\lambda$$

be the gradient of the Lagrangian complete data likelihood with respect to π, and let

$$\lambda(\pi) = \left[H'D_\pi H \right]^{-1} \left[H'D_\pi^{-1}(p - \pi) + h(\pi) \right]$$

be the Lagrange multiplier vector expressed as a function of π. In contrast to the notation in Chapter 2, we now write $l(\cdot | p)$, denoting dependence on p, because in the algorithm p, the 'observed' vector of proportions for the complete table, have to be estimated as well. As discussed in Chapter 2, the ML estimate $\hat{\pi}$ under the model in Eq. 6.11 can be found by solving

$$l(\pi, \lambda(\pi) | p) = 0 .$$

The algorithm proceeds as follows. First, a starting value $(\hat{\pi}_{i\ t}^{ML})^{(s+1)}$ is chosen. Randomly chosen values can be used. Sometimes, however, starting values need to be chosen in such a way that the latent variables are properly identified. The uniform distribution is generally unsuitable as a starting value, because for many models it is a stationary point of the likelihood from which the EM algorithm cannot get away. Then, for iteration $s = 0, 1, 2, \ldots$, the following steps should be repeated until convergence is reached:

$$\text{E-step:} \quad (\hat{p}_{i\ t}^{ML})^{(s)} \quad = p_i^M \times (\hat{\pi}_{t\ i}^{L|M})^{(s)}$$

$$\text{M-step:} \quad (\hat{\pi}_{i\ t}^{ML})^{(s+1)} = (\hat{\pi}_{i\ t}^{ML})^{(s,K_s)}$$

where, in the M-step, $(\hat{\pi}_{i\ t}^{ML})^{(s,K_s)}$ is calculated iteratively (for K_s inner iterations) using the scheme

$$(\hat{\pi}^{ML})^{(s,0)} = (\hat{\pi}^{ML})^{(s)}$$

$$\log(\hat{\pi}^{ML})^{(s,k+1)} = \log(\hat{\pi}^{ML})^{(s,k)} +$$
$$step^{(s)} l \left[(\hat{\pi}^{ML})^{(s,k)}, \lambda \left((\hat{\pi}^{ML})^{(s,k)} \right) \Big| (\hat{p}^{ML})^{(s)} \right]$$

for $k = 0, \ldots, K_s - 1$. Note that the M-step is exactly the algorithm described in Chapters 2 and 3. Since generally no closed-form solutions are available in the M-step, it has to be performed iteratively, with K_s inner iterations in outer iteration s. Full convergence to the complete likelihood ML solution at every step is generally not necessary (Meng & Rubin, 1993). For the models of this chapter K_s between 2 and 10 was used (independently of s). There are no fixed rules for the choice of K_s, but choosing it too large is inefficient, while choosing a value too small can sometimes lead to nonconvergence and may be inefficient as well. Now, even if the number of inner iterations in the M-step can be limited, the number of outer iterations needed may still be very large; typically thousands were needed for the problems of this chapter. In the present applications, the outer iterations were stopped if successive estimates of the expected probabilities changed by a sufficiently small amount, say 10^{-10}. Note that the starting values $(\hat{\pi}_{i\,t}^{ML})^{(0)}$ should not be zero, because the algorithm will fail because logarithms are taken of the estimated probabilities. The same recommendations on the step size as described in Chapter 2 apply to the M-step. In particular, it was found for several of the models of this chapter that the step size should not be too large, e.g., a maximum value of 0.2 was sometimes necessary.

The above algorithm is an extension of the algorithm described by Becker and Yang (1998). They were the first to extend Goodman's (1974b) algorithm to marginal models with latent variables, replacing the IPF procedure in the M-step of Goodman's algorithm by the Lang-Agresti (1994) algorithm. The Becker-Yang algorithm essentially applies to the loglinear marginal models described in Chapter 2, and we have extended this to the broader class of models described in Chapter 3. Additionally, in the M-step we use the algorithm proposed by Bergsma (1997), which is a modification and generalization of the Lang-Agresti algorithm. The modification consists of a reduction of the dimension of the space in which the search is done, which appears to lead to better convergence properties (see Section 2.3.5), and the generalization is to the nonloglinear marginal models of Chapter 3.

Using the provisional version of our computer program, we encountered some numerical problems when fitting the models discussed in this chapter. It appears that latent variable marginal models are more difficult to fit than the manifest variable models of the previous chapters. Fortunately, we were always able to overcome these problems and obtain the ML estimates without too much difficulty, usually by trying out different starting values. Still, this did not always work, and for some models we needed to replace the matrix inverse used in the computation of $\lambda(\pi)$ by the Moore-Penrose inverse. For other models for which the usual inverse gives convergence, using the Moore-Penrose inverse did not yield convergence, a choice between the two had to be made by successively trying out the two methods. How to obtain convergence for all of the models of this chapter is explained on the book's website.

Now even though expressing a model in the constrained form in Eq. 6.11 may be possible, it may not lead to a very efficient algorithm, so in the next two sections more efficient versions of the algorithm are discussed for certain commonly occurring cases. These versions also tend to yield fewer numerical problems than the version of the present section.

6.5.3 ***Marginal Restrictions in Combination with a Loglinear Model for the Complete Table

Many of the models in this chapter are a combination of a loglinear model for the complete table and a marginal model. In such cases, the implementation of EM (in particular, the M-step) of the previous section is usually not very efficient, and we describe a better method below. In practice, it is usually inefficient to write the log-linear model as a constraint as in Eq. 6.11 on π because, for reasonably large tables at least, the number of loglinear parameters in a typical model is small compared to the number of cells in the table. This, in turn, implies a large number of constraints, which can make the computation of H prohibitively expensive (see also the example in Section 2.3.6). It is also easily seen that for saturated or nearly saturated models there may not be a computational gain by using the parametric representation of the loglinear model.

For the complete table, assume a loglinear model of the form

$$\log \pi = W \gamma \tag{6.12}$$

for a design matrix W and a vector of loglinear parameters γ. In addition, assume a marginal model defined by Eq. 6.11. Furthermore, it is assumed that the columns of A are a linear combination of the columns of W — a regularity condition that is almost always satisfied in practice.

The M-step in the algorithm described in the previous section is modified as follows, and as in Section 2.3.6, by using

$$l_W(\pi, \lambda(\pi)|p) = l(\pi, \lambda(\pi)|p) - \frac{p}{\pi} + 1 + W(W'D_\pi W)^{-1}W'(p - \pi),$$

which can be shown to equal the gradient of the Lagrangian complete data likelihood with respect to π. A starting value $\hat{\pi}^{(0)}$ needs to be chosen to *satisfy the loglinear model* (but the uniform distribution should not be chosen). Then, for iteration $s = 0, 1, 2, \ldots$, the following steps should be repeated until convergence is reached:

$$\text{E-step: } (\hat{p}_{i\,t}^{ML})^{(s)} \quad = p_i^M \times (\hat{\pi}_{t\,i}^{L|M})^{(s)}$$

$$\text{M-step: } (\hat{\pi}_{i\,t}^{ML})^{(s+1)} = (\hat{\pi}_{i\,t}^{ML})^{(s,K_s)}$$

where, in the M-step, $(\hat{\pi}_{i\,t}^{ML})^{(s,K_s)}$ is calculated iteratively (for K_s inner iterations) using the scheme

$$(\hat{\pi}^{ML})^{(s,0)} = (\hat{\pi}^{ML})^{(s)}$$

$$\log(\hat{\pi}^{ML})^{(s,k+1)} = \log(\hat{\pi}^{ML})^{(s,k)} +$$
$$\text{step}^{(s)} l_W \left[(\hat{\pi}^{ML})^{(s,k)}, \lambda \left((\hat{\pi}^{ML})^{(s,k)} \right) \middle| (\hat{p}^{ML})^{(s)} \right]$$

for $k = 0, \ldots, K_s - 1$. The same recommendations on the choice of step size apply as in the previous subsection and in Chapter 2.

6.5.4 ***Speeding up of the EM Algorithm for Separable Models

Latent marginal homogeneity models, as described in Section 6.2, have a marginal homogeneity model for the latent variables and a logit model for the manifest variables, given the latent variables. In such cases, the M-step can be simplified considerably, because one only needs to fit the marginal model for the table of latent variables. Specifically, the decomposition of the kernel of the log-likelihood becomes

$$L(\pi) = \sum_{i,t} p_{i\,t}^{ML} \log \pi_{i\,t}^{ML} = \sum_{i,t} p_{i\,t}^{ML} \log \pi_{i}^{M|L}{}_{t} + \sum_{t} p_{t}^{L} \log \pi_{t}^{L}\,.$$

If there is a logit model for $\pi_{i}^{M|L}{}_{t}$ and a marginal model for π_{t}^{L}, this decomposition allows separate fitting of the marginal model. We call such models 'separable models.' The principle can be illustrated as follows.

In the example of Section 6.2, with latent variables Y and Z and manifest variables A, B, C, and D, the loglinear model is $\{YZ, YA, YB, ZC, ZD\}$ and the marginal restriction on the latent variables specifies that $\pi_{t}^{Y} = \pi_{t}^{Z}$. The EM algorithm then goes as follows. First, choose starting values $(\hat{\pi}_{i\,jkl\,t\,u}^{ABCDYZ})^{(0)}$ satisfying $\{YZ, YA, YB, ZC, ZD\}$. Then, for iteration $s = 0, 1, 2, \ldots$, the following steps should be repeated until convergence is reached:

E-step: $(\hat{p}_{i\,jkl\,t\,u}^{ABCDYZ})^{(s)} = p_{i\,jkl}^{ABCD} \times (\hat{\pi}_{t\,u\,i\,jkl}^{YZ|ABCD})^{(s)}$

M-step 1: Compute $(\hat{\pi}_{t\,u}^{YZ})^{(s+1)}$ for model $\pi_{t}^{Y} = \pi_{t}^{Z}$ using $(\hat{p}_{t\,u}^{YZ})^{(s)}$ as 'observed' data with K_s iterations of Bergsma's algorithm

M-step 2: $(\hat{\pi}_{i\,jkl\,t\,u}^{ABCDYZ})^{(s+1)} = (\hat{\pi}_{t\,u}^{YZ})^{(s+1)}(\hat{\pi}_{i\,t}^{A|Y})^{(s)}(\hat{\pi}_{j\,t}^{B|Y})^{(s)}(\hat{\pi}_{k\,u}^{C|Z})^{(s)}(\hat{\pi}_{l\,u}^{D|Z})^{(s)}$

The crucial point is that M-step 1 can be performed very fast because our algorithm is applied to a small table. Furthermore, the E-step and M-step 2 are in closed form and can therefore be performed fast as well. Still, note that in both the E-step and M-step 2 the whole table is involved, so these two steps can become a computational problem for large tables.

6.5.5 ***Asymptotic Distribution of ML Estimates

Under certain regularity conditions, the ML estimates of expected cell probabilities in the complete table for the models of this chapter have a joint asymptotic normal distribution, with asymptotic mean the true probabilities. The regularity conditions involve the assumption that the true probabilities for the complete table are strictly positive, and that the model constraints are sufficiently smooth. The asymptotic covariance matrix of the probabilities and other parameters are given below, for several specifications of the latent class marginal model. The derivations are based on the method of Aitchison and Silvey (1958), who considered general ML estimation with constraints (see also Bergsma, 1997, Appendix A), and Haberman (1977) who considered asymptotics for loglinear and certain nonloglinear models.

Notation is as follows. With π the vector of population probabilities for the complete table, let A_L be a matrix such that $\pi^{L} = A_{L}'\pi$ is the vector of probabilities for the table of latent variables, and let A_M be a matrix such that $\pi^{M} = A_{M}'\pi$ is the vector of probabilities for the table of manifest variables. The vector p consists of observed proportions for the table of manifest variables and the vector Np, with N the sample size, and is assumed to have a multinomial distribution based on the population probability vector π^{M}.

By its definition, each row of A_M contains only a single '1' and the rest 0s, and each column contains the same number of 1s. Therefore it can be verified that $A'_M A_M$ is a diagonal matrix with strictly positive diagonal elements, and that $\hat{\pi}_0 = A_M (A'_M A_M)^{-1} p$ is an estimate of the unrestricted probability vector for the complete table. Since Np is assumed to have a multinomial distribution, p has covariance matrix $D[\pi^M] - \pi^M (\pi^M)'$, and by the delta method, the covariance matrix of $\hat{\pi}_0$ is

$$ V_0 = V(\hat{\pi}_0) = A_M (A'_M A_M)^{-1} \left[D[\pi^M] - \pi^M (\pi^M)' \right] (A'_M A_M)^{-1} A'_M . $$

For other sampling methods, such as stratified sampling, the expression for V_0 has to be modified accordingly by replacing the expression in square brackets by the appropriate covariance matrix for p (see also Chapter 2).

Aitchison and Silvey's method can now be used to find the asymptotic covariance matrix of $\hat{\pi}$, the ML estimate of π under the constraint $h(A'\pi) = 0$. Since the Jacobian of $h(A'\pi)$ with respect to π is AH, it is given by

$$ V(\hat{\pi}) = \frac{1}{N} \left\{ V_0 - V_0 A H \left(H' A' V_0 A H \right)^- H' A' V_0 \right\} , $$

where the '$-$' in the superscript refers to the Moore-Penrose inverse. In some cases, $H'A'V_0AH$ may be nonsingular and the ordinary inverse can be used, but in practice this is often not the case. Remember that, for nonsingular matrices, the Moore-Penrose and ordinary inverse coincide. The delta method leads to the following asymptotic covariance matrix of other estimated parameters:

(i) for the marginals,

$$ V(A'\hat{\pi}) = A'V(\hat{\pi})A ; $$

(ii) for the parameter vector $\hat{\theta} = \theta(A'\hat{\pi})$,

$$ V(\hat{\theta}) = T'A'V(\hat{\pi})AT $$

where $T = T(A'\pi)$ is the Jacobian of θ (see Section 3.3.4);

(iii) for the model parameter vector β in the formulation in Eq. 6.10,

$$ V(\hat{\beta}) = (X'X)^{-1}X'V(\hat{\theta})X(X'X)^{-1} . $$

The asymptotic covariance matrix of the vector of residuals $p - \hat{\pi}^M$ is given as

$$ V(p - \hat{\pi}^M) = \frac{1}{N} A'_M V_0 A H \left(H' A' V_0 A H \right)^- H' A' V_0 A_M , $$

and this expression can be used to calculate the vector of adjusted residuals $(p - \hat{\pi}^M)/\text{se}(p - \hat{\pi}^M)$, which are comparable to a random variable with a standard normal distribution.

Next, consider the case in which a constraint of the form $h(A'\pi) = 0$ is combined with a loglinear model of the form $\log \pi = W\gamma$. Denote by $\hat{\pi}_1$ the ML estimate of π, only under the loglinear model, not subject to the marginal model $h(A'\pi) = 0$. Then its asymptotic covariance matrix is

$$V_1 = V(\hat{\pi}_1) = D_\pi W (W'V_0W)^- W'D_\pi$$

(Haberman, 1977). Again using Aitchison and Silvey's method, the asymptotic co-variance matrix of $\hat{\pi}$, the ML estimate of π under both the loglinear and marginal model, is given as

$$V(\hat{\pi}) = \frac{1}{N} \left\{ V_1 - V_1AH \left(H'A'V_1AH \right)^- H'A'V_1 \right\}$$

Again, the delta method gives us the asymptotic covariance matrices of other estimated parameters:

(i) for the loglinear parameters,

$$V(\hat{\gamma}) = (W'W)^{-1}W'D_\pi V(\hat{\pi})D_\pi W(W'W)^{-1} \; ;$$

(ii) for the marginals,

$$V(A'\hat{\pi}) = A'V(\hat{\pi})A \; ;$$

(iii) for the parameter vector $\hat{\theta}$,

$$V(\hat{\theta}) = T'A'V(\hat{\pi})AT$$

where $T = T(A'\pi)$ is the Jacobian of θ (see Section 3.3.4);

(iv) for the model parameter vector β in the formulation in Eq. 6.10,

$$V(\hat{\beta}) = (X'X)^{-1}X'V(\hat{\theta})X(X'X)^{-1} \; .$$

To compute the adjusted residuals, we also need

$$V(p - \hat{\pi}^M) = \frac{1}{N}A_M'V_1AH \left(H'A'V_1AH \right)^- H'A'V_1A_M \; .$$

7

Conclusions, Extensions, and Applications

As stated in the Preface, this book has been written with quantitatively oriented social and behavioral science researchers in mind. Its main purposes have been to thoroughly explain the methodological issues and principles involved in marginal modeling and to indicate the types of substantive research questions for which marginal modeling is very useful or even the only means of obtaining the appropriate answers. Therefore, a large variety of real-world examples have been presented to help the reader understand the main issues involved. The book does provide a solid statistical basis for marginal modeling, but mathematical statisticians might prefer a more theoretical mathematical approach.

Especially for our intended audience, it is important to have good and accessible computer programs and routines to carry out the marginal analyses. In the last section of this chapter, several computer programs, including our own Mathematica and R routines will be introduced. The reader can find all data sets used in this book and the Mathematica and R routines to carry out the book's analyses on these data sets on the book's website, which will be given in the same section; these routines entail the often complicated matrices with the generalized exp-log notation needed to define the models.

Further, from the practitioner's point of view, a possible limitation of our approach might be that only categorical data have been dealt with. We had good, practical and theoretical reasons for this, as outlined before in the Preface. Nevertheless, how to handle research questions involving continuous data will briefly be indicated in Section 7.1.

When researchers are confronted with dependent observations and clustered data, they have several options for dealing with these dependencies, other than the marginal-modeling approach discussed in this book. Different models, especially random coefficient models, and different estimation procedures such as weighted least squares or GEE procedures might be chosen. These alternatives will briefly be discussed and compared in Section 7.2.

The real-world examples used in this book have been chosen to illustrate and represent as well as possible rather general types of research questions investigators might have. However, marginal models are also very useful to answer a large number

W. Bergsma et al., *Marginal Models: For Dependent, Clustered, and Longitudinal Categorical Data*, Statistics for Social and Behavioral Sciences,
DOI: 10.1007/978-0-387-09610-0_7, © Springer Science+Business Media, LLC 2009

of more specific research questions that one might not so easily link to marginal modeling. Just to illustrate this point and to underline once more the great flexibility and usefulness of the marginal-modeling approach, a few of such more specific research questions will be briefly presented in Section 7.3.

Finally, the present work on marginal modeling can be extended in several directions. Some indications of meaningful possible future developments will be discussed in Section 7.4.

7.1 Marginal Models for Continuous Variables

Until comparatively recently, linear models for multivariate normally distributed data have drawn much more attention from statisticians than models for categorical data. This is also true for marginal models. Many kinds of linear models for continuous data have been developed to deal with dependent observations, some of them widely known and applied, others less well-known, but available. Many of these models are essentially marginal models, although the term 'marginal' is rarely used. Moreover, many linear models for the joint multivariate distribution imply simple and direct inferences for the marginal distributions, unlike the situation for categorical data.

In this section, just a few of these marginal models for continuous data will be mentioned and briefly discussed, mainly to give the reader an idea of what kinds of marginal models for continuous data we are talking about. Many more details can be found in the references provided.

7.1.1 Changes in Means

Research questions involving categorical variables that have to be answered by means of marginal models often make use of data from repeated-measures designs, in which one or more characteristics are measured repeatedly over time on the same sample of subjects. The same is true when continuous variables are involved. Suppose a single continuous characteristic is measured at T successive time points, resulting in T variables. Of interest may be questions such as: does the mean response change over time, or are the changes over time linearly related to time itself? Hypotheses of these kinds can be tested by univariate or multivariate (M)ANOVA procedures for repeated measurements available in the standard statistical packages. Below, just a few main features of these procedures will be mentioned. For overviews of the basic techniques based on the general linear model, see Johnson and Wichern (2007) and Kim and Timm (2007).

The resulting (continuous) longitudinal data and growth curves are often analyzed by means of the univariate ANOVA procedure for repeated measures. This model is generally defined as a mixed model with subjects treated as a random factor and measurement occasions as a fixed factor. This basic model can be further extended by including fixed or random factors representing other independent variables, and their interactions. To obtain exact F-tests for testing null hypotheses about main and interaction effects, the within-subject covariance matrix has to have a specific structure. For a single group analysis, a necessary and sufficient condition for exact F-tests to exist (for p repeated measures) is the circularity condition: there exists a constant matrix C of order $p \times (p-1)$ such that

$$C'\Sigma C = \sigma^2 I$$

where $C'C = I$. The orthonormal matrix C defines a transformation of the repeated measures that yields a homogeneous covariance structure (Kim & Timm, 2007, p. 97). An example of a covariance matrix for the dependencies among the observations that satisfies the circularity condition is a matrix in which the entries on the main diagonal (the variances) are all the same, as are all off-diagonal cells (the covariances). As it was emphasized in this book that marginal modeling is about analyzing marginal tables, but without necessarily having to make restrictive assumptions about the dependency structure in the full table, one might say that this ANOVA model is not a marginal model because it contains explicit and restrictive assumptions about the nature of the dependencies in the data (see the circularity condition). On the other hand, the overall, marginal differences among the means for the T variables can simply and directly be inferred from the mean within-subject differences. For example, considering a research situation in which the simple t-test for matched pairs with husband-wife pairs, the mean within pair difference between husband and wife is equal to the mean difference between husband and wives.

An alternative, more 'marginal' procedure in the strict sense is provided by a particular variant of the multivariate ANOVA (MANOVA) procedure, which is specially useful when the circularity condition does not hold. In this MANOVA variant, the repeated measures of the same characteristic may be measures of different characteristics (although still on the same subjects). It is not necessary to assume that either the variances of the different characteristics (or the T variables concerning the same characteristic) are the same, or that the covariances among the measurements are similar. Hypotheses about the fixed effects (of time) can be tested without restricting the covariance matrix of the measures (Kim & Timm, 2007, Chapter 6). In this particular MANOVA model, the individuals in the sample are treated as exchangeable units drawn from a population in whose global (marginal) characteristics the researcher is interested. In other words, the individual differences only stem from the systematic fixed effects and from individual error terms (as in the standard regression model, for example). More complex MANOVA variants with a random subject factor are available to analyse data from more complex designs, such as arising in particular growth curve studies (Kim & Timm, 2007).

In sum, an important class of models for analyzing repeated continuous measures introduces random effects for individuals or clusters of individuals. In these linear mixed models, hypotheses about change over time can be formulated in terms of fixed effects and of variance components pertaining to the distribution of the random effects. For the class of models assuming multivariate normally distributed continuous variables, marginalizing the outcomes of the random-effect models or setting the (variances of the) random coefficients equal to zero, generally yields estimates that have a straightforward interpretation in terms the differences among the marginal means (but see also Section 7.2.2). In addition to the literature mentioned so far, the interested reader is further referred to Weerahandi (2004) for a comprehensive overview of exact methods for analyzing repeated measurements of continuous characteristics, and to Verbeke and Molenberghs (2000) and Molenberghs and Verbeke (2005, Chapter 4) for a comprehensive review of more general estimation and testing methods.

7.1.2 Changes in Correlation and Regression Coefficients

An investigator may also be interested in the changes in other characteristics than the mean — e.g., in the stability of correlation coefficients in repeated measurement studies. Suppose two characteristics X and Y are measured T times on the same sample of subjects, and let ρ_t be the correlation between X and Y at a particular time point t. A test of the hypothesis that the correlation remains constant over time ($\rho_t = \rho$ for all t) should take into account that the T correlation coefficients are computed on the same sample and, hence, are not independent of each other. Marginal modeling procedures to carry out this test have been proposed by Olkin and Press (1995) and, more recently, by Steiger (2005).

A similar question of stability over time may be stated for the regression coefficient b_{YX} when Y is regressed on X at each point in time. This problem (and the previous problem concerning the correlation coefficients) can also be recast in the form of a Seemingly Unrelated Regressions Model (SUR), first introduced by Zellner (1962). SUR-type models are often estimated by means of generalized least squares procedures (see Srivastava & Giles, 1987). It is less well-known that they can be recast as structural equation or path models with correlated error terms, for which full information maximum likelihood estimation is possible. If there are two measurements on the same subjects ($T = 2$), and if the researcher wants to test the equality of the unstandardized regression coefficients b_1 and b_2, where b_i represents the zero order 'effect' of X_i on Y_i, the SUR model consists of the two equations

$$Y_1 = b_1 X_1 + E_1$$

and

$$Y_2 = b_2 X_2 + E_2 \,.$$

In the original SUR model proposed by Zellner (1962), only the error terms E_1 and E_2 are allowed to correlate. Drton and Richardson (2004) already remarked that this implies that X_1 is conditionally independent of Y_2, given X_2, and that X_2 is conditionally independent of Y_1, given X_1. Although such strong conditional independence relations are probably violated in most real data sets because of omitted common causes, they did not provide a more general model that still captures the essential idea of a SUR model. Such a more general SUR model can be obtained by letting E_1 be correlated with X_2 and E_2 with X_1: each error term E_j is correlated with all exogenous variables that are not in the equation of the dependent variable Y_j corresponding to E_j.

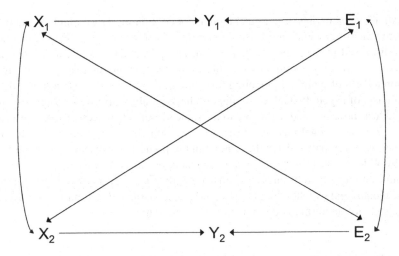

Fig. 7.1. The saturated SUR model

Fig. 7.1 represents this model graphically. The double arrows between E_1 and X_2, and between E_2 and X_1 represent the correlations between the error terms and the exogenous variables not in the relevant equation. Since both error terms are exogenous variables that may be considered as an amalgamation of left-out causes of Y_1 and/or Y_2, they may be correlated with the two other exogenous X_1 and X_2, at least as far as the error term E and the putative cause X pertain to different time periods.

If the two regression coefficients b_1 and b_2 differ, the resulting model is saturated and the hypothesis $H : b_1 = b_2$ can be tested by means of a log likelihood ratio test against this saturated model. An analysis of this kind can be carried out with an available software package for structural equation modeling (LISREL, EQS, AMOS, etc.).

In principle, this procedure can be applied in different equivalent ways since different saturated SUR models can be defined. However, one must make sure that the SUR model still estimates the zero order regression coefficients $b_{Y_1 X_1}$ and $b_{Y_2 X_2}$ that one wants to compare, and not some partial regression coefficient controlling for extra variables, e.g., comparing $b_{Y_1 X_1}$ and $b_{Y_2 X_2.X_1}$. The model obtained by replacing the arrow from X_1 to E_2 by an arrow from X_1 to Y_2, and the arrow from X_2 to E_1 by an arrow from X_2 to Y_1, while still allowing a correlation between E_1 and E_2, constitutes such an alternative, but misleading saturated SUR model. This model does not allow a comparison of $b_{Y_1 X_1}$ and $b_{Y_2 X_2}$, but does allow a comparison of the partial regression coefficients $b_{Y_1 X_1.X_2}$ and $b_{Y_2 X_2.X_1}$.

All above examples of marginal models for continuous data pertain to data from repeated measurement designs. But in principle, each of the general research questions discussed in the previous chapters for categorical data can be translated into similar questions for continuous variables. For example, an interesting psychometric application of marginal modeling for continuous variables arises in the study of parallelism of psychological tests. Suppose that k parallel forms of the same psychological tests have been administered to the same sample of subjects. Parallel tests are assumed to have equal means, equal variances, and equal covariances for all pairs of tests. A (marginal) test for this compound hypothesis was proposed a long time ago by Wilks (1946). Many more specific examples might be given, but as this last example and the previous one show, it is generally not too difficult to find the appropriate models and software for obtaining the correct parameter estimates for marginal models involving normally distributed continuous data.

7.2 Alternative Procedures and Models

7.2.1 Alternative Estimation Procedures: WLS and GEE

All statistical inferences concerning the marginal models in this book start with models that do not make a priori assumptions about the specific nature of the dependencies among the observations, and are based on full maximum likelihood estimation. The proposed estimation and testing procedures only assume that the observed frequencies in the joint table follow a multinomial distribution (see Section 2.3 for a more thorough discussion of these sampling issues).

Although not necessary, it is of course also possible, as described in previous chapters, to also define specific restricted models for the dependencies. For example, when investigating marginal homogeneity for the univariate distributions of A and B in turnover table AB with dimension $I \times J$, one might additionally postulate a loglinear uniform association model for the joint probabilities π_{ij}^{AB}. In this uniform association model, all local odds ratios in the $I \times J$ table are estimated to be the same — that is, all possible adjacent 2×2 tables using cells (i,j), $(i,j+1)$, $(i+1,j)$, and $(i+1,j+1)$ have the same value. As shown in previous chapters, such restricted models may involve rather complex constraints, but they can still be fitted in a straightforward way with our general algorithm.

In our procedure, it is also not necessary, as in some other full maximum likelihood methods, to define the marginal model by specifying the joint distribution of all variables involved in a stepwise, often very cumbersome manner by parameterizing first, second, third (etc.) order distributions (for an overview of this approach, see Molenberghs & Verbeke, 2005, Chapter 7).

In sum, the marginal-modeling approach advocated here is extremely flexible and straightforward, makes no unwanted assumptions about the dependencies, and carries with it all the desirable properties and nice advantages that full maximum likelihood procedures have regarding estimation of parameters and testing models.

However, for some very large problems, computations may be rather time consuming and may require a lot of memory space; sometimes for very large problems even a forbiddingly large amount. Alternative estimation and testing methods may, or even must then be considered. Excellent and very complete overviews of alternative approaches of marginal modeling are presented by Agresti (2002), Diggle et al. (2002), Lindsey (1999), Verbeke and Molenberghs (2000), and Molenberghs and Verbeke (2005). A few of the main conclusions from these works will be discussed below, but the reader is strongly advised to consult these books for many important particulars.

Computationally much simpler to obtain than ML estimates are Weighted Least Squares (WLS) estimates. In the categorical data context, WLS is often called GSK after Grizzle et al. (1969), who were the first to apply WLS systematically to categorical data analysis. They were actually the first to formulate a full-fledged categorical marginal-modeling approach (using WLS). See Koch, Landis, Freeman, and Lehnen (1977) and Reynolds (1977, Section 7.3.8) for an easy to read social science-oriented introduction.

ML and WLS estimates have asymptotically the same properties: they both provide Best Asymptotically Normal (BAN) estimates. WLS estimated parameters can be tested for significance by means of the Wald test, using the estimated variance-covariance matrix of the estimates. Testing the goodness of fit of the whole model is possible by means of a test statistic W, which compares the sample response functions with the model based estimates, weighted by the estimated inverse covariance matrix of the response function. Asymptotically, W has a chi-square distribution (Agresti, 2002, Section 15.1.2). The question may arise: why not simply use WLS rather than ML? The reason is that WLS has several serious disadvantages compared to the ML approach. First, when the observed marginal tables are sparse, the results of WLS tend to be much less reliable than those based on ML, and WLS easily breaks down if there are too many zero observed frequencies in the marginals. Second, nested models cannot be compared directly with each other, since W, unlike G^2, does not partition into separate parts pertaining to separate submodels. Third, WLS results depend on the particular parameterization used for a model — i.e., although this may be rare in practice, it is possible that a hypothesis is rejected if one parameterization is used but not when another one is used. Finally, WLS has not been developed to deal with latent class models, and it is doubtful if it would be advantageous compared to ML. WLS is probably most useful, and has a computational advantage over procedures for obtaining ML estimates, if the original joint table is very large but the marginal tables involved are not sparse.

At this moment, the most popular estimation method for marginal models is the use of Generalized Estimating Equations (GEE) proposed by Liang and Zeger (1986). See also Hardin and Hilbe (2003), who provide a comprehensive introduction to the GEE methodology. GEE estimates are meant to be close approximations of the full ML estimates. Full information ML estimation needs assumptions about the sampling process (e.g., multinomial) to make sure that the kernel of the log likelihood functions is appropriately specified. The ML-estimated frequencies in the joint table follow from the assumed sampling distribution under the imposed marginal restrictions. Roughly speaking, GEE is not based on a statistical model and does not lead to an estimate of the full observed joint distribution of the variables. In this way, many of the important practical problems of 'ordinary' ML estimation can be avoided. In marginal modeling, GEE provides estimates of the marginal means or probabilities under a model that includes the marginal assumptions and a 'working covariance matrix' of the dependence structure of the observations. This working matrix can have a very simple form, even assuming independence among the observations. The GEE estimates for the marginal parameters are consistent whatever the nature of the postulated working matrix. However, the estimated standard errors of the GEE estimates may depart strongly from the true standard errors if the specified working matrix departs from the true situation. Corrections of the GEE standard errors taking the observed data into account, using a so-called sandwich estimator, are necessary and will generally yield much better estimated standard errors. For a discussion of these sandwich estimates, see Agresti, 2002, Section 11.4. In sum, a drawback of GEE is that a working covariance matrix has to be chosen, which may be arbitrary, and may be inaccurate. Furthermore, GEE has not been developed for latent class models.

As documented in the general references given above in this subsection, GEE parameter estimates and the corrected standard errors often approach the ML estimates rather well. However, one has to be careful. There are yet no clear general and practical rules for deciding when the approximation will be satisfactory. Therefore, it is hard to judge for a researcher, whether in his or her particular case GEE will perform well. Further, the examples from the literature in which ML and GEE are compared are almost all rather simple, usually much more simple than the real-world examples discussed in the previous chapters. Moreover, there are some indications in the above literature that when there are strong dependencies in the data that are ignored in the working covariance matrix for the dependencies, the corrections for the standard errors do not work well. Especially in large, complex data sets, very strong dependencies may exist and not enough is known about the behavior of the standard error estimates in these cases. Finally, testing models has to be done by means of testing the significance of the separate parameters. Because GEE is not full maximum likelihood, there is no direct way of comparing nested models with the saturated model, or with each other. Nevertheless, whenever the computational difficulties of full ML estimation are too large to overcome, GEE may be the preferred viable alternative, and should probably be preferred to WLS.

7.2.2 Modeling Dependent Observations: Marginal, Random Effects, and Fixed Effects Models

There are many ways of handling the dependencies in the observations. Extensive comparisons of the several approaches are given by Agresti (2002), Diggle et al. (2002), Verbeke and Molenberghs (2000), and Molenberghs and Verbeke (2005). In addition, Raudenbush and Bryk (2003), Snijders and Bosker (1999), and Hsiao (2003) can be recommended, especially for their treatments of random coefficient models. Again, only a few important highlights from the relevant literature will be presented here and the reader should definitely consult these general references for many important particulars.

The important starting point of our discussion here on the comparison of marginal modeling and alternative procedures is that marginal modeling is about research questions that involve statements about marginal tables. Marginal modeling is essentially about testing restrictions on marginal distributions and estimating marginal parameters under the specified marginal model. Therefore, the focus here will be how alternative approaches handle this basic question compared to marginal modeling.

It is sometimes suggested that models for the joint distributions are always to be preferred above marginal approaches because marginal models do not provide information about the joint table, but models for joint distributions can always be used to investigate their consequences for the marginals. In a way, this statement is true, and in a way, it is not. It is generally true that the processes that go on at the joint level cannot be deduced uniquely from a marginal model, as such. In principle, it is always possible to imagine different models for the joint probabilities that all lead to the same (restricted) marginal distributions. Therefore, marginal models as such can, in principle, not be used for testing models for the joint frequencies. On the other hand, in general, given a particular data set, marginal restrictions do result in unique ML estimates for the joint probabilities. For example, if in turnover table AB, MH is assumed for the marginal distributions A and B, the ML estimates of the joint probabilities $\hat{\pi}_{ij}^{AB}$ under MH will generally be different from the observed p_{ij}^{AB} and the estimated odds ratios estimating the dependencies between A and B based on $\hat{\pi}_{ij}^{AB}$ will generally be different from the observed odds ratios based on p_{ij}^{AB}. In this sense, the MH model tells us what goes on in the joint table under the MH hypothesis and how this differs from the saturated model for the joint table.

Viewed from the other side, a model for the joint probabilities can always be marginalized and investigated regarding its consequences for the marginal tables by summing over the appropriate (cluster) variables (Lee & Nelder, 2004). If in turnover table AB, a linear-by-linear association model is postulated (without further marginal constraints), the estimates $\hat{\pi}_{ij}^{AB}$ can be used to investigate the model's consequences for the marginal probabilities $\hat{\pi}_{i+}^{AB}$ and $\hat{\pi}_{+j}^{AB}$. Sometimes models for the joint probabilities even have explicit implications for the marginal distributions. For example, the symmetry model $\pi_{ij}^{AB} = \pi_{ji}^{AB}$ implies $\pi_{i+}^{AB} = \pi_{+i}^{AB}$. In that sense, such a model for the joint table can be used to test for marginal restrictions, but only under the condition that the model is true in the population. Moreover, the connection with MH is easily seen in the symmetry model. But, in general, it is not so easy (and usually very cumbersome, or next to impossible) to define marginal models in terms of simple restrictions on meaningful parameters for the joint table. In that sense, models for the complete table are difficult to use for testing explicit marginal hypotheses, even if such a joint model is true.

An important exception to this last statement is formed by standard linear models for multivariate normal distributions. For example, as implicated in the discussions in the previous Section 7.1 about the repeated measurement design with continuous data, the fixed effects of Time or other variables in (M)ANOVA models with a random subject effect (a random-intercept model) can also be given a marginal interpretation. For example, if on average within subjects, the growth curve for the dependent variable Y is linear over time with regression coefficient b, then this coefficient b can also be interpreted as the effect on the population level, and the estimated overall sample means will also grow linearly over time with the same coefficient b. This follows from the fact that linear models involve expectations of sums, and the standard rule that the expected value of the sum is the same as the sum of the expected values can be used: $E(v+w) = E(v) + E(w)$. If the model is nonlinear, such transfer of interpretations of the fixed effects is generally not possible. The standard example is the growth curve for a particular response. If the individual growth curves are logistic, their average will not be logistic and the partial individual logistic curve will be different from the marginal logistic growth curve. This has a lot to do with the simple fact that $E \ln(v+w) \neq E \ln(v) + E \ln(w)$. Molenberghs and Verbeke give more precise particulars on the relationship between marginal and random-coefficient models. They also specify under which conditions individual level effects can be given a population-averaged interpretation (Molenberghs & Verbeke, 2005, Chapter 16; see also Agresti, 2002, Section 12.2.2).

How the above general statements work out in practice may be further clarified by means of a simple imaginary example in which the marginal-modeling approach will be compared with random (hierarchical, multilevel) effects models. Imagine a repeated measurement design with five monthly measurements on the same subjects, with the time measurements nested within 200 individuals randomly drawn from some population. In terms of hierarchical modeling, time is here the lower level 1 variable, whereas the individuals define the higher level 2. Imagine further that a campaign to quit smoking was going on in the media during the five months the study lasted. The dependent variable is the dichotomous response variable S - *Smoking* or not, with categories 1 = Respondent does not smoke and 2 = Respondent smokes. The independent or explanatory variables are G (*Gender*) with 1 = Man and 2 = Woman, and T (*Time*) with categories 1 through 5. The researcher is interested in how effective the campaign is in terms of reducing the percentage of smokers in the population over time and whether those changes are the same for men and women.

The data can be summarized in a full table $GS_1S_2S_3S_4S_5$ (with variable S_t representing smoking status at time t), which has dimension 2^6. The relevant collapsed marginal table GTS (with dimension $2 \times 5 \times 2$) contains the marginal distributions of the binary response variable S (*Smoking*) stratified for *Gender G* and *Time T*. A marginal model to study the research questions stated above would start with imposing the restrictions contained in loglinear model $\{GT, TS, GS\}$ for table GTS (where, as done throughout this book, for the representation of the marginal restrictions, this table is treated as if it were a normal table rather than containing dependent marginal distributions of S). Model $\{GT, TS, GS\}$ implies the absence of the three-variable effect GTS: the response growth curves over time are the same for men and women or, what amounts to the same, the men-women response differences are the same at all points in time. In terms of loglinear parameters, the odds ratios in the five conditional 2×2 tables GS, one for each of the five time points, are all the same. However, the observations in these five subtables are not independent of each other because they involve the same respondents. Marginal modeling is needed to estimate and test the model. The interpretations of the effects in this model are on the marginal level. The effects of T in loglinear model $\{GT, TS, GS\}$ estimate how the marginal odds $\pi_{1\,t\,g}^{S|TG} / \pi_{2\,t\,g}^{S|TG}$ change over time within the subpopulation Men and, equally so, given the model, within the subpopulation Women. The effect of G in the same model estimates how different $\pi_{1\,t\,g}^{S|TG} / \pi_{2\,t\,g}^{S|TG}$ is for Men and Women at each point in time, where these differences are the same for all time points. The outcomes of the marginal-modeling approach provide direct information on the research question: how does the percentage of male and female smokers in the population change over time under the influence of the campaign (while taking the dependencies of the observations over time into account).

An alternative approach might be to start with the analogous logit model, but now at the level of the 5×200 observations. Then, one may write *for each individual* i (in a somewhat different notation than used so far to emphasize that individual i is in principle not a category of an ordinary variable I),

$$\ln \left(\frac{\pi_{i1\,t\,g}^{S|TG}}{\pi_{i2\,t\,g}^{S|TG}} \right) = \mu + \beta_1 t + \beta_2 g \, .$$

The left-hand side term in this equation represents the individual specific odds of not smoking versus smoking, given T and G for individual i. However, the five time observations are nested within the individuals and therefore the measurements at the five time points are not independent of each other. This dependency (i.e., the (between) variance that is explained by the cluster variable *Individual*) should be accounted for. Therefore, the model is changed into

$$\ln \left(\frac{\pi_{i1\,t\,g}^{S|TG}}{\pi_{i2\,t\,g}^{S|TG}} \right) = \mu_i + \beta_1 t + \beta_2 g \qquad (7.1)$$

where subscript i in μ_i refers to cluster i, and in this case to individual i. The intercept μ is assumed to be different for each individual. The effects of T and G are assumed to be the same for all individuals and are treated as fixed effects. The term μ_i can be considered either as a fixed effect or as a random effect.

If μ_i is regarded as fixed, cluster variable, *Individual* is essentially treated as a nominal-level variable with, in this case, 200 categories. The term μ_i represents 200 dummy variable effects, one for each individual. Or, adding a constant term μ_0 to the equation, there will be $200 - 1 = 199$ dummy variable effects

$$\mu_i = \mu_* + \sum_1^{199} \beta_{D_i} D_i,$$

where dummy variable D_i equals 1 if the observations belong to individual i and 0 otherwise. If the effects of T or G are also assumed to vary among individuals, fixed interaction effects might be added by multiplying T or G by the dummy variables D.

In a random-effect interpretation of μ_i, the intercept for the individuals is regarded as consisting of a common mean element μ_0, plus for each individual separately an error term — a deviation from the mean in the form of a random draw from an assumed common population distribution (a normal distribution or some other member of the exponential family),

$$\mu_i = \mu_* + u_i,$$

with $E(u_i) = 0$. In the standard application of this model, the random errors u_i are assumed to be independent of the explanatory (exogenous) variables in the model.

If interactions of i with T and G are assumed, the effects of T and G might also be defined in a similar way as random (slope) effects:

$$\ln \left(\frac{\pi_{i1\,t\,g}^{S|TG}}{\pi_{i2\,t\,g}^{S|TG}} \right) = \mu_* + u_i + \beta_{1i}t + \beta_{2i}g,$$

with

$$\beta_{1i} = \beta_1 + \varepsilon_i$$

and

$$\beta_{2i} = \beta_2 + \delta_i .$$

If the model in Eq. 7.1 has a random intercept ($\mu_0 + u_i$), but the effects of T and G are considered fixed, the model as a whole is often called a mixed model.

When should fixed-effect models be used for the clustering, when random-effect models, and how do these two compare to marginal modeling? Random-effect models are especially useful if the clusters are drawn randomly from some population and if the number of clusters is large (say, larger than 20), but the number of observations within the clusters are small (say, less than 6). Fixed-effect models, on the other hand, are most useful if the clusters are not drawn randomly, and when their number is small, but the number of within cluster observations is large.

In principle, random-effect models assume a random sample of clusters and are appropriate for multistage sampling procedures. Fixed-effect models treat the clusters as the mutually exclusive and exhaustive categories of the cluster variable and assume essentially a form of stratified sampling, where the stratifier is the cluster variable. If the clusters are not a random sample, the logic of random effects breaks down, unless some other probability mechanism will be present that mirrors random sampling. Sometimes random measurement error is used as such a mechanism, but then it must be assumed to operate at the cluster level. The discussion of when to use random effects, and when not, has a lot in common with the old discussion about the application of statistical inference procedures to population data.

In the above smoking example, the individuals were randomly drawn. From this point of view, random effects are appropriate here. Moreover, treating the individuals as fixed-effect dummy variables may give rise to a serious statistical problem, because the estimates of some or all structural parameters may no longer be consistent. Some estimates may not improve if more and more individuals are added to the sample, because each added individual adds another parameter, too (this difficulty is sometimes solved by using conditional ML inference procedures — e.g., in Rasch models, see also Molenberghs & Verbeke, 2005, Section 13.3.2).

In the random effects variant, this particular statistical problem is circumvented, because, assuming u_i has a normal distribution, the only parameters to be estimated for the distribution of the individual random effects are its mean μ_0 and variance σ_u^2. In many research situations, however, random-effect models do not make much sense. In most cross-national research, it is hard to imagine an appropriate random mechanism to justify random effects, even when 30 or more countries are being compared. It often makes much more sense to consider the, say, 50 countries as 50 categories of the variable 'Living in country X,' not different from belonging to age category or birth cohort X. Also, the argument that random-effect models are needed to find the explained variance by aggregate characteristics, and the variance at the aggregate level, does not hold. Similar explained variance analyses can be performed at the fixed level, if one is willing to make similar restrictions. Or perhaps formulated a bit more accurately, the seemingly extra potential of random-effect models follow from the extra, generally unverified assumptions that are being introduced in random-effect models. If one would be willing to assume, for example, that the fixed cluster dummy effects are the same for a particular subset of the dummies and that the relationships between the dependent variable and the aggregate level variables are linear, the same possibilities exist for fixed-effect models. Essentially, multilevel analysis is needed for investigating the effects of higher-level variables on lower-level characteristics, but not necessarily in the form of random coefficients.

Further, if there are just a few clusters, even if randomly chosen, random-effect models may not be applicable, simply because the sample size is too small to get robust and accurate estimates of the parameters of their distribution. On the other hand, if for each cluster just a few observations are available, as in the smoking example above, fixed-effect models become problematic because there are not enough observations per category of the cluster variable to yield stable estimates. Random effects usually perform better then, because they make explicit assumptions about (equality of) within cluster variances and restricted models generally yield more robust estimates than less-restricted models (but see also below) (Snijders & Bosker, 1999, p. 43).

Finally, in most standard applications of random-effect models it is assumed that the random components (the latent variable) u_i (and ε_i or δ_i) in the above equation(s) are independent of each other and of the other independent variables in the equation (such as G). In fixed-effect analyses using dummies for the clusters, such an assumption is not made. In that sense, fixed-effect models may capture a larger part of the unobserved heterogeneity (of the cluster dependencies) than random-effect models.

Now how does marginal modeling compare to these two approaches? First, it is good to realize that the marginal-modeling approach advocated in this book also assumes a (multinomial) probability distribution of the joint variables and, therefore, a random sample of clusters (as in all examples in this book). In that respect, it is more similar to random than to fixed effects analysis.

Second, the fixed analysis interpretation of Eq. 7.1 investigates the effects of T and G on average within individuals. As such, the fixed analysis does not answer the research question about the population impact of the campaign against smoking. A seemingly simple and straightforward kind of 'marginal fixed analysis' would be possible by just leaving out the dummy effects for the individual. But then, of course, the dependency of the observations is no longer accounted for. These dependencies get confused with the effects of the other variables and standard statistical procedures that assume independent observations break down.

Also in a mixed model such as in Eq. 7.1, with a random intercept and fixed slopes, the fixed effects for G and T are partial effects on average within individuals (and controlling for T or G, respectively). They are not the intended marginal, estimated population effects, and the random-intercept model also does not answer the research question that started this imaginary investigation directly.

The random-intercept model might be marginalized to investigate the marginal consequences. However, it should be kept in mind that, in general, random-effect models are not saturated models for the dependencies in the data — something that many researchers forget in their routine applications of random-effect models. Unlike marginal models, random-effect models make assumptions about these dependencies, such as the above-mentioned assumption of independence of the random terms of the other variables in the equation. For example, in our simple random-intercept example above, it may be assumed that the autocorrelation pattern over time is caused by the random factor 'Individual' with constant variance σ_u^2. Therefore, the correlation ρ between adjacent time point t and $t+1$ is the same for all values of t; moreover, the same value ρ will be found for t and $t+2$, and for t and $t+3$, as well as t and $t+4$ (cf. the circularity condition in Section 7.1.1). If, however, in reality, an autocorrelation model of lag 1 would apply, the autocorrelations between all t and $t+1$ would be ρ, but for lag 2 (t and $t+2$) it would be ρ^2, etc. A nice feature of random-effect models is that such structures can be defined. But the problem is that the researcher does not know the true dependency structure in the population. If this structure is misspecified, marginalizing can lead to wrong inferences about the marginal distributions. A solution would be to define saturated random-effect models (Lee & Nelder, 2004). But it is certainly not always easily seen how to do this, and if it is possible, it is certainly not simple (and is usually very cumbersome) to specify the further restrictions in such a way that the intended marginal restriction will follow (which was the main focus of the comparison of the alternative techniques).

In sum, marginal restrictions are by far most easily and flexibly handled by marginal models as described in this book, while retaining the possibility to specify a particular joint distribution.

7.3 Specific Applications

As the previous chapters have made abundantly clear, dependencies among observations that have to be reckoned with arise in many research situations, partly because of the research design, partly because of the nature of the research questions. Repeated observations of the same subjects are often a cause of the dependencies, as are several forms of explicit clustering or multi-stage sampling designs (e.g., regions-communities-households-individuals or schools-classes-pupils). But more subtle or implicit varieties in design also exist that can make the observations dependent. For example, in surveys, using face-to-face interviews, the respondents are clustered within interviewers. In almost all analyses of such survey data (including the authors' own substantive analyses), this possible source of dependencies among the observations is completely ignored with possibly serious consequences for statistical inferences (for an early illustration, see Hagenaars & Heinen, 1982). As discussed in the previous section, this clustering effect can and must be taken into account in one of the several indicated ways. Ignoring the problem is not among them, but marginal modeling is.

Below, three other, very different and more specific concrete research situations will be described in which marginal modeling is the appropriate approach: the analysis of data from multiple responses, the analysis of dyadic data, and testing properties of Mokken (cumulative) scales. They are presented here as instances of the many specific research questions where it is not clear how to obtain satisfactory statistical answers using standard techniques, but where awareness of marginal modeling offers an adequate and flexible statistical framework.

7.3.1 Multiple Responses

Multiple responses to a single questionnaire item occur when the respondents are asked to choose an arbitrary number of alternatives (not necessarily just one) from a given list of options. Loughin and Scherer (1998), for instance, analyse data in which livestock farmers were asked to indicate from which sources they obtained their veterinary information: (A) professional consultant, (B) veterinarian, (C) state or local extension service, (D) magazines, and (E) feed companies and representatives. Since farmers may obtain their information from different sources, multiple responses were allowed. In their sample of 262 farmers, 453 responses were recorded, so the mean number of responses per respondent was 1.73.

To investigate whether respondents with a different educational background used different sources, the data can be summarized in a contingency table *Education* × *Info source* using all 453 observed responses. However, the routine application of the traditional chi-square test for independence is inappropriate here, since the 453 observations are not statistically independent because of clustering within farmers. This problem was recognized a long time ago and several solutions have been proposed. For example, Loughin and Scherer (1998) proposed to test the significance of the chi-square value by means of a bootstrap procedure. Decady and Thomas (2000) and Bilder and Loughin (2004) discussed modifications and corrections of the traditional chi-square test to make it appropriate for multiple responses. Agresti and Liu (1999) were the first to apply the full ML marginal-modeling approach to data, including variables with multiple responses.

The analysis of multiple-response data amounts to comparing marginal distributions that are, in our terminology, partially dependent (see the analyses of the NKPS data in Chapters 2, 5, and 6). Agresti and Liu treat a dependent variable Y with c

multiple responses as a set of c binary items with scores equal to zero if the corresponding alternative is not chosen, and equal to one if it is chosen. Each alternative j of Y corresponds to a binary variable J with $J = 1$, if j is chosen and $J = 0$ otherwise. Now suppose that the researcher is interested in the effects of an explanatory variable X with r levels on the (multiple) response. Let $\pi_{1|i}^{J|X}$ denote the probability of selecting alternative j of Y in the subgroup with $X = i$. Then, $\{(\pi_{1|i}^{J|X}, 1 - \pi_{1|i}^{J|X}), J = 1, \cdots, c\}$ are the c marginal univariate distributions for the responses when $X = i$.

Independence of X and Y can now be tested by means of the multiple marginal independence model

$$\log \left(\frac{\pi_{1|i}^{J|X}}{1 - \pi_{1|i}^{J|X}} \right) = \beta_j \,,$$

which states that the probability of choosing alternative j is the same for all values of X. Other less-restrictive, but still nonsaturated marginal models that can be considered in this context are

$$\log \left(\frac{\pi_{1|i}^{J|X}}{1 - \pi_{1|i}^{J|X}} \right) = \alpha_i + \beta_j \,,$$

or, when X has ordered levels

$$\log \left(\frac{\pi_{1|i}^{J|X}}{1 - \pi_{1|i}^{J|X}} \right) = \beta_j + \gamma_j x_i \,.$$

In the latter two models, it is tested whether the (linear) effect of X on each response dummy is the same for all responses.

All these models can be estimated and tested as marginal models defined on the $r \times 2^c$ table, which contains the response patterns for X and the c binary items representing the multiple-response variable Y.

When both X and Y allow multiple responses, two different sets of binary items are defined: r items $I = 1, \ldots, r$ for X, and c items $J = 1, \ldots, c$ for Y. Independence of X and Y can now be tested by a simultaneous test of the $r \times c$ independence hypotheses

$$\pi_{11}^{IJ} = \pi_{1+}^{IJ} \pi_{+1}^{I\,J}$$

for each pair (I, J) of binary items from the X and the Y set. This test is based on the appropriate marginal model for the $2^{(r+c)}$ table, which contains the response patterns for the $r + c$ binary variables representing X and Y.

7.3.2 Categorical Dyadic Data

Dyadic data (Kenny, Kashy, & Cook, 2006) are obtained when the basic unit of analysis is not an individual, but a pair (dyad) of individuals, who interact with each other. Dyadic data abound in social-psychological and sociological research: data collected on husband and wife, parents and children, teacher and students, etc., are just a few examples. A characteristic of dyadic data is that, because of the interaction

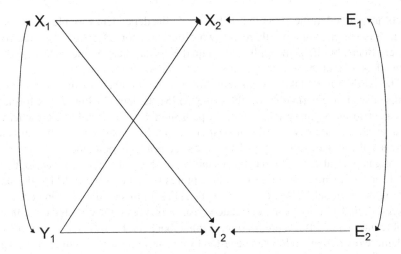

Fig. 7.2. The Actor-Partner Interdependence Model for dyadic data

between the members of a dyad, the observations on them are not statistically independent. Analysis of dyadic data should take this pairwise dependence into account. In a way, and at many places, we have been using dyadic data in this book, such as the above-mentioned parent-child NKPS data in Chapters 2, 5 and 6, but always ignoring, from a substantive point of view, their dyadic character. In this subsection, it will be shown how to handle such data, but now focusing on the dyadic character.

An important distinction for the (marginal) analysis of dyadic data is between situations in which the members of a dyad are clearly distinguishable, and situations in which that is not the case. Husband and wife, and teacher and student, are two examples of dyads with distinguishable members: one member can clearly be identified as the husband (or teacher), the other as the wife (or student). However, in other examples, such as friendship pairs, it is completely arbitrary who is considered the first member of the pair and who is the second one; such dyads consist of indistinguishable members.

An interesting general model for analyzing dyadic data with distinguishable or indistinguishable members is the Actor-Partner Interdependence Model (APIM) described in Kenny et al. (2006). Fig. 7.2 depicts the basic ideas behind this model. For each of the two dyad members, two variables X and Y are observed with X antecedent to Y. The variables pertaining to the first member are denoted by X_1 and Y_1, and those pertaining to the second member by X_2 and Y_2. Actor effects are defined as the effect that X_i has on Y_i, the dependent variable for the same member. Partner effects are defined as the effect that X_i has on Y_j, the dependent variable for the partner. In this simple situation, there are two different actor effects and two different partner effects, but in some applications equal actor or equal partner effects may be assumed. Note that the two error terms are allowed to be correlated.

The APIM model takes on a simple form if the members of the dyad are distinguishable. For continuous variables, the APIM depicted in Fig. 7.2 is formally equivalent to an ordinary two-equations path model (with correlated error terms) that does not require any use of marginal modeling. Its parameters can be estimated by means of standard software packages for structural equation models.

For categorical data summarized in table $X_1X_2Y_1Y_2$, the nonsaturated loglinear model for the distinguishable dyads that captures the core of the APIM is the standard loglinear model $\{X_1X_2, X_1X_2Y_1, X_1X_2Y_2, Y_1Y_2\}$. In this model, the direct relation between Y_1 and Y_2 mirrors the correlated error term, and some possible extra three-variable interaction effects are present. If this model can be accepted for the data at hand, more restrictive models can be tested by constraining some loglinear model parameters in the original model — e.g., setting the loglinear three-variable interaction parameters equal to zero. None of these analyses requires marginal modeling. In general, when the relationship between the members within a dyad is asymmetric, and the members are clearly distinguishable (such as in the teacher-pupil relation), no further marginal constraints have to be imposed on the model parameters of the linear or loglinear model.

This is no longer true when the members of a dyad are indistinguishable. Then, both members of the dyad become exchangeable and the structure of the APIM should be adapted to this exchangeability condition. Although marginal modeling is still not always or strictly required, models incorporating the exchangeability condition are, at the very least, most easily formulated and estimated using marginal-modeling terminology and procedures.

For continuous data, the most common practice to express exchangeability is to equate the means of the two X variables, as well as their variances, and to equate the variances of the two error terms. Moreover, the two actor effects (effects of X_1 on Y_1 and of X_2 on Y_2), and the two partner effects (effects of X_1 on Y_2 and of X_2 on Y_1), should also be set equal. In this way, it does not matter for the distribution of the four random variables who is taken as first and who as the second member of the dyad. Parameters are equated in such a way that the joint distribution f of the four random variables (X_1, X_2, Y_1, Y_2) satisfies the permutation condition

$$f(x_1, x_2, y_1, y_2) = f(x_2, x_1, y_2, y_1) .$$

Equating the means and variances are marginal restrictions. Software for structural equation models that can handle (restrictions on) first and second moments (means and variances) and regression coefficients can be used to estimate all parameters of the APIM models for continuous data from indistinguishable dyads.

For categorical data, when the members of a dyad are indistinguishable, extra constraints have to be imposed on model $\{X_1X_2, X_1X_2Y_1, X_1X_2Y_2, Y_1Y_2\}$ that guarantee that the expected frequencies satisfy the exchangeability condition defined above.

Denoting X_1 and X_2 by A and B, respectively, and Y_1 and Y_2 by C and D, respectively, the joint probabilities must be restricted as follows:

$$\pi_{i\,jk\ell}^{ABCD} = \pi_{ji\,\ell k}^{ABCD}\,.$$

The simplest way to estimate this complete model that consists of the conjunction of the loglinear and the exchangeability constraints is using the marginal-modeling techniques discussed in this book. Because in this particular APIM model the exchangeability restriction amounts to a particular form of symmetry, the complete model can also be formulated in terms of restrictions on the parameters of the assumed loglinear model, as Bishop et al. (1975, Chapter 8) and Hagenaars (1990, Chapters 4, 5, and 6) have shown. However, these kinds of restrictions become rather complex and the model is easily misspecified. For example, the exchangeability condition implies not only restrictions on the first- and second-order parameters, but also on the three-variable interactions in the given example such as

$$u_{i\,jk}^{ABC} = u_{i\,jk}^{ABD}$$
$$u_{i\,jk}^{ABC} = u_{ji\,k}^{ABC}\,.$$

It is much simpler and less prone to mistakes to impose the restrictions directly to the probabilities itself using the marginal-modeling approach. On the other hand, it should be realized that the loglinear model must be compatible with these marginal exchangeability restrictions. This means, in terms of our example, that when the unrestricted three-variable interaction term ABC is present in the model, the term ABD should also be included, as follows from the required parameter restrictions needed for the exchangeability condition indicated above.

7.3.3 Mokken Scale Analysis

Mokken scale analysis (Mokken, 1971; Sijtsma & Molenaar, 2002) is used for scaling items and measuring respondents on an ordinal scale. It can be regarded as a probabilistic version of the deterministic cumulative or Guttman scale. The Mokken scale is based on the monotone homogeneity model, which is a nonparametric item response model. Here, only dichotomous items will be considered — i.e., items with only two responses, one of which can be labeled as the positive response 1 and one as the negative response 0. Let π_1^i denote the probability of a positive response 1 to item i (using a convenient ad hoc notation).

In testing, whether a set of items forms a Mokken scale, several scalability coefficients play a crucial role. Suppose we have a set of J dichotomous items that are ordered with respect to decreasing popularity. Hence, item i precedes item j in this ordering when $\pi_1^i > \pi_1^j$. A *Guttman error* is said to occur if someone solves (or responds positively to) the more difficult item of a pair but not the easier one — i.e., when for an item pair (i, j) with $\pi_1^i > \pi_1^j$, someone answers 1 on item j but 0 on item i. Let F_{ij} be the number of Guttman errors occurring in a sample of n subjects for the pair of items (i, j) with $\pi_1^i > \pi_1^j$, and let E_{ij} be the expected number of such errors under the hypothesis that X_i and X_j are independent. Then three different types of scalability coefficients can be defined.

First, for each item pair (i, j) with $\pi_1^i > \pi_1^j$, the scalability coefficient H_{ij} is

$$H_{ij} = 1 - \frac{F_{ij}}{E_{ij}} .$$

Second, the scalability coefficient for an individual item j is defined as

$$H_j = 1 - \frac{\sum_{i \neq j} F_{ij}}{\sum_{i \neq j} E_{ij}} .$$

Finally, the scalability for the total scale for all items becomes

$$H = 1 - \frac{\sum_{i=1}^{J-1} \sum_{j=i+1}^{J} F_{ij}}{\sum_{i=1}^{J-1} \sum_{j=i+1}^{J} E_{ij}} .$$

Sijtsma and Molenaar (2002) provide more information on the properties and uses of these scalability coefficients.

For investigating the properties of a Mokken scale, tests for several hypotheses about these scalability coefficients are relevant, such as

- Are all or some of the scalability coefficients different from zero?
- Are all or some of the scalability coefficients larger than a scalability criterion c, which is larger than zero?
- Are the scalability coefficients of the different items equal to each other?
- Are the item scalability coefficients equal in different groups?

To test such hypotheses, it must be taken into account that the H coefficients are defined on the same subjects, and therefore involve dependent observations. The marginal-modeling approach can handle these dependencies. Van der Ark, Croon, and Sijtsma (2008a) and Van der Ark, Croon, and Sijtsma (2008b) showed that all scalability coefficients can be written as functions of the response pattern frequencies using the generalized exp-log notation. Marginal modeling can then be applied to test the hypotheses about the H coefficients and to obtain estimates (and standard errors) of these coefficients under various restrictive models. An advantage of using marginal-modeling techniques in this context is that no further simplifying assumptions about the distribution of the H coefficients are needed. More details on data and analyses can be found in Van der Ark et al. (2008a) and Van der Ark et al. (2008b), and on the book's website.

7.4 Problems and Future Developments

In the process of writing this book and carrying out the analyses, a number of difficulties and questions came up that needed answers, but answers that were not available then and not directly available now. These problems will be presented in this section, but not to solve them here and now, not even to promise to work on it, or to try to solve them in the near future. Nothing more is intended here than making the reader aware of some problems that still exist and of possible future developments.

The full ML marginal-modeling approach is based on the asymptotic properties of the estimators, as is often true in statistics. However, researchers work with *finite samples*. More should be known about the behavior of the estimates in finite samples. Although the ML method seems to work very well, even with samples that are not very large and with extremely sparse joint tables, this impression should be further investigated. More needs to be known about how the estimates behave depending on the number of clusters, the number of observations within the clusters, the strength of the dependencies, the sparseness in the joint table, and the sparseness of the marginal table, etc.

Practical problems were encountered with respect to the size of the joint table, the required memory space, and the time-consuming iteration process. With improving computer hardware and software, these problems will no doubt be much lessened, but in the end only to a limited extent because computational complexity increases exponentially with the number of variables involved. What is really needed for very (or extremely) large problems is an improved fitting algorithm that has less computational requirements.

At several places in this book, NKPS data were used, in which the data were clustered within families, and in which one parent and one child was selected from each family. The marginal models for these data nicely took the dependencies in the observations into account. But a problem remains. This data set, because just one child and one parent were chosen, is not representative of the population, because it underrepresents large families. *Weights* are needed to correct for this. This is a common situation in cluster sampling, and the problem can in fact be solved in various ways using marginal-modeling procedures. For example, in the above family problem, a variable *Family size* could be introduced, over which we need to marginalize using appropriate weights. Alternatively, if certain groups are known to be over- or undersampled, the marginal distribution of the grouping variable can be corrected using marginal modeling.

A completely different kind of possible extension came up in discussions about the usefulness of the marginal-modeling approach for the analysis of social mobility tables. In such analyses, use is often made of 'association-2' or *bilinear models*, in which linear restrictions are imposed on the log odds or the log odds ratios, but using *estimated variable scores* rather than fixed ones. These models cannot be simply expressed in terms of zero restrictions on linear combinations of the cell probabilities, as required in our marginal-modeling approach. Perhaps some form of an alternating algorithm in which the variable scores are estimated first, and then the marginal restrictions are applied, and then back again might do the trick.

Finally, in many examples throughout this book, the data sets contained *missing data*. For the time being, this issue was completely ignored. However, given the omnipresence of missing data and its often selective nature, marginal models should take the (partially) missing observations into account. In principle, handling missing data is a form of latent variable analysis, and the procedures discussed in the previous chapter can also be used for missing data problems. Fay has shown how to include missing data in categorical data analysis in a very general way (Fay, 1986, 1989; Hagenaars, 1990, Section 5.5.1; Vermunt, 1997a). Fay introduces response variables to denote whether or not a particular respondent has a score on a particular variable — e.g., $R_A = 0$ indicates that the respondent does not have a score on variable A and $R_A = 1$ means that a score was obtained. In this way, mutually exclusive response groups can be defined. For example, if there are three-variables A, B, and C, response group ABC consists of the respondents that have a score on all three-variables and no missing data. Response group AC consists of all respondents that have scores on A and C, but whose scores on B are missing, etc. For response group AC, the scores on B are latent, because missing. Using the EM algorithm described in the previous chapter, a complete table ABC for response group AC is computed in the E-step, conditional on the observed frequencies n_{ik}^{AC} and using the conditional probabilities $\hat{\pi}_{j|ik}^{B|AC}$ obtained in the previous M step. This is done for all response groups that contain missing data. Next, all complete tables for all response groups are summed to obtain the completed 'observed' table that is then used in the next M-step. In marginal modeling, these response groups have to be defined for each cluster separately, because within a response group, the observations are dependent of each other because of the clustering, but also because the response groups themselves are dependent because one member of a particular cluster might belong to one response group, but another member of the same cluster might belong to another response group. This makes the E-step more complicated than usual. Details have to be worked out. An alternative might be to use Bayesian imputation methods (Schafer, 1997; Rubin, 1987) but then for dependent data.

Probably many other important developments will take (and are taking) place, but the above ones were the most urgently encountered in our own substantive analyses.

7.5 Software, Generalized exp-log Routines, and Website

Marginal modeling can be done in many different ways, by means of WLS, by means of GEE, by means of random-effect models (especially for marginal restrictions in linear models for continuous data), by means of procedures in which the joint table is parameterized in orthogonal sets for successive higher-order distributions (one-way, two-way, three-way marginals, etc.). Agresti (2002), Verbeke and Molenberghs (2000), and Molenberghs and Verbeke (2005) extensively discuss specialized and general purpose software to apply these approaches. SAS is extremely useful in this respect (but Muthen's Mplus and STATA can also be considered).

The specific full ML estimation procedure for categorical data advocated in this book, using the constraint approach with Lagrange multipliers, is as of yet not incorporated in the big standard statistical packages. Nevertheless, a number of routines and programs are available to perform the full ML marginal analyses. Lang has developed R- and S-Plus functions (website: www.stat.uiowa.edu/ jblang/); Vermunt incorporated a rather flexible marginal subroutine in an experimental version of his multipurpose program LEM for loglinear modeling (LEM's webpage: *www.uvt.nl/faculteiten/fsw/organisatie/departementen/mto/software2.html*; for an application, see Vermunt et al., 2001).

All models in this book were estimated by Mathematica procedures, written by the first author. The main advantages of his procedures are that many models can be easily specified with little knowledge about matrix algebra, that latent variables may be included, that it can handle rather large data sets, that it uses the very flexible generalized exp-log notation described in Chapter 3, and that it has some extra provisions to make the algorithm converge faster. Because Mathematica is not available for all users, equivalent R procedures have been developed.

The authors' software, the Mathematica and R procedures, along with the book's data sets, the marginal models, the required matrices, and the outcomes have been put on the book's website. For practical use, especially working out the necessary generalized exp-log notation for defining the models and the restrictions on the relevant (association) coefficients might be (very) cumbersome. That is the main reason they have been put on the website in a general form to be used rather routinely by other researchers. Our programme CMM, which stands for *Categorical Marginal Modeling*, can be found on the book's website that is maintained by the first author:

www.cmm.st

References

Agresti, A. (1984). *Analysis of ordinal categorical data*. New York: Wiley.

Agresti, A. (1990). *Categorical data analysis*. New York: Wiley.

Agresti, A. (1996). *An introduction to categorical data analysis*. New York: Wiley.

Agresti, A. (2002). *Categorical data analysis, (Second edition)*. New York: Wiley.

Agresti, A., & Agresti, B. F. (1977). Statistical analysis of qualitative variation. In K. F. Schuessler (Ed.), *Sociological methodology 1978* (p. 204-237). Washington D.C.: American Sociological Association.

Agresti, A., Booth, J. G., Hobert, J., & Caffo, B. (2000). Random-effects modeling of categorical response data. In M. E. Sobel & M. P. Becker (Eds.), *Sociological methodology 2000* (p. 27-80). Washington, D.C.: American Sociological Association.

Agresti, A., & Coull, B. A. (1998). Approximate is better than "exact" for interval estimation of binomial proportions. *The American Statistician, 52,* 119-126.

Agresti, A., & Liu, I. M. (1999). Modeling a categorical variable alllowing arbitrarily many category choices. *Biometrics, 55,* 936-943.

Aitchison, J., & Silvey, S. D. (1958). Maximum likelihood estimation of parameters subject to restraints. *Annals of Mathematical Statistics, 29,* 813-828.

Aitchison, J., & Silvey, S. D. (1960). Maximum-likelihood estimation procedures and associated tests of significance. *Journal of the Royal Statistical Society, Series B, 22,* 154-171.

Allison, P. D. (1990). Change scores as dependent variables in regression analysis. In C. C. Clogg (Ed.), *Sociological methodology 1990* (p. 93-114). Washington, D.C.: American Sociological Association.

Allison, P. D. (1994). Using panel data to estimate the effects of events. *Soiological Research and Methods, 23,* 174-199.

Alwin, D. F. (2007). *Margins of error: A study in reliability in survey measurement*. New York: Wiley.

Aris, E. M. (2001). *Statistical causal models for categorical data*. Tilburg: Dutch University Press.

Balagtas, C. C., Becker, M. P., & Lang, J. B. (1995). Marginal modelling of categorical data from crossover experiments. *Applied Statistics, 44,* 63-77.

Bartolucci, F., Colombi, R., & Forcina, A. (2007). An extended class of marginal link functions for modelling contingency tables by equality and inequality constraints. *Statistica Sinica*, *17*, 691-711.

Bassi, F., Hagenaars, J. A., Croon, M. A., & Vermunt, J. K. (2000). Estimating true changes when categorical panel data are affected by uncorrelated and correlated errors. *Sociological Methods and Research*, *29*, 230-268.

Becker, M. P. (1994). Analysis of repeated categorical measurements using models for marginal distributions: an application to trends in attitudes on legalized abortion. In P. V. Marsden (Ed.), *Sociological methodology* (Vol. 24, p. 229-265). Oxford: Blackwell.

Becker, M. P., & Yang, I. (1998). Latent class marginal models for cross-classifications of counts. In A. E. Raftery (Ed.), *Sociological methodology* (Vol. 28, p. 293-326). Oxford: Blackwell.

Bekker, M. H. J., Croon, M. A., & Vermaas, S. (2002). Inner body and outward appearance - the relationship between orientation toward outward appearance, body awareness and symptom perception. *Personality and Individual Differences*, *33*, 213-225.

Berger, M. (1985). *Some aspects of the application of the generalized multivariate analysis of variance model*. Unpublished doctoral dissertation, Tilburg University, Tilburg.

Bergsma, W. P. (1997). *Marginal models for categorical data*. Tilburg: Tilburg University Press.

Bergsma, W. P., & Croon, M. A. (2005). Analyzing categorical data by marginal models. In L. A. Van der Ark, M. A. Croon, & K. Sijtsma (Eds.), *New developments in categorical data analysis for the social and behavioral sciences* (p. 83-101). Mahwah, New Jersey: Lawrence Erlbaum Associates.

Bergsma, W. P., & Rudas, T. (2002a). Marginal models for categorical data. *Annals of Statistics*, *30*, 140-159.

Bergsma, W. P., & Rudas, T. (2002b). Modeling conditional and marginal association in contingency tables. *Annales de la Faculté des Sciences de Toulouse*, *XI*, 455-468.

Bergsma, W. P., & Rudas, T. (2002c). Variation independent parameterizations of multivariate categorical distributions. In C. M. Cuadras, J. Fortiana, & J. A. Rodriguez-Lallena (Eds.), *Distributions with given marginals and related topics* (p. 21-28). Amsterdam: Kluwer Academic Publishers.

Berkson, J. (1980). Minimum chi-square, not maximum likelihood! *Annals of Statistics*, *8*, 457-487.

Bilder, C. R., & Loughin, T. M. (2004). Testing for marginal independence between two categorical variables with multiple responses. *Biometrics*, *60*, 241-248.

Bishop, Y. V. V., Fienberg, S. E., & Holland, P. W. (1975). *Discrete multivariate analysis*. Cambridge, MA: MIT Press.

Blair, J., & Lacy, M. G. (2000). Statistics of ordinal variation. *Sociological Methods and Research*, *28*, 251-279.

Bollen, K. A. (1989). *Structural equations with latent variables*. New York: Wiley.

Borsboom, D. (2005). *Measuring the mind: Conceptual issues in contemporary psychology*. Cambridge: Cambridge University Press.

Bradley, R. A., & Terry, M. R. (1952). The rank analysis of incomplete block designs. I. the method of paired comparisons. *Biometrika, 39*, 324-345.

Brown, L. D., Cai, T. T., & Dasgupta, A. (1999). Interval estimation for a binomial proportion. *Statistical Science, 16*, 101–133.

Bryk, A. S., & Weisberg, H. I. (1977). Use of the nonequivalent control group design when subjects are growing. *Psychological Bulletin, 84*, 950-962.

Campbell, D. T., & Clayton, K. N. (1961). Avoiding regression effects in panel studies of communication impact. *Studies in public communication, 3*, 99-118.

Campbell, D. T., & Kenny, D. A. (1999). *A primer on regression artifacts*. New York: Guilford.

Campbell, D. T., & Stanley, J. C. (1963). *Experimental and quasi-experimental designs for research*. Chicago: RandMcNally.

Cartwright, N. (1989). *Nature's capacities and their measurement*. Oxford: Clarendon Press.

Caussinus, H. (1966). Contribution à l'analyse statistique des tablaux de correlation. *Annales de la faculté des sciences de l'Université de Toulouse, 29*, 77-182.

CBS News and the New York Times. (2001). *4/13/2001 2:34:51PM, "CBS News/New York Times Poll: Thomas Hearings Panelback", hdl:1902.4/USC BSNYT1991-OCT91B*. University of Connecticut: Roper Center for Public Opinion Research.

Cliff, N. (1993). Dominance statistics: Ordinal analyses to answer ordinal questions. *Psychological Bulletin, 114*, 494-509.

Cliff, N. (1996). *Ordinal methods for behavioral analysis*. Mahwah, NJ: Lawrence Erlbaum.

Clogg, C. C. (1981a). Latent structure models of mobility. *American Journal of Soiology, 86*, 836-868.

Clogg, C. C. (1981b). New developments in latent structure analysis. In D. J. Jackson & E. F. Borgatta (Eds.), *Factor analysis and measurement in sociological research* (p. 215-246). Beverly Hills, CA: Sage.

Clogg, C. C., & Shihadeh, E. S. (1994). *Statistical models for ordinal variables*. Thousand Oaks, CA: Sage.

Cohen, J. (1960). A coefficient of agreement for nominal scales. *Psychological Bulletin, 70*, 213-220.

Coleman, J. S. (1964). *Introduction to mathematical sociology*. London: Collier.

Cox, D. R., & Wermuth, N. (1996). *Multivariate dependencies: Models, analysis and interpretation*. London: Chapman & Hall.

Croon, M. A. (1990). Latent class analysis with ordered latent classes. *British Journal of Mathematical and Statistical Psychology, 43*, 171-192.

Croon, M. A., Bergsma, W. P., & Hagenaars, J. A. (2000). Analyzing change in categorical variables by generalized log-linear models. *Sociological Methods and Research, 29*, 195-229.

David, H. A. (1968). Gini's mean difference rediscovered. *Biometrika, 55*, 573-575.

David, H. A. (1983). Gini's mean difference. In S. Kotz, N. L. Johnson, & C. B. Read (Eds.), *Encyclopedia of statistical sciences (vol. 3)* (p. 436-437). New York: Wiley.

Dawid, A. P. (1980). Conditional independence for statistical operations. *Annals of Statistics, 8,* 598-617.

Decady, Y. J., & Thomas, D. R. (2000). A simple test of association for contingency tables with multiple column responses. *Biometrics, 56,* 893-896.

Diggle, P. J., Heagerty, P. J., Liang, K. Y., & Zeger, S. L. (2002). *Analysis of longitudinal data (Second edition).* Oxford: Oxford University Press.

Drton, M., & Richardson, T. S. (2004). Multimodality of the likelihood in the bivariate seemingly unrelated regressions model. *Biometrika, 91,* 383-392.

Duncan, O. D. (1972). Unmeasured variables in linear panel models. In H. Costner (Ed.), *Sociological methodology 1972* (p. 36-82). San Francisco, CA: Jossey-Bass.

Duncan, O. D. (1979). Testing key hypotheses in panel analysis. In K. F. Schuessler (Ed.), *Sociological methodology 1980* (p. 279-289). San Francisco, CA: Jossey-Bass.

Duncan, O. D. (1981). Two faces of panel analysis: Parallels with comparative cross-sectional analysis and time-lagged association. In S. Leinhard (Ed.), *Sociological methodology 1981* (p. 281-318). San Francisco, CA: Jossey-Bass.

Durkheim, E. (1930). *Le suicide.* Paris, France: Presses Universitaire de France.

Dykstra, P. A., Kalmijn, M., Knijn, T. C. M., Komter, A. E., Liefboer, A. C., & Mulder, C. H. (2004). *Codebook of the Netherlands Kinship Panel Study: A multi-actor, multi-method panel study on solidarity in family relationships. Wave 1* (Tech. Rep. No. NKPS Working Paper 1). The Hague, NL: Netherlands Interdisciplinary Demographic Institute.

Elliot, D. S., Huizinga, D., & Menard, S. (1989). *Multiple problem youth: Delinquency, substance use, and mental health problems.* New York: Springer-Verlag.

Fay, R. E. (1986). Causal models for patterns of nonresponse. *Journal of the American Statistical Association, 81,* 354-365.

Fay, R. E. (1989). Estimating nonignorable nonresponse in longitudinal surveys through causal modeling. In D. Kasprzyk, G. J. Duncan, G. Kalton, & M. P. Singh (Eds.), *Panel surveys* (p. 375-399). New York: Wiley.

Fienberg, S. E. (1980). *The analysis of cross-classified categorical data.* Cambridge, MA: MIT Press.

Fienberg, S. E., & Larntz, K. (1976). Loglinear representation for paired and multiple comparison models. *Biometrika, 63,* 245-254.

Firebaugh, G. (2008). *Seven rules for social research.* Princeton, NJ: Princeton University Press.

Forthofer, R. N., & Koch, G. G. (1973). An analysis for compounded functions of categorical data. *Biometrics, 29,* 143-157.

Franzoi, S., & Shields, S. A. (1984). The Body Esteem Scale: multidimensional structure and sex differences in a college population. *Journal of Personality Assessment, 48,* 173-178.

Glock, C. Y. (1955). Some applications of the panel method to the study of change. In P. F. Lazarsfeld & M. Rosenberg (Eds.), *The language of social resarch* (p. 242-250). New York: Free Press.

Glonek, G. J. N. (1996). A class of regression models for multivariate categorical responses. *Biometrika, 83*(1), 15-28.

Glonek, G. J. N., & McCullagh, P. (1995). Multivariate logistic models. *Journal of the Royal Statistical Society, Series B, 57*, 533-546.

Goldstein, H. (1979). *The design and analysis of longitudinal studies: Their role in the measurement of change.* London: Academic Press.

Goodman, L. A. (1973). The analysis of multidimensional contingency tables when some variables are posterior to others: A modified path analysis approach. *Biometrika, 60*, 179-192.

Goodman, L. A. (1974a). The analysis of systems of qualitative variables when some of the variables are unobservable. Part I - A modified latent structure approach. *American Journal of Sociology, 79*, 1179-1259.

Goodman, L. A. (1974b). Exploratory latent structure analysis using both identifiable and unidentifiable models. *Biometrika, 61*, 215-231.

Goodman, L. A., & Kruskal, W. H. (1979). *Measures of association for cross classifications.* New York: Springer.

Grizzle, J. E., Starmer, C. F., & Koch, G. G. (1969). Analysis of categorical data by linear models. *Biometrics, 25*, 489-504.

Haber, M. (1985). Maximum likelihood methods for linear and loglinear models in categorical data. *Computational Statistics and Data Analysis, 3*, 1-10.

Haberman, S. J. (1973). Log-linear models for frequency data: sufficient statistics and likelihood equations. *Annals of Statistics, 1*, 617-632.

Haberman, S. J. (1974). *The analysis of frequency data.* Chicago: University of Chicago Press.

Haberman, S. J. (1977). Product models for frequency tables involving indirect observations. *Annals of Statistics, 5*, 1124-1147.

Haberman, S. J. (1979). *Analysis of qualitative data. vol.2: New developments.* New York: Academic Press.

Haberman, S. J. (1988). A stabilized Newton-Raphson algorithm for loglinear models for frequency tables derived by indirect obervation. In C. C. Clogg (Ed.), *Sociological methodology: Vol. 18* (p. 193-211). Washington, D.C.: American Sociological Association.

Hagenaars, J. A. (1986). Symmetry, quasi-symmetry, and marginal homogeneity on the latent level. *Social Science Research, 15*, 241-255.

Hagenaars, J. A. (1988). Latent structure models with direct effects between indicators: Local dependence models. *Sociological Methods and Research, 16*, 379-405.

Hagenaars, J. A. (1990). *Categorical longitudinal data: Log-linear, panel, trend, and cohort analysis.* Newbury Park: Sage.

Hagenaars, J. A. (1992). Analyzing categorical longitudinal data using marginal homogeneity models. *Statistica Applicata, 4*, 763-771.

Hagenaars, J. A. (1993). *Loglinear models with latent variables.* Newbury Park: Sage.

Hagenaars, J. A. (1998). Categorical causal modeling: latent class analysis and directed loglinear models with latent variables. *Sociological Methods and Research, 26,* 436-486.

Hagenaars, J. A. (2002). Directed loglinear modeling with latent variables: Causal models for categorical data with nonsystematic and systematic measurement errors. In J. A. Hagenaars & A. L. McCutcheon (Eds.), *Applied latent class analysis* (p. 234-286). Cambridge: Cambridge University Press.

Hagenaars, J. A. (2005). Misclassification phenomena in categorical data analysis: Regression toward the mean and tendency toward the mode. In L. A. Van der Ark, M. A. Croon, & K. Sijtsma (Eds.), *New developments in categorical data analysis for the social and behavioral sciences* (p. 15-39). Mahwah, NJ: Lawrence Erlbaum.

Hagenaars, J. A., & Heinen, T. G. (1982). Effects of role-independent interviewer characteristics on responses. In W. Dijkstra & J. Van der Zouwen (Eds.), *Response behavior in the survey interview* (p. 91-130). London: Academic Press.

Hagenaars, J. A., & McCutcheon, A. L. (2002). *Applied latent class analysis.* Cambridge: Cambridge University Press.

Halaby, C. N. (2004). Panel models in sociological research: Theory into practice. In R. M. Stolzenberg (Ed.), *Sociological methodology 2004* (p. 507-544). Ames, IA: Blackwell.

Hand, D. (2004). *Measurement: Theory and practice.* London: Arnold.

Hardin, J. W., & Hilbe, J. M. (2003). *Generalized estimating equations.* New York: Chapman & Hall.

Hays, W. L. (1994). *Statistics.* Belmont, CA: Wadsworth.

Heagerty, P. J., & Zeger, S. L. (2000). Marginalized multilevel models and likelihood inference. *Statistical Science, 15,* 1-26.

Heckman, J. J. (2005). The scientific model of causality (with discussion). In R. M. Stolzenberg (Ed.), *Sociological methodology 2005* (p. 1-162). Washington, D.C.: American Sociological Association.

Heinen, T. G. (1996). *Latent class and discrete latent trait models: Similarities and differences.* Thousand Oaks, CA: Sage.

Hosmer, D. W., & Lemeshow, S. (2000). *Applied logistic regression.* New York: Wiley.

Hsiao, C. (2003). *Analysis of panel data.* Cambridge: Cambridge University Press.

Johnson, R. A., & Wichern, D. W. (2007). *Applied multivariate statistical analysis (Sixth edition).* Upper Saddle River, NJ: Pearson.

Jöreskog, K. G., & Sörbom, D. (1996). *Lisrel 8: User's reference manual.* Lincolnwood: Scientific Software International.

Judd, C. M., & Kenny, D. A. (1981). *Estimating the effects of social interventions.* Cambridge: Cambridge University Press.

Kenny, D. A., & Cohen, S. H. (1980). A reexamination of selection and growth processes in the nonequivalent control group design. In K. F. Schuessler (Ed.), *Sociological methodology 1980* (p. 290-313). San Francisco, CA: Jossey-Bass.

Kenny, D. A., Kashy, D. A., & Cook, W. L. (2006). *Dyadic data analysis*. London: Guilford Press.

Kiiveri, H., & Speed, T. P. (1982). Structural analysis of multivariate data: A review. In S. Leinhardt (Ed.), *Sociological methodology 1982* (p. 209-289). San Francisco, CA: Jossey-Bass.

Kim, K., & Timm, N. (2007). *Univariate and multivariate general linear models: Theory and applications with SAS*. Boca Raton, FL: Chapman & Hall/CRC.

Knoke, D., & Burke, P. J. (1980). *Loglinear models*. Beverly Hill, CA: Sage.

Koch, G. G., Landis, J. R., Freeman, D. H., & Lehnen, R. G. (1977). A general methodology for the analysis of experiments with repeated measurements of categorical data. *Biometrics, 33*, 133-158.

Krippendorff, K. (1986). *Information theory: Structural models for qualitative data*. Newbury Park: Sage.

Kritzer, H. M. (1977). Analyzing measures of association derived from contingency tables. *Sociological Methods and Research, 5*, 35-50.

Landis, J. R., & Koch, G. G. (1979). The analysis of categorical data in longitudinal studies of behavioral development. In J. R. Nesselroade & P. B. Baltes (Eds.), *Longitudinal research in the study of behavior and development* (p. 233-262). New York: Academic Press.

Lang, J. B. (1996a). Maximum likelihood methods for a generalized class of log-linear models. *Annals of Statistics, 24*, 726-752.

Lang, J. B. (1996b). On the comparison of multinomial and Poisson log-linear models. *Journal of the Royal Statistical Society, Series B, 58*, 253-266.

Lang, J. B. (2004). Multinomial-Poisson homogeneous models for contingency tables. *Annals of Statistics, 32*, 340-383.

Lang, J. B. (2005). Homogeneous linear predictor models for contingency tables. *Journal of the American Statistical Association, 100*, 121-134.

Lang, J. B. (2008). Score and profile likelihood confidence intervals for contingency table parameters. *Statistics in Medicine, 27*, 5975-5990.

Lang, J. B., & Agresti, A. (1994). Simultaneously modelling the joint and marginal distributions of multivariate categorical responses. *Journal of the American Statistical Association, 89*, 625-632.

Lang, J. B., McDonald, J. W., & Smith, P. W. F. (1999). Association-marginal modeling of multivariate categorical responses: A maximum likelihood approach. *Journal of the American Statistical Association, 94*, 1161-1171.

Lauritzen, S. L. (1996). *Graphical models*. Oxford: Clarendon Press.

Lazarsfeld, P. F. (1950). The logical and mathematical foundation of latent structure analysis. In S. Stouffer (Ed.), *Measurement and prediction* (p. 413-472). Princeton, NJ: Princeton University Press.

Lazarsfeld, P. F. (1955). Interpretation of statistical relations as a research operation. In P. F. Lazarsfeld & M. Rosenberg (Eds.), *The language of social research* (p. 115-124). Glencoe, IL: Free Press.

Lazarsfeld, P. F. (1959). Latent structure analysis. In S. Koch (Ed.), *Psychology: a study of a science. conceptual and systematic. Vol. 3: Formulations of the person and the social context* (p. 476-543). New York: McGrawHill.

Lazarsfeld, P. F. (1972). Mutual effects of statistical variables. In P. F. Lazarsfeld, A. K. Pasanella, & M. Rosenberg (Eds.), *Continuities in the language of social research* (p. 388-398). New York: Free Press.

Lazarsfeld, P. F., Berelson, B., & Gaudet, H. (1948). *The people's choice.* New York: Duell, Sloan and Pearce.

Lazarsfeld, P. F., & Fiske, M. (1938). The 'panel' as a new tool for measuring opinion. *Public Opinion Quarterly, 2,* 596-612.

Lazarsfeld, P. F., & Henry, N. W. (1968). *Latent structure analysis.* Boston, MA: Houghton Mifflin.

Lee, Y., & Nelder, J. A. (2004). Conditional and marginal models: another view. *Statistical Science, 19,* 219-238.

Liang, K. Y., & Zeger, S. L. (1986). Longitudinal data analysis using generalized linear models. *Biometrika, 73,* 13-22.

Lindsey, J. K. (1999). *Models for repeated measurements.* Oxford: Clarendon.

Long, J. S. (1997). *Regression models for categorical and limited dependent variables.* Thousand Oaks, CA: Sage.

Lord, F. M. (1960). Large-sample covariance analysis when the control variable is fallible. *Journal of the American Statistical Association, 55,* 309-321.

Lord, F. M. (1963). Elementary models for measuring change. In C. W. Harris (Ed.), *Problems in measuring change* (p. 21-38). Madison, WI: University of Wisconsin Press.

Lord, F. M. (1967). A paradox in the interpretation of group comparisons. *Psychological Bulletin, 72,* 304-305.

Lord, F. M., & Novick, M. R. (1968). *Statistical theories of mental test scores.* Reading, Massachusets: Addison-Wesley.

Loughin, T. M., & Scherer, P. N. (1998). Testing for association in contingency tables with multiple column responses. *Biometrics, 54,* 630-637.

Luijkx, R. (1994). *Comparative loglinear analyses of social mobility and heterogamy.* Tilburg: Tilburg University Press.

MacLachlan, G. J., & Krishnan, T. (1997). *The EM algorithm and extensions.* New York: Wiley.

McKim, V. R., & Turner, S. (Eds.). (1997). *Causality in crisis: Statistical methods and the search for causal knowledge in the social sciences.* Notre Dame, IN: Notre Dame University Press.

Meng, X. L., & Rubin, D. B. (1993). Maximum likelihood estimation via the ECM algorithm: A general framework. *Biometrika, 80,* 267-278.

Millsap, R. E., & Yun-Tein, J. (2004). Assessing factorial invariance in ordered-categorical measures. *Multivariate Behavioral Research, 39,* 479-515.

Mokken, R. J. (1971). *A theory and procedure of scale analysis.* The Hague, NL: Mouton/De Gruyter.

Molenberghs, G., & Verbeke, G. (2005). *Models for discrete longitudinal data.* New York: Springer-Verlag.

Mooijaart, A., & Van der Heijden, P. G. M. (1991). The EM algorithm for latent class analysis with equality constraints. *Psychometrika, 57,* 261-269.

Morgan, S. L., & Winship, C. (2007). *Counterfactuals and causal inference: Methods and principles for social research (analytical methods for social research)*. Cambridge: Cambridge University Press.

Muthén, L. K., & Muthén, B. O. (2006). *Mplus: Statistical analysis with latent variables. (User's Guide Fourth edition)*. Los Angeles, CA: Muthén and Muthén.

Olkin, I., & Press, S. J. (1995). Correlations redux. *Psychological Bulletin*, 155-164.

Pearl, J. (2000). *Causality: Models, reasoning, and inference*. Cambridge: Cambridge University Press.

Pearl, J. (2001). Direct and indirect effects. In *Proceedings of the seventeenth conference on uncertainty in artificial intelligence* (p. 411-420). San Francisco, CA: Morgan Kaufman.

Plewis, I. (1985). *Analysing change: Measurement and explanation using longitudinal data*. Chichester: Wiley.

Prochaska, J. O., Velicer, W. F., Fava, J. L., Rossi, J. S., & Tosh, J. Y. (2001). Evaluating a population-based recruitment approach and a stage-based expert system intervention for smoking cessation. *Addictive Behaviors*, *26*, 583-602.

Qaqish, B. F., & Ivanova, A. (2006). Multivariate logistic models. *Biometrika*, *93*, 1011-1017.

Raudenbush, S. W., & Bryk, A. S. (2003). *Hierarchical linear models: Applications and data analysis methods*. Thousand Oaks, CA: Sage Publications.

Reynolds, H. T. (1977). *The analysis of cross-classifications*. New York: Free Press.

Robins, J. M. (2003). Semantics of causal DAG models and the identification of direct and indirect effects. In P. J. Green, N. L. Hjort, & S. Richardson (Eds.), *Highly structured stochastic systems* (p. 70-81). Oxford: Oxford University Press.

Rubin, D. B. (1974). Estimating causal effects of treatments in randomized and nonrandomized studies. *Journal of Educational Psychology*, *66*, 688-701.

Rubin, D. B. (1987). *Multiple imputation for nonresponse in surveys*. New York: Wiley.

Rudas, T. (1997). *Odds ratios in the analysis of contingency tables*. Newbury Park: Sage.

Schafer, J. L. (1997). *Analysis of incomplete multivariate data*. London: Chapman and Hall.

Schott, J. R. (1997). *Matrix analysis for statistics*. New York: Wiley.

Searle, S. R. (2006). *Matrix algebra useful for statistics*. New York: Wiley.

Shadish, W. R., Cook, T. D., & Campbell, D. T. (2002). *Experimental and quasi-experimental designs for generalized causal inference*. Boston, MA: Houghton Mifflin.

Sijtsma, K., & Molenaar, I. W. (2002). *Introduction to nonparametric item response theory*. Thousand Oaks, CA: Sage.

Simpson, E. H. (1951). The interpretation of interaction. *Journal of the Royal Statistical Society, Series B*, *13*, 238-241.

Skrondal, A., & Rabe-Hesketh, S. (2004). *Generalized latent variable modeling: Multilevel, longitudinal, and structural equation models*. Boca Raton, FL: Chapman and Hall.

Snijders, T. A. B., & Bosker, R. (1999). *Multilevel analysis: An introduction to basic and advanced data analysis methods*. London: Sage.

Snijders, T. A. B., & Hagenaars, J. A. (2001). Guest editors' introduction to the special issue to causality at work. *Sociological Methods and Research, 30*, 3-10.

Sobel, M. E. (1994). Causal inference in latent variable models. In A. von Eye & C. C. Clogg (Eds.), *Latent variables analysis: Applications for developmental rsearch* (p. 3-35). Newbury, CA: Sage.

Sobel, M. E. (1995). Causal inference in the social and behavioral sciences. In G. Arminger, C. C. Clogg, & M. E. Sobel (Eds.), *Handbook of statistical modeling for the social and behavioral sciences* (p. 1-38). New York: Plenum Press.

Sobel, M. E. (1997). Measurement, causation, and local independence in latent variable models. In M. Berkane (Ed.), *Latent variable modeling and applications to causality* (p. 11-28). New York: Springer-Verlag.

Sobel, M. E. (2000). Causal inference in the social sciences. *Journal of the American Statistical Society, 95*, 647-651.

Srivastava, V. K., & Giles, D. E. A. (1987). *Seemingly unrelated regression equations models*. New York: Marcel Dekker.

Steiger, J. H. (2005). Comparing correlations: Pattern hypothesis tests between and/or within independent samples. In A. Maydeu-Olivaries & J. J. McArdle (Eds.), *Contemporary psychometrics* (p. 377-414). Mahwah, NJ: Lawrence Erlbaum.

Stolzenberg, R. M. (Ed.). (2003). *Sociological methodology 2003*. Washington, D.C.: American Sociological Association.

Suen, H. K., & Ary, D. (1989). *Analyzing behavioral observation data*. Hillsdale, NJ: Lawrence Erlbaum Publishers.

Van der Ark, L. A., Croon, M. A., & Sijtsma, K. (2008a). Mokken scale analysis for dichotomous items using marginal models. *Psychometrika, 73*, 183-208S.

Van der Ark, L. A., Croon, M. A., & Sijtsma, K. (2008b). Possibilities and challenges in mokken scale analysis using marginal models. In K. Shigemasu, A. Okada, T. Imaizuma, & T. Hodhina (Eds.), *New trends in psychometrics* (p. 525-534). Tokyo: Universal Academic Press.

Van der Eijk, C. (1980). *Longitudinaal enquete onderzoek: mogelijkheden en problemen*. Amsterdam: Universiteit van Amsterdam.

Velicer, W. F., Fava, J. L., Prochaska, J. O., Abrams, D. B., Emmons, K. M., & Pierce, J. P. (1995). Distribution of smokers by stage in three representative samples. *Preventive Medicine, 24*, 401-411.

Verbeek, M. (1991). *The design of panel surveys and the treatment of missing observations*. Unpublished doctoral dissertation, Tilburg University, Tilburg.

Verbeke, G., & Molenberghs, G. (2000). *Linear mixed models for longitudinal data*. New York: Springer.

Vermunt, J. K. (1997a). *LEM: A general program for the analysis of categorical data: Users manual* (Tech. Rep.). Tilburg: Tilburg University.

Vermunt, J. K. (1997b). *Log-linear models for event histories*. Thousand Oaks, CA: Sage.

Vermunt, J. K. (1999). A general class of nonparametric models for ordinal categorical data. In M. E. Sobel & M. P. Becker (Eds.), *Sociological methodology 1999, vol. 29* (p. 187-223). Washington D.C.: American Sociological Association.

Vermunt, J. K. (2003). Multilevel latent class analysis. In R. M. Stolzenberg (Ed.), *Sociological methodology 2003* (p. 213-240). Washington, D.C.: American Sociological Association.

Vermunt, J. K., & Hagenaars, J. A. (2004). Ordinal longitudinal data analysis. In R. C. Hauspie, N. Cameron, & L. Molinari (Eds.), *Methods in human growth research* (p. 374-393). Cambridge: Cambridge University Press.

Vermunt, J. K., & Magidson, J. (2004). *Latent gold 4.0 user's guide*. Belmont, MA: Statistical Innovations.

Vermunt, J. K., Rodrigo, M. F., & Ato-Garcia, M. (2001). Modeling joint and marginal distributions in the analysis of categorical panel data. *Sociological Methods and Research, 30*, 170-196.

Wainer, H. (1991). Adjusting for differential base rates: Lord's paradox again. *Psychological Bulletin, 109*, 147-151.

Wedderburn, R. W. M. (1974). Quasi-likelihood functions, generalized linear models and the gaussian method. *Biometrika, 61*, 439-447.

Weerahandi, S. (2004). *Generalized inference in repeated measures: Exact methods in manova and mixed models*. Hoboken, NJ: Wiley.

Whittaker, J. W. (1990). *Graphical models in applied multivariate statistics*. New York: Wiley.

Wiggins, L. M. (1973). *Panel analysis: Latent probability models for attitude and behavior processes*. Amsterdam: Elsevier.

Wilks, S. S. (1946). Sample criteria for testing equality of means, equality of variances, and equality of covariances in a normal multivariate distribution. *Annals of Mathematical Statistics, 17*, 257-281.

Winship, C., & Sobel, M. E. (2004). Causal inference in sociological studies. In M. Hardy (Ed.), *The handbook of data analysis* (p. 481-503). Thousand Oaks, CA: Sage.

Zellner, A. (1962). An efficient method of estimating seemingly unrelated regression equations and tests for aggregation bias. *Journal of the American Statistical Association, 57*, 348-368.

Author Index

Subject Index

Statistical Learning from a Regression Perspective

Richard A. Berk

This book considers statistical learning applications when interest centers on the conditional distribution of the response variable, given a set of predictors, and when it is important to characterize how the predictors are related to the response. Among the statistical learning procedures examined are bagging, random forests, boosting, and support vector machines. Intuitive explanations and visual representations are prominent. All of the analyses included are done in R.

2008. 360 pp. (Springer Series in Statistics) Hardcover
ISBN 978-0-387-77500-5

Studying Human Populations
An Advanced Course in Statistics

Nicholas T. Longford

Studying Human Populations is a textbook for graduate students and research workers in social statistics and related subject areas. It follows a novel curriculum developed around the basic statistical activities of sampling, measurement and inference. The monograph aims to prepare the reader for the career of an independent social statistician and to serve as a reference for methods, ideas for and ways of studying human populations: formulation of the inferential goals, design of studies, search for the sources of relevant information, analysis and presentation of results.

2008. XVI, 474 pp. (Springer Texts in Statistics) Hardcover
ISBN 978-0-387-98735-4

Introduction to Applied Bayesian Statistics and Estimation for Social Scientists

Scott M. Lynch

This book covers the complete process of Bayesian statistical analysis in great detail from the development of a model through the process of making statistical inference. The key feature of this book is that it covers models that are most commonly used in social science research, including the linear regression model, generalized linear models, hierarchical models, and multivariate regression models, and it thoroughly develops each real-data example in painstaking detail.

2007. XXVII, 364 pp. (Statistics for Social and Behavioral Sciences)
Hardcover
ISBN: 978-0-387-71264-2